# Stomatopod Crustacea
# of the Western Atlantic

STUDIES IN TROPICAL OCEANOGRAPHY NO. 8
INSTITUTE OF MARINE SCIENCES, UNIVERSITY OF MIAMI

# Stomatopod Crustacea
# of the Western Atlantic

By RAYMOND B. MANNING

UNIVERSITY OF MIAMI PRESS
Coral Gables, Florida

This volume may be referred to as
Stud. trop. Oceanogr. Miami 8:
(viii + 380 pp.), April, 1969

In 1963 the Institute of Marine Sciences of the University of Miami established a series of publications entitled *Studies in Tropical Oceanography* to accommodate research reports too large for inclusion in regular periodicals. The first volume of this series was Dr. Donald P. deSylva's "Systematics and Life History of the Great Barracuda, *Sphyraena barracuda* (Walbaum)," followed by seven additional numbers covering a diversity of marine topics. As the eighth volume of *Studies in Tropical Oceanography,* the editors now present the results of Dr. Raymond B. Manning's study of the systematics of western Atlantic stomatopod crustaceans.

The Stomatopoda, predaceous crustaceans known to scientists and laymen alike as mantis shrimps, are a characteristic element of tropical and subtropical marine faunas. The adults, mainly burrowers, range in size from a couple of inches to more than a foot in length. They may be exceedingly abundant on a variety of benthic habitats and some species form the basis of commercial fisheries. Larval stomatopods, transparent and sometimes reaching several inches in length, are one of the most conspicuous and typical components of tropical zooplankton. Though predators like the adults, they may form a major part of the diet of pelagic fishes such as tunas.

The present work, the first attempt to summarize existing knowledge of the stomatopod crustaceans of the tropical western Atlantic and the first such monograph on stomatopods for any faunal region in half a century, is a major contribution to the study of this group. More than 60 species, half of which were unknown in the western Atlantic fauna prior to Dr. Manning's studies, are described and illustrated. Synonymies, distributional records and other information are also given for each species. Original keys to families, genera and species facilitate identification and are an important feature of the work.

This monograph is an expansion of a doctoral dissertation by the author. The work was supported initially by research grants from the National Science Foundation and was completed at the Smithsonian Institution. Publication of the volume was made possible by National Science Foundation grant number GN-719 to the Institute of Marine Sciences, University of Miami.

The Editors

# TABLE OF CONTENTS

# STOMATOPOD CRUSTACEA OF THE WESTERN ATLANTIC[1]

RAYMOND B. MANNING[2]
*Institute of Marine Sciences, University of Miami*
and
*Department of Invertebrate Zoology, Smithsonian Institution*

## ABSTRACT

The stomatopod fauna of the western Atlantic is far richer in number of species than previously believed. Sixty-two species, representing 18 genera and four families, are described and illustrated. Twenty-eight species have been added to the western Atlantic fauna, including 24 species described as a result of this study. Ten new species and one new subspecies are described in this report, and four other species, previously unrecorded from the area, are recognized. Complete synonymies are presented, along with keys to American genera and western Atlantic species.

The western Atlantic stomatopods show closest affinities with the faunas of the eastern Atlantic and eastern Pacific areas. Five Indo-West Pacific species also occur in the study area: *Pseudosquilla ciliata* (Fabricius), *P. oculata* (Brullé), *Alima hyalina* Leach, *A. hieroglyphica* (Kemp), and *Odontodactylus brevirostris* (Miers). These are the most widely distributed stomatopods; the species of *Alima* and *Odontodactylus* have been recorded from scattered localities between Hawaii, the Indian Ocean, and the Atlantic. All but *O. brevirostris* occur in the eastern Atlantic, but none occurs in the eastern Pacific. *Heterosquilla mccullochae* (Schmitt), previously known from the Gulf of California, is recorded from Florida and the Virgin Islands. Numerous other eastern Pacific species have closely related but distinct analogues in the western Atlantic.

## INTRODUCTION

The crustacean order Stomatopoda is one of the most primitive of the groups now included in the Malacostraca; although primitive, the members of the order are highly specialized and prominent marine organisms about which relatively little is known. Our knowledge of the taxonomy and biology of the stomatopods is still at the alpha, or descriptive, level. Many species are known from only one or two specimens, and undoubtedly many more remain to be described. In each of several small collections from the Indo-West Pacific area now awaiting study, approximately one-third of the species are undescribed.

My interest in the stomatopods originated with attempts to identify and list the species known from Florida. This interest has been expanded to include studies on the systematics of the group on a world-wide basis, with present emphasis on American species. This report, based on a dissertation submitted for partial fulfillment of the requirements for the Ph.D. degree at the Institute of Marine Sciences, University of Miami,

[1]Contribution No. 991 from the Institute of Marine Sciences, University of Miami.
[2]Present address: Department of Invertebrate Zoology, Smithsonian Institution, U.S. National Museum, Washington, D.C.

1

Florida, considers in detail the western Atlantic fauna, including 62 species in 18 genera. Keys are provided for all American genera and for all western Atlantic species. Each western Atlantic species is described in detail and its diagnostic features are illustrated. Complete synonymies are presented. Wherever applicable, comparisons are made with related species in the eastern Atlantic and eastern Pacific.

The cover illustration is from Marcgrave, 1648. His figure of Tamara guacu *(Lysiosquilla scabricauda)* was the first illustration of a stomatopod from the western Atlantic.

## ACKNOWLEDGEMENTS

A study of this scope could not have been carried out without the aid of many individuals. I am grateful to the following persons for providing literature or for lending materials (addresses given below were in effect at the time of each loan and are not necessarily the current addresses of the persons listed): N. Bahamonde, Centro de Investigaciones Zoologicas, Universidad de Chile; Enrique E. Boschi, Universidad de Buenos Aires; J. Cadenat, Institut Français d'Afrique Noire, Dakar; Jorge Carranza, Instituto Tecnológico de Veracruz, Mexico; Héctor Chapa Saldaña, Instituto Nacional de Investigaciones Biologico-Pesqueras, Mexico City; Eugenie Clark, Cape Haze Marine Laboratory, Sarasota, Florida; Erik Dahl, Zoological Institute, Lund; C. E. Dawson, Gulf Coast Research Laboratory, Ocean Springs, Mississippi; R. Ph. Dollfus, Muséum National d'Histoire Naturelle, Paris; C. H. Edmondson, Bernice P. Bishop Museum, Honolulu; Donald S. Erdman and Carmelo Feliciano, Department of Agriculture and Commerce, Puerto Rico; Armando J. G. Figueira, Museu Municipal do Funchal, Madeira; M. Figueiredo, Instituto de Biologia Marinha, Lisbon; H. A. F. Gohar, University of Cairo; C. B. Goodhart, Museum of Zoology, Cambridge, England; J. R. Grindley, South African Museum, Capetown; H.-E. Grüner, Zoologisches Museums, Berlin; J. Knudsen, Universitetets Zoologiske Museum, Copenhagen; C. V. Kurian, University of Kerala; Alceu Lemos de Castro, Museu Nacional, Rio de Janeiro; C. Bernard Lewis, Science Museum, Institute of Jamaica; H. Lijding, Surinam Fisheries Service; R. H. McConnell, Fisheries Laboratory, Georgetown, British Guiana; L. McCloskey, Duke University Marine Laboratory; R. W. Menzel, Oceanographic Institute, Florida State University; Th. Monod, Institut Français d'Afrique Noire, Dakar; J. E. Randall, Institute of Marine Biology, University of Puerto Rico; L. R. Richardson, Victoria University of Wellington, New Zealand; S. A. Rodrigues, Universidade de São Paulo, Brazil; J. A. Roze, Museo de Biologia, Universidad Central de Venezuela; Carl H. Saloman, U. S. Fish and Wildlife Service, St. Petersburg Beach, Florida; R. Serène, UNESCO, Djakarta; A. E. Smalley, Tulane University; V. G. Springer, Florida State Board of Conservation Marine Laboratory, St. Petersburg, Florida; W. Stephenson, University of Queensland, Brisbane; J. R. Taylor, U. S. Fish

2

and Wildlife Service, St. Petersburg Beach; Sylvia Taylor, Cape Haze Marine Laboratory, Sarasota; K. K. Tiwari, Zoological Survey of India, Calcutta; E. Tortonese, Museo Civico di Storia Naturale, Genoa; S. J. Townsley, University of Hawaii; Marta Vannucci, Instituto Oceanografico, Universidade de São Paulo; H. Vilela, Instituto de Biologia Marinha, Lisbon; Torben Wolff, Universitetets Zoologiske Museum, Copenhagen; V. A. Zullo, Systematics-Ecology Program, Marine Biological Laboratory, Woods Hole.

Special thanks are due Harvey R. Bullis, Jr., U. S. Fish and Wildlife Service, Pascagoula, Mississippi for making available the valuable material collected by the exploratory fishing vessels M/V OREGON and M/V SILVER BAY, and to the following individuals and institutions who provided material on loan and/or made available working space of different times: E. Deichmann and W. A. Newman, Museum of Comparative Zoology, Harvard; T. Abbott, J. E. Böhlke, and R. Robertson, Academy of Natural Sciences, Philadelphia; Dorothy Bliss, American Museum of Natural History, New York; W. D. Hartman, Yale Peabody Museum; Woodhull B. Young, Vanderbilt Marine Museum, Long Island; John S. Garth, Allan Hancock Foundation, University of Southern California, Los Angeles; Juan Rivero, Institute of Marine Biology, University of Puerto Rico; Isabella Gordon and R. W. Ingle, Crustacea Section, British Museum (Natural History); J. Forest, Muséum National d'Histoire Naturelle, Paris; J. Stock, Zoological Museum, Amsterdam; L. B. Holthuis, Rijksmuseum van Natuurlijke Historie, Leiden. Dr. Holthuis has provided technical advice on many problems, and the literature survey for this work could not have been so complete without the use of his unpublished synonymy of the stomatopods.

I am indebted to Fenner A. Chace, Jr., U. S. National Museum, for freely providing working space and specimens before I joined the museum staff in 1963 and for his advice and encouragement throughout the course of the study.

Gilbert L. Voss, Institute of Marine Sciences, University of Miami, has provided direction, encouragement, and technical advice, for which I would like to thank him. It was he who first pointed out to me the need for work on this group.

Grateful thanks also are extended to C. Richard Robins, F. M. Bayer, and A. J. Provenzano, Jr., Institute of Marine Sciences, who often helped with technical advice on this and related problems. Many of my fellow students and staff members at the Institute of Marine Sciences aided me in various ways, for which I would like to thank them. Walter A. Starck, II, made special efforts to obtain material for me, and he succeeded in establishing several new records for Florida. The photographs used in this account were provided by Starck (figs. 7, 76) and Walter R. Courtenay, Jr. (figs. 3, 4, 6, 9, 26, 27, 33, 39, 44, 48, 50, 53, 57a). Most of the line drawings are from the pen of my wife Lilly. Figure 77 has been

3

reproduced with the permission of E. J. Brill, Leiden. Mrs. T. W. McKenney, former librarian of the Institute of Marine Sciences, and Mrs. E. McCormick materially aided in obtaining literature.

The initial phases of the study were supported by the National Science Foundation under grants G-11235, GB-389, and GB-1602, and this support is gratefully acknowledged. This paper is a final report to that organization.

## HISTORICAL RÉSUMÉ

Apparently the first stomatopod to be recorded from the western Atlantic was the species now known as *Lysiosquilla scabricauda* (Lamarck). Marcgrave (1648) included a good figure of this species, which he called "Tamara guacu," in his "Historiae Rerum Naturalium Brasiliae" (see Lemos de Castro, 1962). Another of the more common western Atlantic species, *Gonodactylus oerstedii* Hansen, was first mentioned by Petiver (1712) in "Pterigraphia Americana" as *Mantis marina Barbadensis*. It was not until 1818 that the first Post-Linnaean descriptions of East American stomatopods were published. In that year Thomas Say described *Squilla empusa* from the coasts of Rhode Island and Florida. In the same year Lamarck named *Squilla glabriuscula* and *S. scabricauda;* both are now referred to *Lysiosquilla*. In 1828 Latreille described *Coronis scolopendra* from Brazil. Between 1828 and 1837 Lamarck's and Latreille's species were included in various encyclopedic works on natural history, but no new American species were described.

H. Milne-Edwards (1837) described several American stomatopods, including *Squilla armata, S. dubia,* and *S. vittata*. Few American records of stomatopods were published between 1837 and 1852, although Gibbes (1845, 1850, 1850a) listed the species deposited in American collections and described *Squilla neglecta* from South Carolina. In 1852 Dana recognized two new genera, *Lysiosquilla* and *Pseudosquilla*. In the former he placed *L. inornata,* described from Brazilian material. No East American species known at that time were placed in *Pseudosquilla*. Dana also described *Squilla prasinolineata* and *S. rubrolineata* from Brazil.

Between 1852 and 1880 little new information was added on American stomatopods. Smith (1869) recorded *Gonodactylus chiragra* and *S. scabricauda* from Brazil. Von Martens (1872) reported on the Gundlach collection of crustaceans from Cuba and included *S. rubrolineata, G. chiragra* and *Squilla stylifera*.

In 1880 Miers published the first monograph of stomatopods. Among the East American species mentioned by Miers were *Squilla dubia, S. empusa, S. neglecta, S. prasinolineata, L. scolopendra, S. glabriuscula,* and *L. scabricauda*. Miers also included records for *G. chiragra* and *Pseudosquilla ciliata*. By 1880 nine species of stomatopods had been recorded in the western Atlantic.

Brooks (1886), in his account of the CHALLENGER stomatopods, included a description of *L. excavatrix* from North Carolina, based on his own material. He also included biological notes on *S. empusa*.

R. P. Bigelow was the first carcinologist to study American stomatopods in any detail. In 1893 he published two notes describing new American species. He included descriptions of *Squilla alba* and *Lysiosquilla biminiensis* from the Bahamas in one paper, and in the other, based primarily on material collected by the ALBATROSS, he described four new western Atlantic species, *Gonodactylus (Odontodactylus) havanensis* from Cuba, *Squilla quadridens* from Florida, *S. rugosa* from the Gulf of Mexico, and *S. intermedia* from the Bahamas. In 1894 Bigelow surveyed the stomatopods deposited in the U.S. National Museum and provided illustrations for some of the species described in 1893. Although he added no new species, he recognized the subgenus, *Odontodactylus,* as a full genus. In 1895 Hansen pointed out the distinction between the western Atlantic *G. oerstedii* and the Indo-West Pacific *G. chiragra.*

Most of the references to western Atlantic stomatopods between 1894 and the present time are included in faunal papers, only a few of which need be mentioned here. Berg (1900), in the only report to include the stomatopods in the National Museum at Buenos Aires, described *L. platensis* from Argentina. Bigelow (1901) reported on the stomatopods of Puerto Rico and described *L. plumata* and *L. maiaguesensis,* two species of uncertain position, which are now placed in *Eurysquilla.* Moreira (1903), in one of several papers on the crustaceans of Brazil, described *Pseudosquilla braziliensis.* In 1917 Calman described *S. brasiliensis,* also from Brazil. He also pointed out the value of the thoracic epipods and position of the bifurcation of the median carina of the carapace as characters which can help to distinguish various species of *Squilla.* In two faunal papers, Schmitt (1924, 1924a) described *Gonodactylus curacaoensis* and *G. spinulosus* as varieties of *G. oerstedii.*

Lunz published three papers on stomatopods between 1933 and 1937; in the last he described *Chloridella edentata* from the Gulf of Mexico. Chace (1939) described *Chloridella heptacantha* from deep water off the northern coast of Cuba and in 1942 described *Odontodactylus nigricaudatus* from the Gulf of Campeche. Balss (1938) published a comprehensive survey of the Stomatopoda, and recognized the affinities of the stomatopod faunas separated by the isthmus of Panama. He listed four pairs of closely related species from the Atlantic and Pacific coasts of America. In the first complete treatment of the eastern Pacific stomatopod fauna, Schmitt (1940) described *S. hildebrandi* from Panama; although included in the Pacific report, the species is Atlantic, as the type-locality, Fort Sherman, is on the Atlantic coast. Holthuis (1941) described *S. tricarinata* from the Antilles and gave almost complete synonymies for seven American species. Chace (1954) briefly summarized the stomatopod fauna of the Gulf of Mexico, but no new species were described.

5

Three papers have recently been published on the stomatopods of Brazil. Andrade Ramos (1951) gave a preliminary survey of that fauna, and Lemos de Castro (1955) covered the area in more detail. In the latter paper, most of the known species from Brazil were illustrated and a new species, *Squilla schmitti,* was described. In 1966 Manning published a report on South American stomatopods collected by the CALYPSO; one new species, *Gonodactylus lacunatus,* was described.

Chace (1958) described a small species, *Lysiosquilla grayi,* from Massachusetts, and included a key to the eight American species of *Lysiosquilla* known at that time. Holthuis (1959) reported on the stomatopods of Surinam and described three species of *Squilla;* none of the six species treated by Holthuis had been included in Lemos de Castro's previous account of the Brazilian stomatopods, which included 13 species.

The present author has contributed several papers on the stomatopods of the northwestern Atlantic. In 1959 a list of the species from Florida and the Gulf of Mexico was compiled; 19 species were included, among which were several new records. In 1961 a deep-water species, *Lysiosquilla microps,* was described from South Florida. In the same year a report on the stomatopods from northern South America was published, in which one new genus and species, *Parasquilla meridionalis,* and two new species of *Lysiosquilla, L. antillensis* and *L. hancocki,* were described. Sixteen pairs of analagous species on either side of the isthmus of Panama were recognized. In 1962 eight new species were described, including three species of *Squilla* and four of *Lysiosquilla* in one paper, and, in another paper, a second western Atlantic species of *Parasquilla* was described. In 1963 a new *Lysiosquilla, L. insolita,* was described from Florida.

In 1963 I also published a preliminary revision of the genera *Pseudosquilla* and *Lysiosquilla* in which six new genera were recognized. The distinctness of these two genera and their relatives was emphasized. In two subsequent papers (Manning, 1967, 1968) I reviewed the family Squillidae and showed that it comprised four families, of which one, Bathysquillidae, was newly named. The genus *Squilla* was shown to be a complex of genera; two genera were removed from its synonymy and eight were described as new. The 1968 paper included keys to the families and to all genera in each family. At the present time, 62 species are known from the western Atlantic; 10 of these are described as new herein.

The western Atlantic species previously recorded in the literature are listed in chronological order in Table 1; all nomenclatural changes have been incorporated into this table.

## METHODS

*Collection of material.*—A considerable portion of the material reported herein was collected with a bottom trawl or dredge by exploratory fishing

6

## TABLE 1

A Chronological List of Stomatopods Reported from the Western Atlantic, with a Summary of Name Changes[3]

| Original record | Current name |
|---|---|
| *Squilla empusa* Say, 1818 | same |
| *Squilla glabriuscula* Lamarck, 1818 | *Lysiosquilla glabriuscula* |
| *Squilla scabricauda* Lamarck, 1818 | *Lysiosquilla scabricauda* |
| *Coronis scolopendra* Latreille, 1828 | same |
| *Squilla dubia* H. Milne-Edwards, 1837 | *Cloridopsis dubia* |
| *Squilla vittata* H. Milne-Edwards, 1837 | *Lysiosquilla glabriuscula* |
| *Squilla neglecta* Gibbes, 1850 | same |
| *Lysiosquilla inornata* Dana, 1852 | *L. scabricauda* |
| *Squilla prasinolineata* Dana, 1852 | same |
| *Squilla rubrolineata* Dana, 1852 | *Cloridopsis dubia* |
| [*Gonodactylus chiragra:* Smith, 1869] | *Gonodactylus lacunatus* |
| [*Squilla stylifera:* von Martens, 1872] | *Pseudosquilla ciliata* |
| *Lysiosquilla armata* Smith, 1881 | *Heterosquilla armata* |
| *Gonodactylus minutus* Brooks, 1886 | *G. minutus* |
| *Lysiosquilla (Coronis) excavatrix* Brooks, 1886 | *Coronis excavatrix* |
| [*Squilla armata:* A. Milne-Edwards, 1891] | *Pterygosquilla armata* |
| *Squilla alba* Bigelow, 1893 | *Alima hyalina* |
| *Lysiosquilla biminiensis* Bigelow, 1893 | *Acanthosquilla biminiensis* |
| *Gonodactylus havanensis* Bigelow, 1893 | *Odontodactylus brevirostris* |
| *Squilla quadridens* Bigelow, 1893 | *Meiosquilla quadridens* |
| *Squilla intermedia* Bigelow, 1893 | same |
| *Squilla rugosa* Bigelow, 1893 | same |
| *Gonodactylus oerstedii* Hansen, 1895 | same |
| [*Pseudosquilla oculata:* Rathbun, 1900] | same |
| *Lysiosquilla platensis* Berg, 1900 | *Heterosquilla platensis* |
| *Lysiosquilla maiaguesensis* Bigelow, 1901 | *Eurysquilla maiaguesensis* |
| *Lysiosquilla plumata* Bigelow, 1901 | *Eurysquilla plumata* |
| *Pseudosquilla braziliensis* Moreira, 1903 | *Hemisquilla braziliensis* |
| *Squilla brasiliensis* Calman, 1917 | same |
| *Gonodactylus oerstedii* var. *curacaoensis* Schmitt, 1924 | *Gonodactylus curacaoensis* |
| *Gonodactylus oerstedii* var. *spinulosus* Schmitt, 1924 | *Gonodactylus spinulosus* |
| *Chloridella edentata* Lunz, 1937 | *Squilla edentata* |
| *Chloridella rugosa* var. *pinensis* Lunz, 1937 | *Squilla rugosa* |
| *Chloridella heptacantha* Chace, 1939 | *Squilla heptacantha* |
| *Squilla hildebrandi* Schmitt, 1940 | *Alima hieroglyphica* |
| *Squilla tricarinata* Holthuis, 1941 | *Meiosquilla tricarinata* |
| *Odontodactylus nigricaudatus* Chace, 1942 | *Odontodactylus brevirostris* |
| *Alima lebouri* Gurney, 1946 | *Meiosquilla lebouri* |
| *Squilla schmitti* Lemos de Castro, 1955 | *Meiosquilla schmitti* |
| *Lysiosquilla grayi* Chace, 1958 | *Nannosquilla grayi* |
| [*Lysiosquilla polydactyla:* Chace, 1958] | *Heterosquilla polydactyla* |
| *Squilla lijdingi* Holthuis, 1959 | same |
| *Squilla obtusa* Holthuis, 1959 | same |
| *Squilla surinamica* Holthuis, 1959 | same |
| *Lysiosquilla microps* Manning, 1961 | *Bathysquilla microps* |
| *Lysiosquilla antillensis* Manning, 1961 | *Nannosquilla antillensis* |

[3]Brackets indicate extra-limital species recorded for the first time from the western Atlantic Ocean.

| | |
|---|---|
| *Lysiosquilla hancocki* Manning, 1961 | *Nannosquilla hancocki* |
| *Parasquilla meridionalis* Manning, 1961 | same |
| *Parasquilla coccinea* Manning, 1962 | same |
| *Squilla chydaea* Manning, 1962 | same |
| *Squilla discors* Manning, 1962 | same |
| *Squilla randalli* Manning, 1962 | *Meiosquilla randalli* |
| *Lysiosquilla campechiensis* Manning, 1962 | same |
| *Lysiosquilla floridensis* Manning, 1962 | *Acanthosquilla floridensis* |
| *Lysiosquilla enodis* Manning, 1962 | *Platysquilla enodis* |
| *Lysiosquilla schmitti* Manning, 1962 | *Nannosquilla schmitti* |
| *Lysiosquilla insolita* Manning, 1963 | *Heterosquilla insolita* |
| *Gonodactylus lacunatus* Manning, 1966 | same |

vessels, including the ALBATROSS, FISH HAWK, SILVER BAY, OREGON, and COQUETTE. The otter trawl, of virtually any size, is an extremely useful piece of gear for obtaining stomatopods and other bottom-dwelling invertebrates. This study would be much less complete without the material taken by these fishing vessels. Stomatopods occasionally occur in dredge hauls, but dredges usually yield fewer specimens, most of which are damaged. Dredges are perhaps most efficient over a rough substrate which provides burrows for such forms as *Gonodactylus.*

In shallow water a pushnet can be used for collecting some species of *Squilla* and *Pseudosquilla.* Night stations are particularly productive. Designs for two types of pushnets are found in papers by Allen & Inglis (1958) and Strawn (1954). Hand collecting in shallow tropical waters will usually yield specimens of *Gonodactylus:* beds of *Porites,* coralline algae clumps, *Phragmatopoma* clumps, large, porous pieces of coralline rock, and sponges are all suitable habitats for species of *Gonodactylus.*

Brooks (1886) has described how the secretive *Coronis excavatrix* can be enticed with a small piece of fish to the mouth of its burrow where its retreat can be cut off with a trowel. Dahl (1954) noted that *Nannosquilla chilensis* (Dahl) was collected by digging and sifting sand. Most species are cryptic, and commercial poisons have proved very effective as a collecting method. Manning (1960) described the use of one such poison for collecting crustaceans in general. The poisons are perhaps most effective in the coral reef habitat of the tropics, where the reef provides many inaccessible hiding places. Use of poison at one station in Florida waters has yielded six new records in a five-year period.

*Identification.*—Identification of stomatopod genera is based on many features, including qualitative characters, such as general body shape, whether the body is depressed and loosely articulated (as in *Lysiosquilla*) or compressed and compact (as in *Pseudosquilla* and *Pseudosquillopsis*). The presence of body carination, a characteristic of *Squilla,* and its extent is also important. Meristic features, including numbers of teeth on the raptorial claw, number of epipods, and number of denticles on the telson, are very important for identification of species. Qualitative characters, such as shape of eye, rostral plate, and lateral processes of the exposed thoracic somites are no less important. Finally, color patterns are diagnostic in several genera. Species of *Pseudosquilla* can be distinguished on the basis of color alone. In some species of *Squilla,* color patterns may be visible 100 years after specimens were preserved. The most important characters which vary from family to family are discussed separately under the introductory remarks for each family.

As in many invertebrate groups, stomatopods are rich in differentiating characters. There are so many characters that can be used at the species level that there is usually no need for detailed statistical analysis to show distinctness of species. Most of the recognized species can be differentiated immediately by the use of one or two characters in combination with color pattern. However, future work on the species problem in *Gonodactylus* may well have to be accompanied by detailed statistical analyses. There are very few characters that can be used to distinguish species in that genus, and most of the few that are available are qualitative features of telson morphology.

*Measurements.*—All measurements are in millimeters (mm).

Total Length (TL) is measured along the dorsal midline, from the anterior margin of the rostral plate to a line between the apices of the submedian teeth of the telson. Total length, which varies considerably according to the relative contraction or expansion of the body at preservation, can only be used as a gross estimate of overall size; measurements of total length may vary as much as 5 percent on the same specimen.

Carapace Length (CL) is measured on the midline of the carapace, from the base of the movable rostral plate to the posterior median margin. The carapace length can be measured with accuracy and should be used as a base for indices or ratios.

Cornea Width is the greatest width of the cornea. This is not a difficult measurement to make on species in which the cornea is bilobed. In some species of the Gonodactylidae, particularly in the genera *Hemisquilla* and *Odontodactylus,* the cornea is subglobular. All of these forms have a line of ocelli dividing the cornea into two portions; in such cases the width of the cornea is measured at right angles to the line of cells.

Eye Length is measured from the base of the stalk to the distal margin of the cornea. In those species with a bilobed cornea the measurement is made to the division of the lobes. In those species with a subglobular cornea the measurement is made from the base of the stalk to the anteriormost edge of the longitudinal line dividing the cornea.

Cornea Length has been measured in species of *Hemisquilla* and *Odontodactylus.* In some species of these genera the Cornea Width-Cornea Length relationship is an important character. Cornea Length is the greatest length of the cornea measured along the line of cells dividing the cornea into two portions.

Rostral Plate Length is always measured on the midline. Rostral Plate Width is measured at the widest portion of the plate, whether it is basal or in advance of the base as in many species with a cordiform plate.

Antennal Scale Length is measured from its articulation to its distal margin; Antennal Scale Width is the greatest width. These measurements have been used only in *Lysiosquilla*.

The length of the propodus of the claw (Propodus Length) is measured on its pectinate (superior) margin, from the edge articulating with the carpus to the distal edge with the dactylus folded. The depth of the propodus is its greatest depth, measured across the longitudinal axis of the segment. Measurements of the propodus have been used only in *Lysiosquilla* and *Pseudosquilla*.

The width of the abdomen is always measured at the fifth abdominal somite, on a line just anterior to the articulation of the uropods.

Telson Width is the greatest width. Telson Length is measured two ways: in those species in which the submedian teeth are long and movable, as in *Pseudosquilla*, length is measured from the anterior margin to the bases of the movable submedian teeth; in other forms Telson Length is measured to a line drawn between the apices of the submedian teeth.

Measurements of the uropodal exopod are taken on the midline of each segment; these measurements have been used only in *Odontodactylus* and *Pseudosquilla*.

The length of the uropodal endopod is measured from the inner, proximal angle or fold to the distal edge; width of the endopod is the greatest width. Measurements of the endopod are used primarily in *Lysiosquilla*.

*Indices.*—Several indices have been found to be of value in species recognition, and the following indices have been employed here to different degrees in different taxa:

Abdominal Width-Carapace Length Index (AWCLI): Calculated by dividing the abdominal width by the carapace length and multiplying by 100. This can be used in *Pseudosquilla* and *Gonodactylus* as an expression of the relative slenderness of the body.

Corneal Index (CI): This is obtained by dividing the carapace length by the cornea width and multiplying by 100. This index is particularly useful in distinguishing some species in different genera of the Squillidae, including *Squilla* and *Oratosquilla*. The corneal index used by Kemp (1913) was a simple ratio. The index changes with age, for the eye is comparatively larger in small specimens, so that the index can be compared only in specimens of similar size.

Cornea Length-Width Index (CLWI): This is obtained by dividing the cornea length by the cornea width and multiplying by 100. It is a good expression of the shape of the cornea in forms such as *Hemisquilla* with a subglobular cornea.

Eye Length-Width Index (ELWI): This is obtained by dividing the eye length by the cornea width and multiplying by 100. It can be used to express the relative inflation of the cornea in some species of *Pseudosquilla*.

Propodal Index (PI): This is obtained by dividing the carapace length by the propodus length and multiplying by 100. It has been used in *Pseudosquilla* and *Lysiosquilla* as an expression of the relative size of the raptorial claw.

Propodus Length-Depth Index (PLDI): This is obtained by dividing the propodus length by the propodus depth and multiplying by 100. This has been used in *Pseudosquilla* as an expression of the slenderness of the propodus.

Three simple ratios have also been used, two in the course of this study. In *Lysiosquilla* a length-width ratio of the antennal scale and the uropodal endopod have been used herein. In another paper (Manning, 1966a), the telson width-median carina height ratio was used to show sexual dimorphism in the shape of the median carina of the telson in one species of *Odontodactylus*.

*Terminology.*—Most of the technical terms used herein have been in common use by students of the stomatopods. Some of the less common terms and methods of expressing meristic features are listed below. Generalized terms used in most species accounts are shown in Figure 1; more specialized terms are used only for some families or genera and these are discussed separately in the introduction to each of those taxa.

Throughout the descriptive portion of the text arabic numerals are used for meristic features.

One of the characteristic features of the stomatopods is the presence of two movable body segments or *somites* anterior to the carapace. The anteriormost is the *ophthalmic somite,* bearing the paired, stalked eyes. On the dorsal surface of the ophthalmic somite are paired, erect processes, the *ocular scales.* Their shape, position, and whether or not they are fused are important characters.

The antennular somite, the dorsal portion of which is usually covered by the rostral plate, bears the paired, three-segmented *antennules* and, dorsolaterally, the *antennular processes.* The latter may be blunt lobes or slender spines. The antennules terminate in three flagella, with the ventral two fused basally.

The *antennal peduncle* (endopod) and *antennal scale* (exopod) are situated at the anterolateral portion of the carapace. The peduncle is two-segmented and provided with a distal flagellum; there are no important characters to be found on the peduncle. The shape and setation of the scale can be important. In the Lysiosquillidae the *antennal protopod* may

11

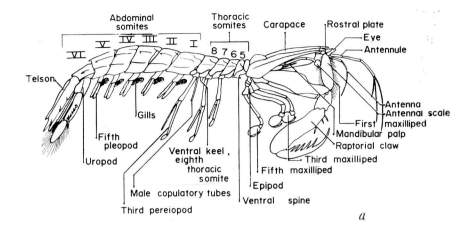

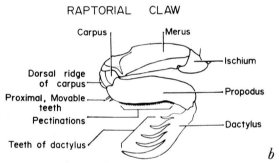

FIGURE 1. Diagrammatic sketches of *a,* a squillid, and *b,* a raptorial claw, illustrating terms used in the descriptive accounts.

be provided with *dorsal, mesial,* or *ventral papillae;* in that family the number and position of papillae are important generic characters.

The *rostral plate* is the median articulated plate on the carapace. Its shape, spination, and carination are often important characters.

The carapace is a dorsal cuticular fold covering the remainder of the cephalic as well as the first thoracic somites. The carapace typically bears three grooves, a transverse *cervical groove* on the posterior third, and parallel, longitudinal *gastric grooves* separating the median portion from the lateral plates. The carapace may be variously carinate or smooth; terminology of the carinae is illustrated under the account of the Squillidae. The *anterolateral angles* are armed in many species, unarmed in all of the Lysiosquillidae. In most species of Squillidae there is a *median carina* on the carapace, which often bifurcates anteriorly. All species have a median pore or *dorsal pit,* and it is conspicuous in those species with a median carina. It lies between the cervical groove and the *anterior bifurcation of the median carina.*

There are eight pairs of thoracic appendages, of which the first five are uniramous *maxillipeds* and the last three are biramous *walking legs.* The maxillipeds consist of eight segments, *precoxa* and *coxa* (fused), *basis, ischium, merus, carpus, propodus,* and *dactylus;* all maxillipeds are subchelate, with the dactylus folded upon the propodus. The second maxilliped is the largest of the five, and is the *claw* or *raptorial claw.* The *dactylus* is often provided with slender, serrate teeth; the count of these teeth given in the descriptions always includes the terminal tooth. In some species the dactylus is unarmed. The *outer* or *inferior margin* of the dactylus may be notched or swollen basally, evenly rounded or sinuous. The *upper margin* or *pectinate margin* of the propodus bears a longitudinal groove, often slotted, into which the dactylus fits when the claw is folded. The outer edge of this groove is completely or partially pectinate or, in the genera *Harpiosquilla* and *Bathysquilla,* is provided with erect spines of varying length. The inner proximal edge of this groove is usually provided with three or four movable teeth. The distal edge of the propodus may be armed with a slender spine. The carpus is a short segment with a prominent *dorsal ridge;* the ridge may be even, undivided and terminate bluntly or may be ornamented with erect tubercles and terminate in one or more spines. The merus is a stout, large segment, with a partial *(Gonodactylus* and *Odontodactylus)* or complete (remainder of genera) groove for reception of the propodus. In some species the *inferodistal angle* of the outer surface of the merus is produced into a spine.

The merus and ischium of the claw are often inconspicuous; in some species of *Squilla* the merus is armed with a ventrally-projecting spine. The *ischiomeral articulation* is usually terminal, but it is subterminal in *Gonodactylus* and *Odontodactylus.*

The last three maxillipeds appear to exhibit no important specific characters, although the shape of the propodus, whether broadened or elongate, is a family character. In the Lysiosquillidae the propodi of the third and fourth thoracic appendages are beaded ventrally. In all stomatopods the propodus of the last thoracic appendage is provided with a stiff brush of bristles which is used to clean the body.

Each of the maxillipeds may be provided with a coxal appendage, the *epipod.*

The mandible is often provided with a three-segmented *mandibular palp.* The palp, visible at the base of the mandible when the claw is deflexed, may be two-segmented or absent.

The last four thoracic somites are visible behind the carapace in dorsal view and are referred to as the *exposed thoracic somites.* These may be provided with dorsal carinae (although the fifth somite is rarely carinate)

13

and variously-shaped *lateral processes*. These lateral processes provide important generic and specific characters. A pair of ventrolateral spines may be present on the fifth thoracic somite, and the lateral process of the fifth somite, when present, may be single or bilobed. The last three thoracic somites each bear one pair of *walking legs,* with a one-or two-segmented endopod and a two-segmented exopod. The shape and number of segments on the endopod often is used as a generic or specific feature. The presence or absence and shape of the *median keel* on the eighth thoracic somite is also a specific character.

Males can be distinguished by the presence of paired *copulatory tubes* arising from the base of the last walking legs. In general, the buds of the male appendages are visible in first stage juveniles. In females the genital openings are paired submedian structures on the ventral surface of the sixth thoracic somite.

There are six flattened abdominal somites which in many species are armed with longitudinal carinae; usually eight are present, paired *submedian, intermediate, lateral,* and *marginal carinae.* In many species of the Squillidae the carinae are armed posteriorly; this armature is expressed in the following formula: submedian, 5-6; intermediate, (1-2) 3-6; lateral, 1-6; marginal, 1-5. This shows that the submedian carinae of the last two somites are armed and the intermediate carinae of the third to sixth, occasionally first and second, are armed, etc. The spine formula refers only to adults; one of the characteristics of juveniles is the reduction of the number of spines. Most species have paired *anterolateral plates* on the abdomen; these plates are usually articulated. The carination of the last abdominal somite may be reduced; on those species of Squillidae with eight carinae on the first five somites, there are usually only six carinae on the last, the marginals being absent.

The appendages of the first five abdominal somites are the biramous *pleopods* which bear the gills. The endopod of the first pleopod of the male undergoes secondary sexual modification; there does not seem to be any diagnostic value in the shape of this organ, but Ingle (1963) believes it may be of value at the generic level. It has not been used as a diagnostic character in this study.

The paired appendages of the sixth abdominal somite are the *uropods,* with a one-jointed protopod or *basal segment,* a two-segmented exopod, and a one-segmented endopod. The *basal prolongation of the uropod* is a flattened, usually bifurcate ventral projection affording several characters at the generic and specific levels. The proximal segment of the exopod is provided with a series of graded, movable spines of various shapes. The shape and setation of the endopod is an important character, particularly in the Gonodactylidae and Lysiosquillidae.

The *telson* is the flattened terminal body segment. A *dorsal median carina* or crest is present except in the Lysiosquillidae; other carinae and projections may be present. In families other than the Lysiosquillidae there

14

are three or four pairs of marginal teeth, *submedians,* which may have movable apices, *intermediates, laterals,* and *marginals.* Only in the Bathysquillidae are all of the marginal teeth provided with movable apices. Marginal *denticles* are present between the marginal teeth. A *denticular formula* of 3-4, (5) 6-10, 1 indicates that on each side of the midline there are three to four submedian denticles, six to ten (rarely five to ten) intermediate denticles, and one lateral denticle. In the Gonodactylidae there are never more than two intermediate denticles; in the Squillidae there are always four or more. The ventral surface of the telson is smooth but may be provided with spines or carinae. The *anal pore* is situated on the ventral surface, and a *postanal carina* may be present.

*Organization of the text.*—Genera appear in the text in the order in which they fall out of the key. A detailed definition is provided for each of the genera occurring in the western Atlantic. In general, characters outlined in the generic definition are not repeated in the species description.

Most of the subheadings used in the species accounts are self-explanatory. The synonymies are complete to the extent that all references that have come to my attention are included; some of the entries are accompanied by my comments in brackets. A question mark next to an entry in the synonymy indicates that that record should be regarded with caution, pending examination of the material reported therein. If the synonymy is incomplete, as in cases of the species which occur also in other geographic areas, it will be preceded by the notation "Restricted Synonymy," which indicates that only original references (for extra-limital species) or western Atlantic references are included.

The section entitled "Previous Records" summarizes distribution records in the literature. Entries in this section which are accompanied by a question mark need to be verified by reëxamination of the material; in some cases more than one species is presumed to have been recorded by the author in question.

In the sections on "Previous Records" and "Material" the entries are from north to south, beginning at Bermuda, Maine through Georgia, the Bahamas, Florida, through the Gulf of Mexico and Caribbean Sea to Panama, then the Antilles from north to south, and South America from Colombia to Argentina. In general, a glance at the beginning and end of these sections will indicate to the reader the northern and southern distributional limits of the species.

The number following the number of specimens in the section on "Material" is usually total length (TL); occasionally carapace length (CL) is given. Localities, except for those marked with a question mark, have been verified on one of several major atlases. The National Geographic Atlas of the World (1963 edition) was used to verify most localities. Localities for specimens collected by the ALBATROSS, COMBAT, COQUETTE, FISH HAWK, OREGON, and SILVER BAY are general; station data for each of these collections have been assembled in an appendix (p. 341). In four cases specimens are listed from broad, general localities; these are: (1) Long Island Sound, (2) Straits of Florida, (3) Tortugas shrimp grounds (the commercial shrimping grounds between Key West and Dry Tortugas, Florida), and (4) off the Mississippi Delta. All depths are in meters (fathoms x 1.82).

In most cases all pertinent data relating to the specimens are included. In the case of *Squilla empusa, Pseudosquilla ciliata, Gonodactylus oerstedii,* and *G. bredini,* the volume of material has been so great that only number of specimens, general locality, and repository have been given.

All stomatopods collected by the Smithsonian-Hartford Expedition of 1937 and the Smithsonian-Bredin Expeditions to the Caribbean in 1956, 1958, 1959, and 1960 (Bredin Sta.) are included herein. In general, station data for material from these expeditions has been abbreviated. Full data for these stations is on file in the Division of Crustacea, U. S. National Museum.

The initials at the end of each entry in the material section indicate the repository for that lot. The following abbreviations have been used:

AHF—Allan Hancock Foundation, Los Angeles.
ANSP—Academy of Natural Sciences, Philadelphia.
BMNH—British Museum (Natural History).
BPBM—Bernice P. Bishop Museum, Honolulu.
CHML—Cape Haze Marine Laboratory, Sarasota, Florida. (now Mote Marine Laboratory).
CPP—Centro de Pesquisas de Pesca, Florianopolis, Brazil.
DUML—Duke University Marine Laboratory, Beaufort, N. Carolina.
FSBCML—Florida State Board of Conservation Marine Laboratory, St. Petersburg, Florida.
GCRL—Gulf Coast Research Laboratory, Ocean Springs, Mississippi.
IM—Indian Museum (Zoological Survey of India), Calcutta.
IMB—Institute of Marine Biology, University of Puerto Rico, Mayaguez.
IOSP—Instituto Oceanografico, Universidade de São Paulo, Brazil.
ITV—Instituto Tecnólogico de Veracruz, Mexico.
LB—Instituto Nacional de Investigaciones Biologico-Pesqueras, Mexico City.
MCSN—Museo Civico di Storia Naturale, Genoa.
MCZ—Museum of Comparative Zoology, Harvard.
MMF—Museu Municipal do Funchal, Madeira.
MNRJ—Museu Nacional, Rio de Janeiro.
MNHNP—Muséum National d'Histoire Naturelle, Paris.
RMNH—Rijksmuseum van Natuurlijke Historie, Leiden.
SEP—Systematics-Ecology Program, Marine Biological Laboratory, Woods Hole, Massachusetts.
SMIJ—Science Museum, Institute of Jamaica.

16

SRF—Sapelo Island Research Foundation, Sapelo Island, Georgia.
UCVMB—Museo de Biologia, Universidad Central de Venezuela, Caracas.
UHML—Marine Laboratory, University of Hawaii.
UMML—Reference collection, Institute of Marine Sciences, University of Miami.
USNM—U. S. National Museum.
USP—Universidade de São Paulo, Brazil.
UZM—Universitetets Zoologiske Museum, Copenhagen.
VMM—Vanderbilt Marine Museum, Huntington, L. I., N. Y.
YPM—Peabody Museum, Yale University, New Haven.
ZMA—Zoological Museum, Amsterdam.
ZMB—Zoological Museum, Berlin.

A brief diagnosis and a more detailed description are usually given for each species. The diagnoses refer primarily to features that will distinguish the species under discussion from other species in the same genus. In those cases where most of the body features are found in all species, particularly in *Coronis, Gonodactylus,* and *Pseudosquilla,* only a diagnosis is given. More detailed descriptions are provided for the remainder of the species. The diagnoses and descriptions are composites and include observed variation in major characters of each species. Both descriptive accounts refer to features of adults; major differences between adults and juveniles are also noted.

Notes given in the section on "Color" usually refer to color in alcohol. Where available, observations on color in life are presented. The size range given under "Size" is based on material examined. The measurements in that section should provide a general idea of the body proportions of the species.

The "Discussion" section includes comments on nomenclature, the major characteristics of the species being treated, and comparisons with similar species. Comments of a more general nature are given under "Remarks." The paragraph on "Ontogeny" gives only available references to the larval forms of each species. The paragraphs on "Type" and "Type-locality" are self-explanatory. The final paragraph in each species account, "Distribution," summarizes the known vertical and horizontal distribution of the species.

## ORDER STOMATOPODA Latreille

Stomapodes Latreille, 1817, p. 40.
Stomatopoda Voigt, in Cuvier, 1836, p. 188 [correction of Stomapodes].—Balss, 1938, p. 1.

*Remarks.*—Until recent studies by myself and others the Order Stomatopoda was considered to comprise the single family, Squillidae Latreille; fossil species were assigned to that family or to the exclusively fossil family, Sculdidae Dames. Manning (1967, 1968) pointed out that evidence

17

derived from the morphology of adult and larval stomatopods supported the recognition of four recent families, Bathysquillidae Manning, Gonodactylidae Giesbrecht, Lysiosquillidae Giesbrecht, and Squillidae. These families can be distinguished according to the key given below; representatives of each family occur in the western Atlantic. The reader is referred to Balss (1938), Holthuis & Manning (in press), and Manning (1968) for background on the order.

## KEY TO RECENT FAMILIES OF STOMATOPODA

1. Telson lacking sharp median carina; propodi of last 3 maxillipeds broad, beaded or ribbed ventrally ............... Lysiosquillidae, p. 18.
1. Telson with sharp median carina; propodi of last 3 maxillipeds slender, not beaded or ribbed ventrally ............................. 2

2. All marginal teeth of telson with movable apices..................
..................................... Bathysquillidae, p. 94.
2. At most submedian marginal teeth with movable apices .......... 3

3. More than 4 intermediate denticles present on telson.............
........................................ Squillidae, p. 99.
3. No more than 2 intermediate denticles on telson.................
................................. Gonodactylidae, p. 239.

## Family LYSIOSQUILLIDAE Giesbrecht, 1910

Erichthidae White, 1847, p. 82 [larvae; type-genus suppressed by ICZN].
Lysiosquillinae Giesbrecht, 1910, p. 148.—Manning, 1963a, p. 324.
Lysiosquillidae Manning, 1967, p. 238 [key]; 1968, p. 109 [discussion and references].

*Definition.*—Propodi of third and fourth thoracic appendages as broad as or broader than long and beaded ventrally, propodus of fourth almost twice as broad as that of fifth appendage; telson without distinct median carina, number of marginal teeth variable; submedian teeth of telson usually movable, submedian denticles present.

*Type-genus.*—*Lysiosquilla* Dana, 1852.

*Number of genera.*—Nine. The seven American genera can be distinguished by the key given below.

*Ontogeny.*—Early larva an antizoea, with biramous appendages on first 5 thoracic somites; abdomen partially segmented, without appendages; later larva an erichthus, with 1 intermediate denticle on the telson; pleopods appear in succession from front to back.

*Remarks.*—There are several distinctive features that will help to recognize members of this family in addition to the main features cited above. In general the body is loosely articulated and the exoskeleton is "softer" than in the members of the Squillidae and the Gonodactylidae; there are rarely

18

any longitudinal carinae on the body, never any on the carapace. All of the species that I have examined have the propodus of the third and fourth maxillipeds strongly beaded or ribbed ventrally (inferiorly) and all have more setae on the outer margin of the distal segment of the uropodal exopod than on the inner.

The close relationship of the genera included here was pointed out by Manning (1963a), who surveyed previous ideas of relationship based upon both the adults and the larvae. It must be pointed out that our knowledge of the larvae of this family is still fragmentary, for no species has been hatched from the egg and the later larvae of only five species, two in *Lysiosquilla* and three in *Acanthosquilla,* have been studied (Alikunhi, 1952). The characters cited for the larvae are given only to show that, as far as we know, the distinctness of this family, based primarily on the structure of the adults, can be shown even for larvae. Additional work on the larvae would be most interesting.

Several characters were used by Manning (1963a) in his preliminary characterization of the genera of the Lysiosquillidae, and several others were mentioned but not stressed. These features, which have not generally been used in the past, are of importance in recognizing genera and species:

1. Presence and position of antennal papillae. These papillae are totally lacking in *Nannosquilla,* present in varying numbers in other genera. Up to four papillae may be found on the inner and ventral surfaces of the antennal protopod. Schmitt (1940) was among the first to point out their importance.

2. Mandibular palp. In most genera the presence or absence of a palp is a specific character, but it is found in none of the species of *Nannosquilla.*

3. Epipods. The epipod of the last maxilliped is lacking in *Nannosquilla* and in one species of *Austrosquilla,* but is usually present in the other genera.

4. Shape of propodi of last three thoracic appendages. This seems to be a familial rather than a specific or generic character.

5. Endopods of walking legs. These are apparently two-segmented in all members of the family. *Lysiosquilla, Heterosquilla,* and *Coronida* have elongate distal sements; the remainder of the genera have ovate or subcircular distal segments. See also notes below.

6. Ventral keel on eighth thoracic somite. *Lysiosquilla* has a prominent longitudinal keel on the last thoracic somite which may even be spined posteriorly in some species. Most species of *Heterosquilla* also have a keel, not as prominent as in any *Lysiosquilla.* In *Coronis, Platysquilla,* and *Hadrosquilla* the keel is reduced to a low rounded prominence, and it is lacking in *Nannosquilla* and *Acanthosquilla.*

7. Uropod structure. Two general types can be found in the Lysiosquillidae. The first, characterized by the presence of a strong fold on the outer, proximal margin of the uropodal endopod, is found in *Coronis,*

19

*Platysquilla, Nannosquilla, Hadrosquilla, Austrosquilla,* and *Acantho-squilla.* The second, lacking the strong fold on the uropodal endopod, is found in *Lysiosquilla, Heterosquilla* and *Coronida.*

The last feature, structure of the uropod, seems to be linked with two others, the shape of the endopod of the walking legs and the size and shape of the spine or process of the sixth abdominal somite which overhangs the articulation of the uropod. In general, in those genera which have the strong fold on the endopod of the uropod, the endopod of each walking leg is broad, with the distal segment ovate or subcircular, and there is a conspicuous, ventrally-directed process on the sixth abdominal somite which overhangs the articulation of the uropod. In the forms lacking the broad fold, the endopod of the walking legs is elongate, even strap-shaped, and there is at most a small lateral spine in front of the articulation of the uropod. These features still need to be verified in some species of *Coronida* and *Heterosquilla.*

The exact position of *Coronida* is not clear. There can be little doubt of its close relationship to the other members of the Lysiosquillidae, but none of the other genera in this family have the inflated dactylus of the raptorial claw.

There is a wide range of variation throughout the genera placed in the family; Chace (1958) commented on the telson structure of the western Atlantic species then placed in *Lysiosquilla.* In *Lysiosquilla s.s.* the marginal armature of the telson is typically masked by fusion of the submedian denticles and suppression of the movable submedian teeth; movable submedians are present in only one species, *L. aulacorhynchus.* The dorsal surface of the telson of *Lysiosquilla* is usually smooth, with a low triangular median boss, but in some species there may be dorsal and lateral spinules, granules, and pits. *Heterosquilla* includes a variety of species, now assigned to two subgenera, both of which may have to be recognized as distinct genera. The basic telson pattern is similar to that of *Lysiosquilla,* but the median boss is raised and usually spined or lobed posteriorly, sometimes carinate laterally, and there may be, on either side of the midline, lateral carinae or rows of spines. The marginal armature consists of a series of small, distinct submedian denticles, a movable submedian tooth, two or four fixed intermediate denticles, a fixed intermediate tooth, a lateral denticle, and a lateral tooth. In *Coronis* and *Platysquilla* the telson is smooth dorsally, with a raised triangular or rounded median boss, above the marginal amature; movable submedian marginal teeth are present in both genera, but *Coronis* has but one pair of fixed teeth lateral to the sub-medians, *Platysquilla* has four. *Acanthosquilla* can be recognized by the fan-shaped series of five or more dorsal teeth on the telson, *Austrosquilla* by the single dorsal projection, and *Nannosquilla* and *Hadrosquilla* by the unadorned dorsal surface which is produced into a false eave overhanging the true marginal armature. The telson of *Coronida* is very ornate, with many spines, tubercles, or carinae. The marginal armature is difficult to characterize, but movable submedian marginal teeth are present.

In the descriptions given below, the arrangement of marginal teeth and denticles is given for each side of the telson. The fusion and suppression of teeth and denticles in different genera limits the use of a fixed nomenclature for the marginal armature.

### KEY TO AMERICAN GENERA OF THE FAMILY LYSIOSQUILLIDAE

1. Distal segment of endopod of first 2 walking legs elongate; proximal portion of outer margin of uropodal endopod at most angled inward, not folded . . . . . . . . . . . . . . . . . . . . . . . . . . . . . . . . . . . . . . . . . . 2
1. Distal segment of endopod of first 2 walking legs ovate or subcircular; proximal portion of outer margin of uropodal endopod folded over . . 4

2. Dactylus of raptorial claw inflated basally; propodus of claw pectinate proximally only; rostral plate rounded or subrectangular. . . . . . . . . . . . .
 . . . . . . . . . . . . . . . . . . . . . . . . [*Coronida* Brooks, 1886; East Pacific].
2. Dactylus of raptorial claw not inflated basally; propodus fully pectinate; rostral plate cordiform or triangular . . . . . . . . . . . . . . . . . . . . . . . . . 3

3. Median dorsal surface of telson with at most a low triangular boss; marginal teeth of telson usually fused, movable submedian teeth rarely present . . . . . . . . . . . . . . . . . . . . . . . . . . . . . . . . . *Lysiosquilla,* p. 21.
3. Median dorsal surface of telson with raised median projection, lobed or spined posteriorly; movable submedian marginal teeth of telson always present, remainder of teeth and denticles distinct, not fused . . . . . . . . . . . . . . . . . . . . . . . . . . . . . . . . *Heterosquilla,* p. 42.

4. Dorsal surface of telson with fan-shaped series of 5 or more spines. . .
 . . . . . . . . . . . . . . . . . . . . . . . . . . . . . . . . . . . . *Acanthosquilla,* p. 61.
4. Dorsal surface of telson unarmed or with at most a single median projection . . . . . . . . . . . . . . . . . . . . . . . . . . . . . . . . . . . . . . . . . . . . . 5

5. Posterior margin of dorsal surface of telson produced into false eave overhanging true posterior margin . . . . . . . . . . . *Nannosquilla,* p. 69.
5. Posterior margin of telson with single median projection, not produced into a false eave . . . . . . . . . . . . . . . . . . . . . . . . . . . . . . . . . . . . . . . 6

6. Mandibular palp present; telson with 1 pair of fixed marginal teeth . . . . . . . . . . . . . . . . . . . . . . . . . . . . . . . . . . . . . . . . *Coronis,* p. 83.
6. Mandibular palp absent; telson with 4 pairs of fixed marginal teeth . . . . . . . . . . . . . . . . . . . . . . . . . . . . . . . . . . . . . . *Platysquilla,* p. 90.

### Genus *Lysiosquilla* Dana, 1852

*Erichthus* Latreille, 1817, p. 43.—Holthuis, 1951a, p. 83.—Hemming, 1954, p. 145 [suppressed by ICZN; name on Official Index, no. 121].
*Lysiosquilla* Dana, 1852, p. 615.—Balss, 1938, p. 129.—Hemming, 1954, p. 145.—Manning, 1963a, p. 317; 1968, p. 110 [key] [ICZN Official List no. 730].

21

*Definition.*—Size large, TL to 300 mm or more; body smooth, depressed, loosely articulated; eyes large, T-shaped, cornea bilobed, set obliquely on stalk; rostral plate cordiform, usually with short anterior median carina; antennal protopod with 1 mesial and 2 ventral papillae; carapace narrowed anteriorly, without carinae or spines, cervical groove indicated on lateral plates only; thoracic somites without dorsal carinae, lateral margins rounded; eighth thoracic somite with prominent median ventral keel; 5 epipods present; mandibular palp present; propodus of fifth maxilliped as broad as long, with ventral brush of setae; raptorial claw slender, large, dactylus armed with 5 or more teeth; outer margin of dactylus with faint basal notch; propodus fully pectinate, with 4 movable spines at base, proximal largest, distal smallest; dorsal ridge of carpus terminating in strong tooth; ischiomeral articulation terminal, merus about 2 times as long as ischium; endopods of walking legs two-segmented, strap-shaped; abdomen depressed, without carinae, loosely articulated, anterolateral plates obscurely articulated; last 2 abdominal somites with dorsal and posterior spinules in some species; sixth somite without strong postero-lateral spines but with ventrolateral spine in front of each uropod; telson broad, smooth or ornamented with dorsal spines, tubercles, or corruga-tions; dorsal surface with low, triangular median boss, low submedian bosses occasionally present; marginal teeth and denticles usually fused, movable submedian teeth discernible in some species only; basal segment of uropod with inner and outer carina, inner terminating in distal tooth, dorsal surface otherwise smooth or with erect spinules; distal segment of exopod longer than proximal; proximal segment with small, spatulate spines on outer margin, inner margin setose; endopod elongate or ovate, proximal portion of outer margin at most angled inward; spines of basal prolongation slender, triangular in cross-section, inner much the longer.

*Type-species.*—*Lysiosquilla inornata* Dana, 1852, by subsequent designa-by Fowler, 1912, p. 539.

*Gender.*—Feminine.

*Number of species.*—Nine, of which three occur in the western Atlantic. A list of the species was given by Manning (1963a). *L. hoevenii* (Herk-lots, 1851), from West Africa, here tentatively synonymized with *L. scabricauda,* may prove to be a distinct species.

*Nomenclature.*—Five of the names cited as generic synonyms by Holthuis & Manning (in press) *Erichthus, Smerdis, Pontiobius, Erichthoidina,* and *Lysioerichthus,* were originally used for larval forms. Two of the names, erichthus and lysioerichthus, are in current use as category names for stomatopod larvae. Only some of the larvae for which the names were used have been definitely identified with an adult. Retention of *Erichthus* and *Smerdis* would result in suppression of the well-known *Lysiosquilla,* which has been in use since 1852 for the characteristic, large species now placed in it. *Erichthus* has been suppressed by the International Commission on Zoological Nomenclature (ICZN) and is no. 121 on the Official Index. Suppression of *Smerdis* has been requested (Holthuis & Manning, 1964). *Lysiosquilla* is name no. 730 on the ICZN Official List.

*Remarks.*—Manning (1963a) restricted this genus to eight species and one subspecies. Four of the species now placed here, *L. maculata* (Fabricius, 1793), *L. tredecimdentata* Holthuis, 1941, *L. capensis* Hansen, 1895, and *L. sulcirostris* Kemp, 1913, are found primarily, if not exclusively, in the Indo-West Pacific. *L. maculata* and *L. sulcirostris* have been reported from West Africa but these records need to be verified. *L. aulacorhynchus* Cadenat, 1957 is known from West Africa, where *L. maculata* and *L. scabricauda* (Lamarck, 1818) may also occur. *L. desaussurei* Stimpson, 1857 is a West American species; *L. maculata* also has been reported in the eastern Pacific. *L. glabriuscula* (Lamarck, 1818), *L. scabricauda,* and *L. campechiensis* Manning are all found in the western Atlantic; the first two have extensive ranges, the last is not known outside of the Gulf of Campeche.

Examination of the large series of western Atlantic specimens of *Lysiosquilla* for this study has shown that the three characters listed below can be very helpful in species recognition:

1. Shape of the keel on the ventral surface of the eighth thoracic somite. In *L. campechiensis* this keel is rounded; in *L. glabriuscula* its margin is parallel to the body line and the posterior edge is truncate; and in *L. scabricauda* the entire keel is triangular, with the posterior margin higher than the anterior.

2. Shape of the antennal scale. This may be ovate or elongate, and its shape is characteristic for each species. A simple length/width ratio can be used to express differences in overall shape of the scale.

3. Shape of the uropodal endopod. Again, this may be ovate or elongate; in *L. glabriuscula* the uropodal endopod is twice as long as broad, whereas in *L. scabricauda* it is two-and-a-half times as long as broad. A simple length/width ratio can also be used to compare the shape of the appendage between species.

*Lysiosquilla* includes some of the largest known stomatopods. The Indo-West Pacific *L. maculata* may attain a total length in excess of 300 mm; the Atlantic *L. scabricauda* may get as large as 275 mm. All of the species are conspicuously banded with dark pigment.

The larvae of two species, *L. maculata* and *L. sulcirostris,* have been identified, and Alikunhi (1952) has shown that the larvae of *Lysiosquilla* can be distinguished from those of *Acanthosquilla.* Larvae have been attributed to the Atlantic *L. scabricauda* and *L. glabriuscula* (see below under each species), but these identifications have not been definitely established.

Several of the species of *Lysiosquilla* show a marked sexual dimorphism.

23

Kemp (1913), Holthuis (1941a), and Serène (1954) have commented on this dimorphism in *L. maculata,* in which the adult females have the raptorial claw reduced in size, variable in shape, with the dactylar teeth reduced in size and number. A similar dimorphism has been recorded for *L. sulcirostris* by Serène (1954) and for *L. glabriuscula* (see below). In *L. scabricauda* sexual dimorphism is reflected in reduction in size of the claw, but not in the reduction in size and number of dactylar teeth, and the sculpture and the shape of the telson can be modified (see below). In each of these four species the proximal portion of the propodus and the carpus of the claw are ornamented with numerous setae in adult females.

*Affinities.*—*Lysiosquilla* seems to be more closely related to *Heterosquilla* than any other genus in the Lysiosquillidae. With *Heterosquilla* it shares the slender endopod of the walking legs, the uropodal endopod lacking a prominent proximal fold, and a reduced spine in front of the articulation of the uropod. Some species of *Heterosquilla,* particularly *H. platensis* and *H. polydactyla,* also have dark bands on the body. *Lysiosquilla* differs from *Heterosquilla* in having the sculpture and armature of the telson reduced. Only one species of *Lysiosquilla, L. aulacorhynchus,* has movable submedian teeth on the telson.

### KEY TO WESTERN ATLANTIC SPECIES OF *Lysiosquilla*

1. Posterior margin of last 2 abdominal somites and dorsal surface of telson spinulose . . . . . . . . . . . . . . . . . . . . . . . . . . . .*scabricauda,* p. 24.
1. Posterior margin of last 2 abdominal somites unarmed; dorsal surface of telson not spinulose . . . . . . . . . . . . . . . . . . . . . . . . . . . . . . . . . 2

2. Sixth abdominal somite and telson smooth dorsally and laterally . . . . . . . . . . . . . . . . . . . . . . . . . . . . . . . . . . . . . . *glabriuscula,* p. 34.
2. Sixth abdominal somite and telson wrinkled and rough laterally . . . . . . . . . . . . . . . . . . . . . . . . . . . . . . . . . *campechiensis,* p. 40.

### *Lysiosquilla scabricauda* (Lamarck, 1818)
### Figures 2-4, 5a-b

Tamara guacu Marcgrave, 1648, p. 186, upper text-fig. on p. 187 [1942 reprint and translation of 1648 ed.].—Sawaya, 1942, p. lxiii [discussion].—Lemos de Castro, 1962, p. 41, pl. 4, fig. 24.
?*Astacus vitreus* Fabricius, 1775, p. 417 [larva].
?*Smerdis vulgaris* Leach, in Tuckey, 1817, fig. on unnumbered plate [larva].
*Squilla scabricauda* Lamarck, 1818, p. 188.—Latreille, 1818, p. 6, pl. 325, fig. 1.—Desmarest, 1823, p. 341, unnumbered pl. in Atlas; 1825, p. 251, pl. 42.—Latreille, 1828, p. 470.—Voigt, in Cuvier, 1836, p. 193.—H. Milne-Edwards, 1837, p. 519.—H. Milne-Edwards, in Lamarck, 1838, p. 323 [listed]; 1839, p. 376 [listed].—Gibbes, 1845, p. 70; in Tuomey, 1848, p. xvi [listed].—White, 1847, p. 83.—Gibbes, 1850, p. 199 [p. 35 on separate; listed].—Guérin-Méneville, in Sagra, 1855, p. xxii; in Sagra, 1857, p. lxvi.—Herklots, 1861, p. 152 [p. 39 on separate; listed].—Smith, 1869, p. 41 [listed].—Howard, 1883, p. 294 [listed].—Guppy, 1894, p. 115 [listed].—von Ihering, 1897, p. 156 [discussion].—Young, 1900, p. 502.—

Torralbas, 1917, p. 620, fig. 68 [p. 80 on separate].—Sawaya, 1942, p. lxiii [discussion].—Manning, 1963a, p. 317 [listed].
?*Squilla maculata:* Latreille, 1828, p. 470.—Osorio, 1888, p. 189.
?*Erichthus Leachii* Eydoux & Souleyet, 1842, p. 258, pl. 5, figs. 26-31 [larva].
?*Squilla hoevenii* Herklots, 1851, p. 17, pl. 1, fig. 11.—Osorio, 1887, pp. 223, 231; 1888, p. 189; 1889, p. 130; 1898, p. 194 [possibly a distinct species].
?*Erichthus vestitus* Dana, 1852, p. 627; Atlas, 1855, pl. 41, fig. 7 [larva].
*Lysiosquilla inornata* Dana, 1852, p. 616, pl. 41, figs. la-e.—Smith, 1869, p. 41 [listed].—Ernst, 1880, p. 436.—Fowler, 1912, p. 539 [designation of *L. inornata* as type-species of *Lysiosquilla*].—Holthuis, 1951a, pp. 83, 84 [petition to place *Lysiosquilla* on Official List].—Manning, 1961b, p. 106 [designation of lectotype].
not *Squilla scabricauda* Saussure, 1853, p. 367 [= *L. desaussurei* Stimpson, 1857].
"A Species of Squilla": Ernst, 1870, p. 3.
*Lysiosquilla scabricauda:* Miers, 1880, p. 7.—Preudhomme de Borre, 1882, p. cxi.—(?)Osorio, 1889, p. 138 [West Africa].—Sharp, 1893, p. 106.— Bigelow, 1893, p. 101; 1894, p. 508.—Faxon, 1895, p. 237 [listed].— (?)Osorio, 1898, p. 194 [West Africa].—Moreira, 1901, pp. 1, 20.—Rathbun, 1905, p. 29 [?larva].—Kemp, 1913, p. 204 [listed].—Sumner, Osburn & Cole, 1913, p. 662 [listed].—(?)Balss, 1916, p. 51 [West Africa].— Luederwaldt, 1919, p. 429.—Parisi, 1922, p. 110.—(?)Monod, 1925, pp. 91, 92 [West Africa].—Luederwaldt, 1929, p. 52.—Glassell, 1934, p. 454.—Pratt, 1935, p. 446.—Lunz, 1935, p. 154, text-fig. 3; 1937, p. 7. —Balss, 1938, pp. 139-141 [listed; discussion].—Oliveira, 1940, p. 145.— Schmitt, 1940, p. 180 [discussion].—(?)Holthuis, 1941, p. 36 [West Africa].—Reed, 1941, p. 46 [discussion].—Anonymous, 1942, p. 5.— Lemos de Castro, 1945, p. 29, fig. [popular account].—Behre, 1950, p. 19.—Dennell, 1950, p. 63.—(?)Dartevelle, 1951, p. 1034 [West Africa]. —(?)Monod, 1951, p. 139, text-figs. 1-3 [West Africa].—Andrade Ramos, 1951, pp. 142, 148, pl. 1, fig. 3.—Hildebrand, 1954, p. 261 [listed].— Chace, 1954, p. 449 [listed].—Hildebrand, 1955, p. 189 [listed].—Lemos de Castro, 1955, p. 34, text-fig. 24, pl. 10, fig. 42, pl. 17, fig. 53 [*Sysiosquilla* under text-fig. 24].—Menzel, 1956, p. 46 [listed].—(?)Cadenat, 1957, p. 133 [West Africa; key, discussion].—Chace, 1958, p. 145 [key].— (?)Longhurst, 1958, p. 86 [West Africa].—Manning, 1959, p. 17; 1961, p. 30 [key]; 1961b, p. 101 *et seq.*, text-figs. 1-2.—Lemos de Castro, 1962, p. 41, pl. 4, figs. 24, 25.—Dawson, 1963, p. 7 [listed]; 1965, p. 13 [listed]. —Bullis & Thompson, 1965, p. 13 [listed].—Boss, 1965, p. 2 [discussion of commensal].—Manning, 1966, p. 381; 1967b, p. 104.
?*Lysiosquilla Hoevenii:* Büttikofer, 1890, pp. 466, 487.—Johnston, 1906, p. 862 [both West Africa].
*Lysiosquilla maculata:* Stebbing, 1904, p. 407.—Boone, 1930, p. 29, pl. 3.— (?)Vilela, 1949, p. 67, fig. 17 [West Africa] [misidentified; not *L. maculata* (Fabricius, 1793)].
?*Lysierichthus vitreus:* Gurney, 1946, p. 167 [larva; older references and synonyms].
?*Lysiosquilla maculata* var. typica: Cadenat, 1950, p. 26 [West Africa].

*Previous records.*—BERMUDA: Dennell, 1950.—SOUTH CAROLINA: Gibbes, in Tuomey, 1848; Gibbes, 1850; Howard, 1883; Lunz, 1935.— FLORIDA: Bigelow, 1893; Sharp, 1893; Boone, 1930; Lunz, 1937; Chace, 1954; Manning, 1959, 1961a, 1967b.—MISSISSIPPI: Dawson,

1965.—LOUISIANA: Anonymous, 1942; Behre, 1950; Chace, 1954; Manning, 1959; Bullis & Thompson, 1965.—TEXAS: Bigelow, 1894; Reed, 1941; Chace, 1954; Hildebrand, 1954; Manning, 1959.—MEXICO: Preudhomme de Borre, 1882; Hildebrand, 1954, 1955; Manning, 1959.— PANAMA: Boss, 1965.—CUBA: Guérin-Méneville, 1855, 1857; Torralbas, 1917.—WEST INDIES AND ANTILLES: H. Milne-Edwards, 1837; Sharp, 1893; Young, 1900. — ANTIGUA: Stebbing, 1904. — VENEZUELA: Ernst, 1870, 1880; Guppy, 1894.—FRENCH GUIANA: Latreille, 1828.—BRAZIL: Marcgrave, 1648 (1942 reprint); Latreille, 1828; White, 1847; Dana, 1852; Miers, 1880; Sharp, 1893; Bigelow, 1894; Moreira, 1901; Parisi, 1922; Luederwaldt, 1919, 1929; Oliveira, 1940; Lemos de Castro, 1945, 1955, 1962; Andrade Ramos, 1951; Manning, 1966. — (?)WEST AFRICA: Gambia to Angola (Monod, 1951).

*Material.*—BERMUDA: 1 ♂, 254.0; Hamilton Harbour; September 1948; R. Dennell; BMNH 49.XI.5.1.—BAHAMAS: 1 postlarva; San Salvador; surface; ALBATROSS; 1886; USNM 11549.—FLORIDA: 1 ♀, 127.0; Indian River; 15 October 1957; E. Klima; UMML 32.1163.— 1 ♂, 158.0; N bridge, Fort Pierce; 18 July 1956; D. Tabb; USNM 111158.—1 ♂, 156.0; Palm Beach; C. E. Boone; AMNH 6187.— 1 ♂; Haulover Beach, Miami; 7-13 May 1961; A. Pfleuger, Jr.; UZM.— 1 damaged ♀, 123.0; 61st St., Miami Beach; 9 March 1962; Mrs. Grady; USNM 111162.— 1 ♀, 161.0; Norris Cut, Miami; 10 May 1960; M. Balanfont; IM.— 1 ♀, 192.0; North Bay Island, Biscayne Bay; 5 October 1948; UMML 32.1166.— 1 ♂, 210.0; Biscayne Bay, Miami Beach; 10 July 1946; UMML 32.164.— 1 ♂, 275.0; Pier 1, Miami; UMML 32.1161. — 1 ♂, 250.0; flats W of Seaquarium, Virginia Key, Miami; 20 October 1960; W. S. Bitler; USNM 111165.— 1 ♂, 210.0; Dinner Key Yacht Basin, Miami; 19 May 1959; L. Thomas; USNM 111166.— 1 ♂; Florida Keys; March 1924; VMM 168.— 1 ♂, ca. 250.0; 1 ♀, ca. 250.0; Key West; MCZ.— 2 dry ♀, ca. 175.0; Key West; S. Ashmead; ANSP 20, 21. — 1 ♂, 210.0; Key West; USNM 111164.— 1 ♀, 220.0; near the Aquarium, Key West; February 1948; Aquarium manager; USNM 111163.— 1 ♂, 235.0; Key West; A. Volpe; 7 November 1958; UMML 32.1173.— 1 ♂, 233.0; Key West; ca. 1896; J. W. Curry; USNM 24812.— 1 ♂, 167.0; same; December 1883; D. S. Jordan; USNM 14112.— 1 ♂, 52.0; same; ALBATROSS; USNM 21490.—1 ♀, 200.0; same; USNM 69594.— 1 ♂, 245.0; 30 mi NW of Key West; 22 March 1951; R. Siebenaler; UMML 32.854.— 1 ♂, 266.0; Fort Jefferson, Tortugas; 22 April 1861; J. B. Holder; MCZ.— 1 ♂, 156.0; Tortugas; J. E. Mills; MCZ 562.— 1 brk. ♀; Captiva Key, Charlotte Harbor; R. Wurdemann; MCZ 575.— 1 ♂, 235.0; Lemon Bay; Summer 1953; M. Rathrock; CHML.— 1 ♀, 235.0; Placida; G. A. Casterling; 3 December 1957; CHML.— 1 ♀, 200.0; Gasparilla Sound; 11 May 1955; E. Peckham; CHML.— 1 ♂,

206.0; Sarasota Bay; 21 August 1949; USNM 111160.— 1 ♀, 207.0; Long Boat Key, Long Beach; D. Nicol; USNM 91221.— 1 ♂, 89.0; 1 ♀, 171.0; John's Pass; January 1884; H. Hemphill; USNM 6471.— 2 ♂, 194.0-213.0; off Cape St. George; USNM 111161.— 1 ♀, 261.0, Grand Lagoon, St. Andrews Bay; 11 January 1959; M. Jones; USNM 104266.— MISSISSIPPI: 1 ♂, 257.0; Dog Keys Pass; on hook and line; 18 August 1959; GCRL 160.21.— LOUISIANA: 1 ♀, 154.0; near Grand Isle; 1-20 November 1939; shrimp boat GOLDEN MEADOW; USNM 110857.— 1 ♀, 166.0; Timbalier; 17 July 1934; USNM 76017.— 1 ♂, 141.0; Breton Id.; S. Springer; USNM 64240.— 2 damaged ♂, ca. 187.0-214.0; Louisiana; AMNH.— TEXAS: 1 ♀, 147.0; PELICAN Sta. 54-1; USNM 110356.— 1 ♀, 165.0; Galveston; M. Wallace; USNM 2268.— MEXICO: 1 ♂, 217.5; Veracruz; 14 February 1964; R. Barron; USNM 113324.— 1 ♂, ca. 200.0; W coast of Yucatan, Bay of Campeche; 19°47'N, 91°58'W; 49 m; Galveston Red Snapper Co.; March 1900; USNM 23448.— 1 ♂, 225.0; Las Barrancas, between Punta Cayal and Soldado Grande; 24 November 1960; V. Rendon; ITV.— HONDURAS: 1 ♂, 178.0; Tela; 1927; T. Barbour; MCZ.— NICARAGUA: 1 ♂, 193.0; off Nicaraguan coast; June 1927; Lt. B. W. Harris; USNM 60609.— COSTA RICA: 1 ♂, 223.0; Limon; 25 September 1962; A. Esna; USNM 119149.— PANAMA: 1 ♂, 241.0; Cristobal, Canal Zone; caught on hook and line; 23 August 1934; C. Hall; AMNH 6884.— CUBA: 1 ♂, 168.0; C. T. Hackett; USNM 110865.— WEST INDIES: 1 brk. ♂, ca. 190.0; no other data; ANSP 45.— COLOMBIA: 1 ♀, 104.6; OREGON Sta. 4866; USNM 119150.— TRINIDAD: 1 ♂, 145.0; Port of Spain; BMNH 1926.9.22.1.— 1 ♀, 146.0; Trinidad; BMNH 1940.VII.8.8.— BRAZIL: 1 brk. claw; 18°20'S, 41°08.5'W; 38m; CALYPSO Sta. 89; MNHNP.— 1 dry ♂, ca. 250.0; no other data; ANSP 22.— 2 ♂, 218.0-235.0; Guanabara Bay, Rio de Janeiro; MNRJ 552.— 1 ♂, 142.0; Rio de Janeiro; Thayer Exped.; MCZ 7903.— 1 ♂, 264.0; same; MCZ 7894.— 2 ♂ (1 brk.), ca. 240.0; same; January 1872; HASSLER Exped.; MCZ 7896.— 1 ♂, 215.0; same; M. J. Heade; MCZ.— 1 ♂, 256.0; same; J. D. Dana; U. S. Exploring Exped.; lectotype of L. inornata Dana; USNM 2115. —1 ♂, 179.0; Ilha São Sebastião, S. Paulo; 1915; E. Garbe; USNM 50680.— 1 ♂, ca. 168.0; Ponta da Pria, Santos; 13 February 1949; C. de Jesus; IOSP.— 1 ♂, 164.0; 2 mi. from Isla Moela, off Santos; 24°03'S, 46°16'W; Inst. Oceanografico, São Paulo; RMNH 403.

*Diagnosis.*—Dactylus of raptorial claw with 8-11 teeth, usually 9-10; fifth abdominal somite with posterior spinules; sixth abdominal somite and telson roughened dorsally, with numerous spinules and tubercles, irregularly arranged; endopod of uropod elongate, length about 2.5 times greatest width.

*Description.*—Eye very large, T-shaped, cornea bilobed, set obliquely on stalk; eyes not extending past end of second segment of antennular

27

peduncle; ocular scales separate, triangular, apices acute, inclined anteriorly; anterior margin of ophthalmic somite produced into sharp spine.

Antennular peduncle short, little more than half as long as carapace; antennular processes produced into broad spines, apices directed anteriorly.

Antennal scale slender, curved, elongate, about 3 times as long as broad (Table 4).

Dactylus of raptorial claw armed with 8-11 teeth, usually 9-10; outer margin of dactylus sinuate, with faint basal notch; propodus longer than carapace in males, relatively shorter in females (Tables 2-3, fig. 4); claw reduced in size, propodus and carpus ornamented with stiff setae in adult females; carpus with strong dorsal spine.

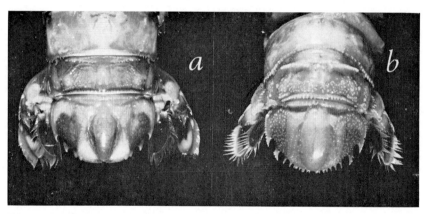

FIGURE 2. *Lysiosquilla scabricauda* (Lamarck), from Florida, telson in dorsal view. *a*, male; *b*, female.

Ventral keel on eighth thoracic somite high, elongate, triangular, posterior apex rounded or sharp.

First 4 abdominal somites smooth, unarmed; fifth somite smooth dorsally, occasionally wrinkled laterally, posterior margin armed with numerous small spinules, usually absent along midline; sixth somite roughened, variously pitted and tuberculate, usually with anterior and posterior transverse rows of spinules; anterior row of spinules may fork toward midline; sixth somite also with longitudinal groove flanked by mesial swelling parallel to lateral margin, swelling not longitudinally sulcate; sixth somite with triangular spine in front of articulation of uropod.

Telson broader than long, rectangular or ovate, with faintly raised, smooth median boss, tapered posteriorly; lateral surfaces of telson variously tuberculate or pitted; lateral margin smooth or with series of erect spinules, with or without 1 or 2 distinct parallel lateral carinae; transverse anterior carina usually spinulose; marginal teeth either completely fused along

TABLE 2
SUMMARY OF CORNEAL (CI) AND PROPODAL (PI) INDICES FOR MALE
*L. scabricauda.*

| TL | CL | CI | PI |
|----|----|----|----|
| 52 | 9.5 | 216 | 065 |
| 141 | 23.5 | 326 | 068 |
| 142 | 24.1 | 287 | 071 |
| 158 | 28.4 | 322 | 070 |
| 178 | 30.0 | 323 | 065 |
| 210 | 33.9 | 336 | 074 |
| 206 | 35.7 | 372 | 071 |
| 218 | 36.2 | 345 | 063 |
| 194 | 37.9 | 351 | 059 |
| 215 | 39.8 | 343 | 059 |
| 210 | 40.2 | 346 | 062 |
| 241 | 42.2 | 370 | 058 |
| 245 | 44.4 | 338 | 060 |
| 275 | 45.1 | 378 | 072 |
| 250 | 47.6 | 406 | 061 |
| 266 | 47.7 | 384 | 066 |

TABLE 3
SUMMARY OF CORNEAL (CI) AND PROPODAL (PI) INDICES FOR FEMALE
*L. scabricauda.*

| TL | CL | CI | PI |
|----|----|----|----|
| 127 | 24.1 | 286 | 072 |
| 165 | 26.5 | 323 | 082 |
| 171 | 27.0 | 329 | 079 |
| 166 | 28.0 | 368 | 084 |
| 161 | 28.7 | 329 | 078 |
| 192 | 31.4 | 341 | 077 |
| 200 | 33.0 | 367 | 078 |
| 207 | 34.1 | 367 | 080 |
| 220 | 35.8 | 372 | 096 |
| 200 | 36.7 | — | 105 |
| 227 | 38.5 | 493 | 111 |
| 235 | 39.5 | — | 090 |
| 261 | 40.2 | 490 | 111 |

midline, armature reduced to 3-4 pairs of projections, or with 8-10 pairs of marginal projections distinct; marginal teeth occasionally with erect dorsal spinules; shape and sculpture of telson sexually dimorphic (figs. 2-3) (see Discussion).

Basal segment of uropod usually ornamented with dorsal spinules, rarely smooth; inner carina of basal segment usually spinulose; proximal segment of exopod shorter than distal, inflated, usually spinulose distally, with graded series of 7-9 blunt movable spines on outer margin, last not extending to midlength of distal segment; inner margin of proximal segment sinuous, setose proximally and distally; endopod elongate, curved, more than 2 times as long as broad, often with longitudinal dorsal row of spinules (Table 4).

*Color.*—Antennal scale outlined in dark pigment, covered with numerous small dark chromatophores; carapace with three broad bands of dark pigment; each body segment with broad anterior and contracted posterior dark line, posterior band the darker, pigment not interrupted along midline, telson with single mesial and two lateral dark spots; distal portions of uropod dark, apex of endopod usually light.

*Size.*—Males, TL 52.0-275.0 mm; females, TL 127.0-261.0 mm. Other measurements of male, TL 256.0 (lectotype of *L. inornata*): carapace length, 41.5; cornea width, 11.5; rostral plate length, 10.4, width, 10.2; fifth abdominal somite width, 48.4; telson length, 35.6, width, 45.5. This is the largest species of stomatopod found in the Americas.

*Discussion.*—The scabrous sixth abdominal somite and telson as well as the large number of teeth on the dactylus of the raptorial claw will separate *L. scabricauda* from the other western Atlantic species of *Lysiosquilla*. The color pattern of *L. scabricauda* differs from that of *L. glabriuscula,* as noted under the Discussion of that species.

The eyes of *L. scabricauda* are larger than those of either *L. campechiensis* or *L. glabriuscula,* and they are smaller in adult females of *L. scabricauda* than in the males. Corneal Indices for selected specimens are listed in Tables 2 and 3.

Florida and Gulf of Mexico specimens of *L. scabricauda* show a well-marked sexual dimorphism, described in detail by Manning (1961b). The dimorphism is apparent in specimens between TL 150 and 200 mm. In males from Florida the telson is relatively smooth, flattened, with the posterior margin truncate and the lateral margins smooth dorsally. In the females the telson is very spinous, strongly arched dorsally, convex posteriorly, and the lateral margins are usually lined with erect spinules. The females also have a smaller raptorial claw, with the carpus and proximal portion of the propodus heavily setose; there is no apparent reduction in the number of teeth on the claw. The reduction in claw size is reflected by the Propodal Index (PI) which is given for males and females of various sizes in Tables 2 and 3. Lateral views of the claw and dorsal and lateral views of the telson of both males and females are shown in figures 2-4.

Caribbean and Brazilian males resemble Florida females in telson structure. Too few females from either of these areas have thus far been examined to determine whether or not the dimorphism is completely reversed in the two areas. No other major differences are evident. The ventral keel on the eighth thoracic somite, in general, is lower and sharper posteriorly in the southern forms than in the northern. There are no noticeable differences in the Corneal Indices, Propodal Indices, or in the length/width ratios of the antennal scale and the uropodal endopod.

30

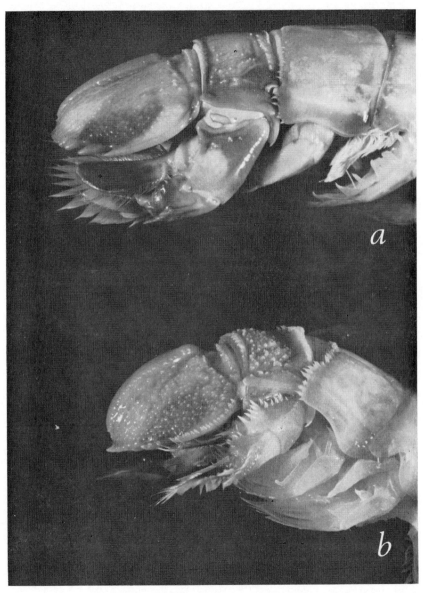

FIGURE 3. *Lysiosquilla scabricauda* (Lamarck), from Florida, posterior portion of body in lateral view. *a*, male; *b*, female.

31

Two specimens of the eastern Pacific *L. desaussurei* Stimpson have been compared with *L. scabricauda*. Both are males and both resemble Florida females in telson sculpture. The ventral process on the eighth thoracic somite is sharper and more erect than in any western Atlantic specimen. The largest specimen of *L. desaussurei,* a male, TL ca. 210 mm, has a smaller eye (CI 402) than an Atlantic male of the same size (CI 345), and the antennal scale is more slender (L/W ratio 3.71) than that of any Atlantic specimen. Finally, the longitudinal boss that parallels the lateral margin of the sixth abdominal somite in *L. desaussurei* is longitudinally sulcate; it is never noticeably grooved in *L. scabricauda.*

The West African *Squilla hoevenii* Herklots is tentatively synonymized with *L. scabricauda;* it may prove to be a distinct species or subspecies. It seems possible that *L. hoevenii, L. desaussurei,* and the northern and southern populations of *L. scabricauda* should be recognized as four subspecies of the latter.

Latreille's *Squilla maculata* (1828) from St. Vincent is tentatively placed in the synonymy of *L. scabricauda.* The exact locality was not given; his specimen could be from Australia or the West Indies. Osorio's *S. maculata* could be *L. scabricauda, L. maculata,* or *L. aulacorhynchus.* The records given by Stebbing (1904), Boone (1930) and Vilela (1949) as *L. maculata* are based on misidentifications.

*Remarks.*—Commensal bivalve mollusks have been observed on at least two specimens of *L. scabricauda.* Donald R. Moore, Institute of Marine Sciences, University of Miami, examined a specimen from Mississippi which had bivalves attached to the underside of the body. The specimen from Panama listed above (AMNH 6884) also had many small bivalves attached to the swimmerets as well as the ventral surface of the body. These mollusks were studied by Boss (1965) who assigned them to a new genus and species of erycinid, *Parabornia squillina.*

During the spring of 1963 (exact date not recorded) several reports of actively swimming specimens of *L. scabricauda* were brought to my attention. In each case these were associated with evening "runs" of penaeid shrimps on outgoing tides in Biscayne Bay, Miami, Florida. Several of the reports were in the form of telephone calls to the Institute of Marine Sciences requesting information on edibility of the stomatopods; one caller noted that had she known they were edible she could have netted in excess of two dozen in one evening. These observations are included here because of the general paucity of observations on the biology of *Lysiosquilla* and the general belief that species of this genus are more or less restricted to their burrows.

Several of the specimens taken at night-lights were brought to the Institute in fresh condition for identification. They proved to be as tasty as the Florida lobster.

This species is known in Florida as the queen mantis shrimp.

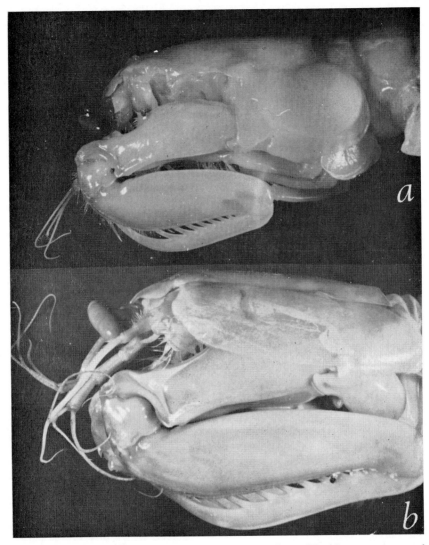

FIGURE 4. *Lysiosquilla scabricauda* (Lamarck), from Florida, carapace and raptorial claw in lateral view. *a*, female (carapace damaged); *b*, male.

*L. desaussurei* is the eastern Pacific analogue of *L. scabricauda*.

*Ontogeny.*—Several larval forms, including *Astacus vitreus* Fabricius, (now *Lysierichthus*), *Smerdis vulgaris* Leach, *Erichthus leachii* Eydoux and Souleyet, and *E. vestitus* Dana, have been identified with *Lysiosquilla scabricauda*. As all of these identifications are tentative, the names are

33

accompanied by a question-mark in the synonymy. The references to the larvae given above are not complete; Gurney (1946) included a list of references which should be consulted for further information. A study of the late pelagic larvae and postlarvae of *L. scabricauda* is in progress.

*Type.*—The type may be in the Museum of Natural History, Geneva. A dry specimen in that collection bearing the label "scabricauda Lamr. k, cette, M. Leneer" was examined there by H. H. Hobbs, II in 1965. Other portions of Lamarck's collection have been located at Geneva, so it is possible that this is the type. I did not find the type in the collection at Paris in 1964.

*Type-locality.*—L'Ocean Indien.

*Distribution.*—This species has been recorded from localities on both sides of the Atlantic. In the western Atlantic it is known from Bermuda, the Bahamas, South Carolina, the Gulf of Mexico, through the Caribbean to southern Brazil. Off West Africa the species has been recorded at several localities between Cape Verde and Gambia in the north to Angola (Monod, 1951).

TABLE 4

SUMMARY OF LENGTH/WIDTH RATIOS FOR ANTENNAL SCALE AND UROPODAL ENDOPOD OF THE WESTERN ATLANTIC SPECIES OF *Lysiosquilla*

|  | *L. campechiensis* | *L. glabriuscula* | *L. scabricauda* |
|---|---|---|---|
| Antennal Scale |  |  |  |
| L/W ratio: range | 333-405 | 224-249 | 287-363 |
| mean | 354 | 235 | 327 |
| N | 5 | 9 | 14 |
| Uropodal Endopod |  |  |  |
| L/W ratio: range | 207-263 | 182-219 | 234-289 |
| mean | 248 | 199 | 280 |
| N | 6 | 11 | 17 |

*Lysiosquilla glabriuscula* (Lamarck, 1818)
Figures 5c-d, 6

*Squilla glabriuscula* Lamarck, 1818, p. 188.—Latreille, 1828, p. 470.— H. Milne-Edwards, 1837, p. 519.—H. Milne-Edwards, in Lamarck, 1838, p. 323 [listed]; 1839, p. 376 [listed].—Miers, 1880, p. 7.—Manning, 1963a, p. 317 [listed].

*Squilla vittata* H. Milne-Edwards, 1837, p. 519.—White, 1847, p. 83.— Gibbes, 1850, p. 199 [p. 35 on separate; listed].—Herklots, 1861, p. 152 [p. 39 on separate; listed].—Young, 1900, p. 502.

?*Erichthus edwardsii* Eydoux & Souleyet, 1842, p. 260, pl. 5, figs. 39-54 [larvae; see Gurney, 1946, for other references].—Gurney, 1946, p. 163 [listed].

*Coronis glabriusculus:* Rathbun, 1883, pp. 121, 130.

*Lysiosquilla glabriuscula:* Sharp, 1893, p. 106.—Bigelow, 1894, p. 508.— Hansen, 1895, p. 75 [larvae].—Rathbun, 1899, p. 628.—Kemp, 1913, p. 203 [listed].—Calman, 1917, p. 143 [larva].—Boone, 1934, p. 27

34

[discussion].—Schmitt, 1940, p. 180 [listed].—Holthuis, 1941, p. 36.—
Gurney, 1946, p. 169 [larvae].—Chace, 1954, p. 449 [listed].—Lemos de
Castro, 1955, p. 37, text-figs. 25-26, pl. 11, fig. 33, pl. 17, fig. 54.—
Cadenat, 1957, p. 133 [key].—Chace, 1958, p. 145 [key].—Manning,
1959, p. 17; 1961, p. 31 [key].—Bullis & Thompson, 1965, p. 13 [listed].
?*Lysiosquilla maculata:* Bouvier & Lesne, 1901, p. 14 [see Remarks].
?*Lysierichthus edwardsii:* Gurney, 1946, p. 165 [larvae; older references].

*Previous records.*—SOUTH CAROLINA: Sharp, 1893; Bullis & Thomp-
son, 1965.—FLORIDA: Rathbun, 1883; Sharp, 1893; Bigelow, 1894;
Boone, 1934; Manning, 1959; Bullis & Thompson, 1965.—GULF OF
MEXICO: Chace, 1954.—WEST INDIES: Young, 1900; Holthuis, 1941.
—JAMAICA: Rathbun, 1899.—ST. VINCENT: White, 1847; Miers,
1880.—BRAZIL: Lemos de Castro, 1955.—?WEST AFRICA: Bouvier
& Lesne, 1901.

*Material.*—SOUTH CAROLINA: 1 brk. ♂, 172.0; 1 brk. ♀, CL 29.7;
from stomach of *Dasyatis centroura;* SILVER BAY Sta. 3655; USNM
111117.— 1♀, ca. 188.0; Hilton Head; J. J. Craven; ANSP 37.—BA-
HAMAS: 1 postlarval ♀, 22.0; Lyford Cay, Nassau; found swimming at
night; 1 August 1964; T. Pederson; USNM 113312.—FLORIDA: 2♂, 1
brk., 130.0-139.0; Lake Worth; 21 July, 7 August 1945; W. G. Van Name;
AMNH.—1♂, 151.0; same; June 1945; A. H. Verrill; AMNH 9791.—
1♂, 132.6; same; 1945; A. H. Verrill; IM.—1♂, 86.4; Crandon Beach,
Key Biscayne, Miami; 22 October 1957; E. Dodge; USNM 111118.— 1
soft ♀, 187.8; off Fowey Light, Dade Co.; shallow water, rocky; J. B. Henderson; USNM 46125.— 1♀, 76.5; Soldier Key, Biscayne Bay; 19 September 1960; E. Boschi; USNM 111115.— 1♀, 160.0; Virginia Key, Miami; suction dredge; 15 April 1954; UMML 32.1165.— 1♂, 156.0; dock
at Institute of Marine Sciences, Virginia Key; 27 May 1965; Chuensri;
UMML.— 1♂, 191.0; south Florida; UMML.— 1 soft ♀, 60.6; off
Florida Keys; SILVER BAY Sta. 2366; USNM 111114.— 1♂, ca. 209.0;
Key West; MCZ 7899.— 1 dry ♀, 120.0; same; S. Ashmead; ANSP 18.—
1♀, 132.8; same; 1901; B. A. Bean, W. H. King; USNM 24861.— 2
postlarvae (1♂, 1♀), both 25.0; 200 yds SE of Sands Key Light, Key
West; 1-5 m; 21 August 1961; W. A. Starck, II, *et al.;* USNM 111113.—
1 soft ♂, 173.5; 2 soft ♀, 105.5-214.1; Garden Key, Tortugas; Dr.
Whitehurst; USNM 2052.— 1♂, 206.2; same; YPM 1428.— 1♀, ca.
140.0; same; MCZ 1530.— 1♂, 135.0; Gulf coast; 1955; CHML.—
MEXICO: 1♂, 178.0; Gulf of Campeche, Yucatan; P. Fuentes; MCZ
12361.— 1♂, 93.3; same; MCZ.— HONDURAS: 1♀, 145.0; H. B.
Newham; BMNH 1923.10.4.1.— WEST INDIES: 3♂, 185.5-190.5; no
other data; RMNH 42.— DOMINICAN REPUBLIC: 1♂, 138.9; Bara-
hona; G. Hamer; USNM 82991.— PUERTO RICO: 1♂, 208.0; lagoon,
Cayo de Lawal, Parguera; 27 October 1960; IMB.—1♂, 181.2; Montalva,
Guanica; 27 April 1948; V. Balnes; IMB.—1♂, 186.5; Parguera Bay,
Lajas; near shore; 22 October 1957; J. Ramos; IMB.—1 fragmented ♂,

35

CL 12.2; Puerto Rico; from stomach of lane snapper; 13 January 1962; J. Randall; UMML 32.2135.—VIRGIN ISLANDS: 1 ♂, 102.6; Salt Pond Bay, St. John; 6 m; 27 January 1961; R. E. Schroeder; UMML 32.2065.—1 fragmented ♂; St. John; from stomach of *Lutjanus analis;* 29 March 1961; J. Randall; UMML.—1 soft ♂, 182.4; St. Thomas; 1923; E. Sebastien; USNM 57690.—LESSER ANTILLES: 1 brk. ♀, CL 28.0; Pointe-à-Pitre, Guadeloupe; Smithsonian-Bredin Sta. 68-56; 30 March 1956; USNM 111116. — 1 ♂, 139.0; Saint Vincent; BMNH 707. — CURAÇAO: 1 brk. ♂, CL 38.3; Piscadera Bay; 6 February 1957; L. B. Holthuis; RMNH 339.

*Diagnosis.*—Dactylus of raptorial claw with 6-7 teeth; last 2 abdominal somites and telson smooth dorsally, lacking spinules, tubercles, or corrugations; endopod of uropod ovate, length about twice greatest width.

*Description.*—Eye large, cornea strongly bilobed, set obliquely on stalk; ocular scales separate, triangular, small; eyes not extending past end of second segment of antennular peduncle; anterior margin of ophthalmic somite produced into sharp spine.

Antennular peduncle short, not half as long as carapace; antennular processes produced into broad, anteriorly-directed spines.

Antennal scale broad, ovate, less than 3 times as long as broad (fig. 5c; Table 4).

TABLE 5

Summary of Corneal (CI) and Propodal (PI) Indices for Selected Specimens of *L. glabriuscula*

|  | TL | CL | CI | PI |
|---|---|---|---|---|
| ♂ ♂ | 86.4 | 16.2 | 312 | 086 |
|  | 93.3 | 18.6 | — | 075 |
|  | 102.6 | 18.2 | 303 | 083 |
|  | 132.6 | 23.7 | 370 | 083 |
|  | 178.0 | 33.4 | 407 | 073 |
|  | 191.0 | 33.5 | 418 | 077 |
| ♀ ♀ | 60.6 | 11.9 | 233 | — |
|  | 76.5 | 13.4 | 257 | 081 |
|  | 132.8 | 23.8 | 413 | 096 |
|  | brk | 28.0 | 491 | 122 |
|  | 213.0 | 37.7 | 509 | 122 |

Dactylus of raptorial claw with 6-7 teeth, reduced to small denticles in adult female; entire claw reduced in size in adult female (Table 5), propodus and carpus of female ornamented with numerous stiff setae; propodus longer than carapace in adult male, shorter than carapace in female; carpus with small distal tooth on dorsal surface.

36

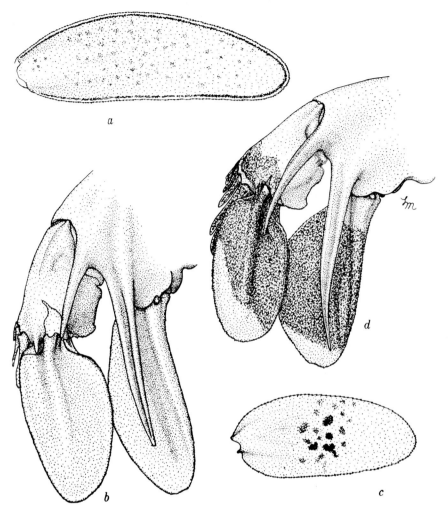

FIGURE 5. *Lysiosquilla scabricauda* (Lamarck), male, TL 158.0, Fort Pierce, Florida. *a*, antennal scale; *b*, uropod, ventral view. *Lysiosquilla glabriuscula* (Lamarck), female, TL 76.5, Soldier Key, Biscayne Bay, Florida. *c*, antennal scale; *d*, uropod, ventral view (setae omitted).

Ventral keel on eighth thoracic somite low, posterior margin angled, occasionally sharp.

Abdomen smooth, without trace of spinules, tubercles, or irregular wrinkles.

Telson broader than long, smooth, posterior margin with 2 pairs of blunt prominences, submedian denticles smoothly fused.

Basal segment of uropod without dorsal wrinkles or spinules; proximal segment of exopod with 7-8 blunt, movable spines, last not extending past mid-length of distal segment; inner margin of proximal segment fully setose, distal setae larger than proximal; endopod curved, broad, about 2 times as long as broad (Table 4).

*Color.*—Antennal scale with median spotted area; carapace with 3 dark bands, anterior 2 diffuse on lateral plates, appearing speckled; body segments each with 2 dark bands, anterior diffuse, usually broken into median and lateral spots, posterior sharper, usually interrupted along midline; telson with 3 large black spots, one medial, 2 lateral; distal portion of uropod dark, apex of exopod clear; endopod usually dark.

*Size.*—Males, TL 25.0-208.0 mm; females, TL 25.0-214.1 mm; postlarva, TL 22.0 mm. Lemos de Castro (1955) reported two males measuring 220-230 mm. Other measurements of male, TL 138.9: carapace length, 27.0; cornea width, 6.8; rostral plate length, 6.4, width, 7.7; fifth abdominal somite width, 38.3; telson length 16.2, width, 26.5.

*Discussion.*—The completely smooth telson of this species will immediately separate it from both *L. campechiensis* and *L. scabricauda.* That feature, in combination with the small number of teeth (6-7) on the raptorial claw, will distinguish *L. glabriuscula* from all other species in the genus.

*L. glabriuscula* can also be distinguished from *L. scabricauda* by the color pattern. The antennal scale of the latter species is outlined in dark pigment (fig. 5a); in *L. glabriuscula* the dark outline is absent and the central portion of the scale is covered by a dark spot (fig. 5c). The bands on the body segments are usually interrupted in *L. glabriuscula,* entire in *L. scabricauda.*

The shapes of the antennal scale and the endopod of the uropod apparently have not been used as specific characters in this genus. Length-width ratios for these appendages in each of the western Atlantic species are summarized in Table 4. The appendages are broader and more ovate in *L. glabriuscula* than in either *L. campechiensis* or *L. scabricauda;* only in *L. glabriuscula* is the endopod of the uropod only two times longer than broad.

Bouvier & Lesne (1901) identify this species with *L. maculata.* It seems possible that their West African specimen was actually *L. maculata* or possibly even *L. aulacorhynchus,* for other than their record, this species is known only from the western Atlantic.

In her discussion of *L. scabricauda,* Boone (1930) suggested that *L. maculata, L. scabricauda,* and *L. glabriuscula* should be considered as a single species. There does not seem to be any real basis for this idea, for specimens of each species are readily distinguishable.

*Remarks.*—Bigelow (1894) briefly commented upon the sexual differences shown by this species. Adult females have a small raptorial claw, with the

38

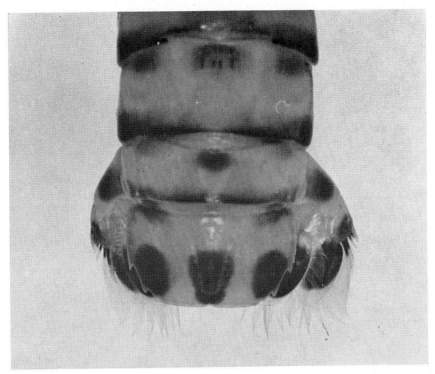

FIGURE 6. *Lysiosquilla glabriuscula* (Lamarck), male, TL 102.6, St. John, V.I., posterior portion of body.

dactylar teeth reduced in number and size and with the carpus and propodus of the claw ornamented with numerous stiff setae. In *L. scabricauda* large females also have the dactylus reduced in size and ornamented with setae, but there is no accompanying reduction in number of teeth on the claw. Holthuis (1941a) has summarized sexual differences found in the Pacific *L. maculata* which are similar to those of *L. glabriuscula*. The size at which these differences appear in *glabriuscula* is between 130-150 mm TL. Reduction in size of the claw is reflected in the Propodal Index (PI), which is summarized in Table 5 for a selected series of male and female *L. glabriuscula*. Normally the propodus is longer than the carapace in males and small females, whereas in large females it is considerably shorter than the carapace.

The eyes also are smaller in large females than in males and are smaller in general than those of *L. scabricauda;* this is shown in the larger Corneal Index (CI) also summarized in Table 5.

*L. maculata* is the eastern Pacific analogue of *L. glabriuscula*.

39

*Ontogeny.*—Although the larval form *Erichthus edwardsii* has been identified with *L. glabriuscula,* there is no definite evidence for this identification. Inasmuch as *edwardsii* and other larval forms which have been synonymized with it have been reported from the Pacific and Indian Oceans as well as the Atlantic it seems possible that larvae of several species are being confused. Gurney (1946) summarized the literature on the larval forms.

*Type.*—Probably not extant. It could not be located at Paris by me in 1964 nor at Geneva by H. H. Hobbs, II, in 1965.

*Type-locality.*—L'Ocean Indien. The species is not known with certainty outside of the western Atlantic.

*Distribution.*—Western Atlantic, from South Carolina, Florida, the Gulf of Mexico, and scattered localities through the Caribbean and northern South America to Brazil. It is apparently more common in the northern part of its range. It has not been recorded previously from Yucatan, Honduras, the Dominican Republic, Puerto Rico, the Virgin Islands, Guadeloupe, or Curaçao.

As Lemos de Castro (1955) pointed out, Holthuis's reference to this species in Luederwaldt (1929) was in error. Lemos de Castro was the first to record the species from Brazil.

The record from West Africa based on Bouvier & Lesne's (1901) identification of *S. vittata* and *S. maculata* needs to be verified.

### *Lysiosquilla campechiensis* Manning, 1962
### Figure 7

*Lysiosquilla campechiensis* Manning, 1962c, p. 219; 1963a, p. 317 [listed]. —Bullis & Thompson, 1965, p. 13 [listed].

*Previous records.*—MEXICO: Gulf of Campeche (Manning, 1962c; Bullis & Thompson, 1965).

*Material.*—MEXICO: 1 ♂, 100.4; Gulf of Campeche; OREGON Sta. 411; holotype; USNM 92651.—2 ♂, 72.1-ca. 110.0; Gulf of Campeche; MCZ. —3 ♂, 64.0-132.0; same; MCZ.

*Diagnosis.*—Dactylus of raptorial claw with 7 teeth; last 2 abdominal somites and telson corrugated and wrinkled laterally; endopod of uropod ovate, length 2.5 times greatest width.

*Description.*—Eye large, cornea bilobed, set obliquely on stalk; eyes extending about to end of second segment of antennular peduncle; ocular scales separate, triangular, apices directed anteriorly; anterior margin of ophthalmic somite produced into sharp spine.

Antennular peduncle short, slightly more than one-half as long as carapace; antennular processes produced into broad, sharp spines, apices directed anteriorly.

Antennal scale slender, elongate, over 3 times as long as broad (Table 4).

Dactylus of raptorial claw armed with 7 teeth, outer margin of dactylus with basal notch; propodus short, but longer than carapace (Table 6), inflated in largest male; dorsal spine of carpus sharp.

Ventral keel on eighth thoracic somite low, rounded.

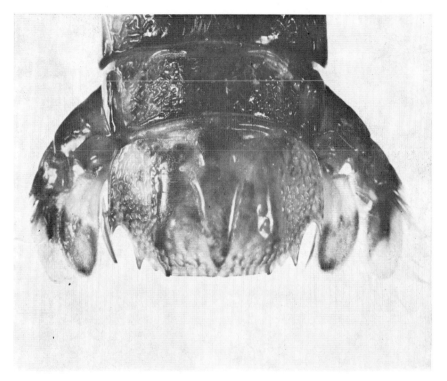

FIGURE 7. *Lysiosquilla campechiensis* Manning, male holotype, TL 100.4, Gulf of Campeche, posterior portion of body.

Abdomen without dorsal tubercles or posterior spinules; last 2 somites pitted or wrinkled laterally, both somites with shallow lateral grooves paralleling margin.

Telson with 3 smooth longitudinal dorsal bosses, lateral and posterior area deeply pitted and wrinkled but not spinulose; longitudinal carina present on either side near and parallel to lateral margin; posterior margin pitted; 4 pairs of distinct posterior projections present.

41

Dorsal surface of basal segment of uropod pitted; outer margin of proximal segment of exopod with 8 movable spines, last not extending to midlength of distal segment; inner margin of proximal segment of exopod bilobed, fully setose, distal setae longer than proximal; endopod about 2.5 times as long as broad (Table 4).

TABLE 6

SUMMARY OF CORNEAL (CI) AND PROPODAL (PI) INDICES FOR MALES OF
*L. campechiensis*

| TL | CL | CI | PI |
|---|---|---|---|
| 64.0 | 12.2 | 321 | 081 |
| 72.1 | 15.7 | 301 | 076 |
| 100.4 | 17.4 | 348 | 075 |
| 97.0 | 20.5 | 347 | 079 |
| 132.0 | 26.1 | 421 | 081 |

*Color.*—Carapace with 3 broad, dark bands; body segments each with 2 dark bands, anterior broad, diffuse, posterior sharper, neither interrupted along midline; telson with median and submedian patches of dark pigment, area bordering articulation of sixth somite dark; uropod with anterior dark spot on basal segment, exopod with dark patch covering distal portion of proximal segment and proximal portion of distal segment; most of distal segment light; endopod with distal half dark.

*Size.*—Males only known, TL 64.0-132.0 mm. Other measurements of holotype, TL 100.4 mm: carapace length, 17.4; cornea width, 5.0; rostral plate length, 4.3, width, 5.8; fifth abdominal somite width, 20.6; telson length, 12.8, width, 19.1.

*Discussion.*—This species resembles *L. glabriuscula* in having less than 8 teeth on the dactylus of the raptorial claw and in lacking tubercles or spinules on the posterior abdominal somites and telson; in these respects it differs from *L. scabricauda*. The lateral corrugations of the posterior abdominal somites and telson will separate *L. campechiensis* from *L. glabriuscula*. The shape of the antennal scale and uropodal endopod of *L. campechiensis* more closely resembles that of *L. scabricauda* than that of *L. glabriuscula* (Table 4).

*Ontogeny.*—Unknown.

*Type.*—U. S. National Museum.

*Type-locality.*—Gulf of Campeche.

*Distribution.*—Gulf of Campeche, in 62-66 m.

Genus *Heterosquilla* Manning, 1963
*Heterosquilla* Manning, 1963a, p. 320; 1966a, p. 118; 1968, p. 110 [key].

*Definition.*—Size moderate, maximum TL around 100 mm; body smooth, depressed, compact; cornea usually bilobed or expanded; rostral plate usually cordiform, with apical spine; antennal protopod with 1 or more papillae; carapace narrowed anteriorly, carinae and spines absent, cervical groove indicated on lateral plates only; exposed thoracic somites without longitudinal carinae; eighth thoracic somite usually with low, median ventral keel; at least 4 epipods present; mandibular palp present or absent; propodus of fifth maxilliped as broad as long, ventral surface with stiff brush of setae; shape of raptorial claw variable, usually slender, dactylus armed with 4 or more teeth; propodus fully pectinate, with 4 movable basal spines; dorsal ridge of carpus undivided, terminating in sharp spine; ischiomeral articulation terminal, merus longer than ischium; endopods of walking legs two-segmented, distal segment slender on all legs or ovate on first 2, slender on third; abdomen depressed but compact, usually without spines except at posterolateral angles of sixth somite; anterolateral plates articulated; sixth somite with small spine or process on each side in front of articulation of uropod; telson broader than long, with upraised median projection, often armed or lobed posteriorly, on dorsal surface, submedian dorsal projections or carinae also present; marginal armature consisting of a row of slender submedian denticles, a movable submedian tooth, 2 or 4 intermediate denticles, a fixed tooth, a denticle, and a fixed lateral tooth; uropod structure variable, basal segment with dorsal spine; proximal segment of exopod with sharp movable spines on outer margin, inner margin setose, distal setae larger; endopod without strong fold on basal portion of outer edge; basal prolongation produced into two spines, length variable.

*Type-species.*—*Lysiosquilla platensis* Berg, 1900, p. 230, by original designation.

*Gender.*—Feminine.

*Number of species.*—Ten, of which five occur in the western Atlantic.

*Remarks.*—Two of the species originally placed in *Heterosquilla, Squilla eusebia* Risso and *Lysiosquilla enodis* Manning, were transferred to a new genus, *Platysquilla,* by Manning (1967). A third species, *L. perpasta* Hale, was transferred to *Hadrosquilla* by Manning (1966a).

In that report I showed that *H. brazieri* (Miers) was the Australian analogue of the Japanese *H. latifrons* (de Haan) and that *H. tricarinata* (Claus) from New Zealand was distinct from *H. spinosa* (Wood-Mason) from the Andamans. The Australian-New Zealand members of the genus *Heterosquilla* were redescribed in detail and were assigned to three subgenera, one of which, *Austrosquilla,* containing *H. vercoi* (Hale) and *H. osculans* (Hale), was recognized as a full genus by me in 1968. *Austrosquilla* was discussed in some detail by me in 1966a and will not be treated further here.

43

The two subgenera recognized in *Heterosquilla* are the nominate sub-genus, *Heterosquilla* Manning, 1963, and the subgenus *Heterosquilloides* Manning, 1966. These two taxa are briefly discussed below.

## Subgenus *Heterosquilla* Manning, 1963

*Diagnosis.*—Cornea small, subglobular; rostral plate subcordiform; merus of raptorial claw without inferodistal spine; 2 intermediate denticles on telson; endopod of uropod lacking strong proximal fold on outer margin; outer spine of basal prolongation of uropod the stronger.

*Remarks.*—This subgenus includes two American species, *H. polydactyla* and *H. platensis* (the type-species), and two species from the Indo-West Pacific, *H. spinosa* (Wood-Mason, 1895) and *H. tricarinata* (Claus, 1871).

Members of this subgenus seem to be more similar in basic facies to *Lysiosquilla s.s.* than any of the other members of the Lysiosquillidae. The body is fairly compact, often ornamented with broad bands of dark color, and the dactylus of the claw is usually armed with numerous teeth.

## Subgenus *Heterosquilloides* Manning, 1966

*Diagnosis.*—Cornea large, bilobed; rostral plate triangular or subcordiform; merus of raptorial claw without inferodistal spine; 4 intermediate denticles on telson; endopod of uropod without strong proximal fold on outer margin; inner spine of basal prolongation of uropod usually the longer.

*Remarks.*—Six species have been assigned to this subgenus, including three from the Americas. These are *H. armata, H. insolita* (the type-species), and *H. mccullochae.* The Indo-West Pacific representatives of the subgenus are *H. brazieri* (Miers, 1880), *H. insignis* (Kemp, 1911), and *H. latifrons* (de Haan, 1844).

The members of this subgenus have the most ornate telsons within the family; *H. mccullochae,* for example, has 13 spinous propections on the telson. The species assigned herein are relatively small, lack the broad bands of dark color on the body, but otherwise also resemble *Lysiosquilla* in having slender endopods on the walking legs and in lacking a strong fold on the uropodal endopod.

There is an undescribed species of *Heterosquilla* in the western Atlantic which closely resembles *H. insolita;* this undescribed species, from off the east coast of Florida, is represented by a fragment of the body and cannot be adequately characterized.

*Affinities.*—*Heterosquilla* seems to be nearer to *Lysiosquilla* than any of the other genera of the Lysiosquillidae. With *Lysiosquilla* it shares the elongate endopod on the last walking leg, the uropod articulation, in which the prominent curved ventral process in front of each uropod is reduced to a small spine, and the endopod of the uropod usually lacks

44

a broad fold on the proximal portion of the outer margin. Members of the subgenus *Heterosquilla* have the prominent dark bands on the body that are characteristic of *Lysiosquilla*. *Lysiosquilla* differs in having a more prominent ventral keel on the eighth thoracic somite and in lacking a median dorsal prominence or submedian carinae on the telson.

### KEY TO AMERICAN SPECIES OF *Heterosquilla*

1. Telson with 2 intermediate marginal denticles (subgenus *Heterosquilla*) .......................................... 2
1. Telson with 4 intermediate marginal denticles (subgenus *Heterosquilloides*) ....................................... 3

2. Rostral plate longer than broad ............... *polydactyla*, p. 45.
2. Rostral plate broader than long ................. *platensis*, p. 48.

3. Posterior margin of last 3 abdominal somites with spinules ..........
.......................................... *armata*, p. 52.
3. Posterior margin of fourth and fifth abdominal somites unarmed .... 4

4. Sixth abdominal somite unarmed dorsally; claw with 4 teeth .........
.......................................... *mccullochae*, p. 55.
4. Sixth abdominal somite with 3 pairs of spines; claw with 8 teeth ......
.......................................... *insolita*, p. 58.

### *Heterosquilla (Heterosquilla) polydactyla* (von Martens, 1881)
### Figure 8

*Lysiosquilla polydactyla* von Martens, 1881, p. 92.—Bigelow, 1894, p. 504 [key].—Rathbun, 1910, p. 608 [listed].—Doflein & Balss, 1912, p. 40.— Kemp, 1913, p. 203 [listed].—Schmitt, 1940, p. 187, text-fig. 19.—Dahl, 1954, pp. 10, 11 [discussion].—Bahamonde, 1957, p. 119, text-figs. 1-3.— Chace, 1958, p. 145 [key].—Manning, 1963a, p. 321 [listed].

*Previous records.*—TIERRA DEL FUEGO: Orange Bay, Hoste Id. (Doflein & Balss, 1912).—CHILE: Maullin Id. (Bahamonde, 1957).

*Material.*—ARGENTINA: 1 ♂, 36.2; Puerto Piramides, Golfo Nuevo, Chubut Province; 7 June 1920; USNM 110001.—1 ♂, 65.3; Puerto Madryn, Golfo Nuevo, Chubut Province; 13 January 1920; USNM 110002.—CHILE: 1 ♀, ca. 89.0; Valparaiso; March 1920; E. Reed; USNM 109999.—1 brk. ♀, CL 7.2; Bahia de San Vincente, Concepcion; 3 May 1960; G. Hartman; RMNH 396.

*Diagnosis.*—Rostral plate longer than broad; antennal protopod with 2 ventral papillae; dactylus of raptorial claw with 17-20 teeth; telson with truncate or obscurely bilobed median dorsal projection, non-carinate submedian projections also present; 2 intermediate denticles present on each side of telson.

45

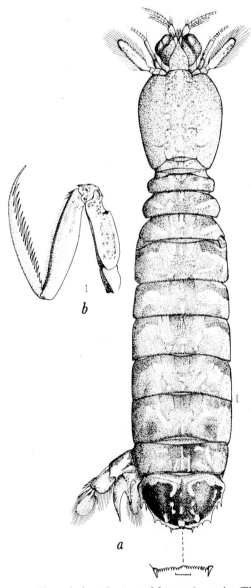

FIGURE 8. *Heterosquilla polydactyla* (von Martens), male, TL 69.5, probably Orange Bay, Hoste Id., Tierra del Fuego. *a*, specimen in dorsal view; *b*, raptorial claw. Insert at bottom shows submedian denticles (from Schmitt, 1940).

*Description.*—Eye large, cornea faintly bilobed, expanded laterally beyond stalk; outer margin of stalk with rounded projection; ocular scales fused, apices curved forward; eyes extending to or slightly beyond end of second segment of antennular peduncle.

Antennular peduncle short, less than half as long as carapace; antennular processes produced into blunt spines directed anterolaterally.

Antennal scale slender, size moderate, about half as long as carapace; antennal protopod with 2 ventral papillae.

Rostral plate subcordiform, longer than broad; anterolateral angles obtusely rounded, anterolateral margins sloping to acute apex.

Raptorial claw slender, dactylus armed with 17-20 teeth, outer margin lacking prominent basal notch; 4 movable spines at base of propodus, first longest, second shortest; dorsal ridge of carpus ending in prominent spine; ischiomeral articulation terminal, merus about half again as long as ischium.

Mandibular palp and 5 epipods present.

Lateral process of fifth thoracic somite an obscure, rounded ventral lobe; lateral margins of last thoracic somite rounded; basal segment of each walking leg with triangular posterior prominence, acute in smallest specimen; distal segment of endopods of walking legs ovate, slender on last leg; eighth thoracic somite with low, rounded ventral keel.

Abdomen depressed but compact, unarmed except at posterolateral angles of sixth somite; each somite with lateral groove, more pronounced posteriorly, flanked by mesial swelling on sixth somite; sixth somite lacking spine in front of articulation of uropod.

Telson broader than long, almost smooth dorsally, with 2 rounded lateral prominences near anterior margin; median prominence low, truncate or obscurely bilobed posteriorly, flanked laterally by submedian obtuse projection; marginal armature consisting of a transverse row of 9-13 slender submedian denticles, a movable submedian tooth, 2 intermediate denticles, an intermediate tooth, a lateral denticle, and a lateral tooth.

Uropod broad, basal segment with inner carina not extending to distal spine; proximal segment of exopod with 5-7 slender movable spines on outer margin, last extending at least to midlength of distal segment; segments of exopod subequal in length; inner margin of proximal segment with distal setae longer than proximal; endopod ovate, curved; spines of basal prolongation broad, flattened, triangular in cross-section, outer the longer.

*Color.*—Faded in present material. Bahamonde's (1957) figures show that the body is banded, as in *H. platensis,* and that there are three broad, dark areas on the telson, the median with a longitudinal clear stripe.

*Size.*—Males, TL 36.2-95.0 mm; only known female, 89.0 mm. Other measurements of male, TL 65.3: carapace length, 11.3; cornea width, 2.7;

rostral plate length, 3.3, width, 3.0; fifth abdominal somite width, 13.8; telson length, 7.5, width, 11.8.

*Discussion.*—This species seems most closely related to *H. platensis,* but differs in having a longer rostral plate, more teeth on the raptorial claw, and poorly developed dorsal sculpture on the telson. The dorsal submedian carinae present in *H. platensis* are completely lacking in this species.

*H. polydactyla* also shows some similarity to *H. tricarinata,* but that species has a more globular cornea, a shorter rostral plate, and a triangular median projection flanked by carinate submedian projections on the telson.

*Ontogeny.*—Unknown.

*Type.*—Not traced.

*Type-locality.*—Questionable, presumably somewhere off the west coast of South America. The type was found unlabelled among a group of Chilean and Peruvian species (Schmitt, 1940).

*Distribution.*—Southern South America, from Maullin Id. (near Chiloe Id.), Chile, to Golfo Nuevo, Argentina.

*Heterosquilla (Heterosquilla) platensis* (Berg, 1900)
Figures 9-10

> *Lysiosquilla platensis* Berg, 1900, p. 230, fig. on p. 231.—Parisi, 1922, p. 108, text-fig. 4.—Chace, 1958, p. 145 [key].—Manning, 1963a, p. 321 [listed].

*Previous records.*—ARGENTINA: Mar del Plata (Berg, 1900; Parisi, 1922).

*Material.*—URUGUAY: 1 dry ♂, 83.5; Rio del Plata; F. Felippone; USNM 65057.—ARGENTINA: 1 dry ♂, 69.5; Mar del Plata; F. Felippone; USNM 54623.—1 ♀, 101.9; Mar del Plata; 18 October 1920; USNM 110000.—3 ♀, 79.2-109.4; Mar del Plata, Cabo Corrientes, Buenos Aires Province; January 1961; E. Boschi; USNM 111027 (smallest ♀ to IM).—6 ♀, 70.3-96.7; Bajo de los Huesos, Rawson, Chubut Province; 29 January 1962; E. Boschi; USNM 111026.—2 ♀, 96.4-99.0; same; 24-29 February 1962; E. Boschi; USNM 111025.—1 ♂, 68.4; same; February 1962; E. Boschi; USNM 111024.

*Diagnosis.*—Rostral plate as broad as or broader than long; antennal protopod with 2 ventral papillae; dactylus of raptorial claw with 13-15 teeth; telson with broad median projection, carinate laterally, flanked laterally by carinate submedian projections; 2 intermediate denticles present on each side of telson.

*Description.*—Eye large, cornea faintly bilobed, expanded laterally beyond

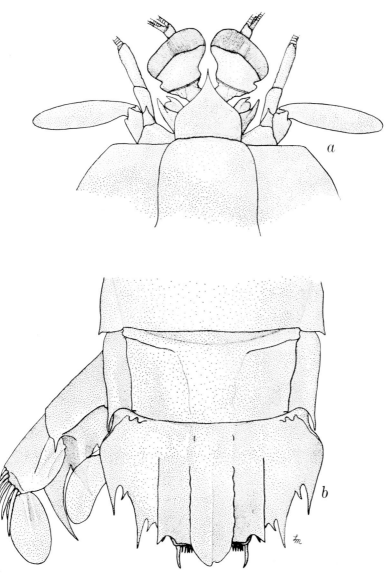

FIGURE 9. *Heterosquilla platensis* (Berg), female, TL 70.3, Rawson, Chubut, Argentina. *a,* anterior portion of body; *b,* last abdominal somite, telson, and uropod (setae omitted).

stalk; stalk with rounded lateral projection; ocular scales small, fused mesially; eyes extending beyond end of second segment of antennular peduncle.

Antennule short, peduncle about half as long as carapace; antennular processes produced into strong spines, directed anterolaterally.

Antennal scale small, about half as long as carapace; antennal protopod with 2 ventral papillae.

Raptorial claw slender, dactylus armed with 13-15 teeth, outer margin of dactylus without prominent basal notch; propodus with small fixed spine proximally on dorsal surface, 4 movable spines on inner margin, first longest, last smallest; dorsal ridge of carpus terminating in strong tooth; ischiomeral articulation terminal, merus longer than ischium.

Mandibular palp and 5 epipods present.

Rostral plate subcordiform, as broad as or broader than long; antero-lateral margins sloping to acute apex.

Lateral process of fifth thoracic somite a rounded, ventral lobe; lateral processes of next 2 somites rounded, broadest posteriorly; endopods of walking legs ovate on first 2 legs, slenderest on last leg.

Abdomen compact, first 5 somites without carinae or spines; sixth somite with posterolateral spines and 1 pair of lateral carinae parallel to and separated from lateral margin by deep groove; sixth somite without ventral spines in front of articulation of uropod.

Telson broader than long, with broad, raised median projection, carinate laterally and posteriorly, apex often divided into blunt lobes; median area flanked by dorsal submedian carinae each terminating in a strong tooth; marginal armature consisting of a transverse row of 10-12 slender sub-median denticles, a movable submedian tooth, 2 intermediate denticles, an intermediate tooth, a lateral denticle, and a lateral tooth.

Uropod broad, basal segment with inner dorsal carina terminating in strong spine; proximal segment of exopod of uropod slightly longer than distal, with 6-7 slender, curved spines on outer margin, last extending past midlength of distal segment; endopod ovate; basal prolongation produced into two triangular but flattened spines, outer the longer, inner with small proximal spine on inner margin at origin of endopod.

*Color.*—Background color light; antennal scale and upper margin of raptorial propodus outlined in dark pigment; each body segment with diffuse anterior and sharper posterior band of dark color; telson with 4 dark patches, 2 submedian, 2 lateral; distal portion of uropods dark.

*Size.*—Males, TL 68.4-83.5 mm; females, TL 70.3-109.4 mm. Other measurements of female, TL 96.4: carapace length, 19.0; cornea width, 4.6; rostral plate length, 4.6, width, 5.0; fifth abdominal somite width, 21.6; telson length, 12.2, width 20.0.

*Discussion.*—H. *platensis* closely resembles H. *polydactyla* but differs in

FIGURE 10. *Heterosquilla platensis* (Berg), female, TL 96.7, Rawson, Chubut, Argentina, dorsal view.

having a broader rostral plate, fewer teeth (13-15 instead of 17-20) on the raptorial claw, and more prominent median and submedian dorsal projections on the telson.

*H. tricarinata* from the Indo-West Pacific also has submedian longitudinal carinae on the telson but the median prominence in that species is reduced to a circular knob; the Indo-West Pacific species also has smaller eyes, with the cornea subglobular instead of faintly bilobed or expanded. Other differences were discussed by Manning (1966a).

*Remarks.*—*H. platensis* is a south temperate species, apparently restricted to the area around the Rio del Plata of Argentina and Uruguay. To the south it is replaced by *H. polydactyla*.

This is one of the few species omitted by Kemp (1913) in his list of the Atlantic and eastern Pacific stomatopods.

*Ontogeny.*—Unknown.

*Type.*—Not traced.

*Type-locality.*—Mar del Plata, Argentina.

*Distribution.*—Rio del Plata of Uruguay and Argentina, south to Chubut Province, Argentina.

## *Heterosquilla (Heterosquilloides) armata* (Smith, 1881)
### Figure 11

*Lysiosquilla armata* Smith, 1881, p. 446.—Verrill, 1885, p. 558 [listed].—Rathbun, 1905, p. 29.—Fowler, 1912, p. 539 [listed].—Sumner, Osborn & Cole, 1913, p. 662 [listed].—Kemp, 1913, p. 202 [listed].—Allee, 1923, p. 180 [listed].—?Verrill, 1923, p. 195, pl. 52, figs. 1-4, pl. 54, fig. 1, text-fig. 2 [larvae].—Fish, 1925, p. 154 [larvae; discussion].—Schmitt, 1940, pp. 180, 184 [discussion].—?Coventry, 1944, p. 544 [larvae].—Gurney, 1946, p. 167 [larvae; listed].—Deevey, 1952, p. 96 [larvae].—Chace, 1958, p. 145 [key].—Manning, 1963a, p. 321 [listed].
not *Lysiosquilla armata*: Bigelow, 1894, p. 507 [=*Platysquilla enodis* (Manning)].

*Previous records.*—NEW ENGLAND: Smith, 1881.

*Material.*—NEW ENGLAND: 1 ♂, 38.0; off Martha's Vineyard, Massachusetts; Fish Hawk Stats. 865-867; lectotype; USNM 35398.—1 damaged ♀, 47.3; off New England; from stomach of tilefish; Fish Hawk Sta. 876; paralectotype; YPM 4435.—2 ♂, ca. 22.1-24.0; 1 ♀, ca. 25.0; off Martha's Vineyard; Fish Hawk Sta. 949; YPM 4425.—2 ♂ (1 brk.), 20.7; 2 ♀, 20.2-25.0; same; USNM 21491.—1 ♂, 66.8; off New England; Fish Hawk Sta. 1110; YPM 4436.—1 damaged ♀; off New England; Albatross Sta. 2248; USNM 7199.

*Diagnosis.*—Rostral plate almost as long as broad; antennal protopod with 2 ventral papillae; dactylus of raptorial claw with 10-11 teeth; last 3

52

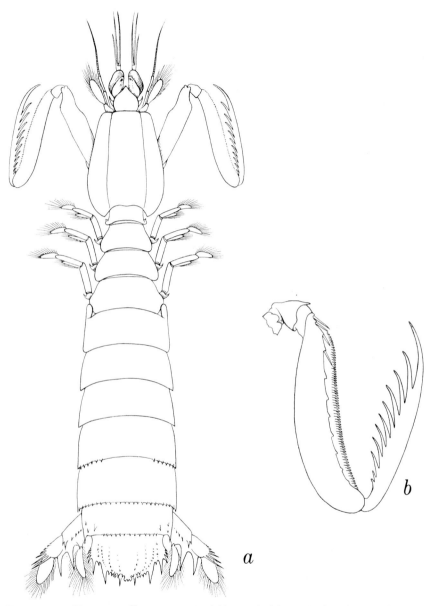

FIGURE 11. *Heterosquilla armata* (Smith), probably male lectotype, TL 38.0. *a*, dorsal view; *b*, raptorial claw. Illustration by R. P. Bigelow.

abdominal somites with posterior spinules; median projection of telson trilobed; 4 intermediate denticles present on each side of telson.

*Description.*—Eye large, cornea faintly bilobed, set obliquely on and expanded laterally beyond stalk; ocular scales separate, rounded; eyes not extending past end of second segment of antennular peduncle.

Antennular peduncle short, slightly more than half as long as carapace; antennular processes produced into slender, anteriorly-directed spines.

Antennal scale short, ovate, not half as long as carapace; antennal peduncle with 2 ventral papillae.

Rostral plate cordiform, length and width subequal or width slightly greater; anterolateral margins rounded, anterior margins sloping to sharp apical spine.

Raptorial claw slender, dactylus armed with 10-11 teeth; outer margin of dactylus with basal notch, proximal flanking lobe of notch more prominent than distal; proximal movable spine on propodus the longest; dorsal ridge of carpus terminating in strong spine; ischiomeral articulation terminal, merus elongate, over twice as long as ischium.

Mandibular palp and 5 epipods present.

Lateral process of fifth thoracic somite produced into sharp, ventrally-directed spine, with or without supplementary spine on outer margin; lateral processes of next 2 somites rounded, broadening posteriorly; basal segment of each walking leg with sharp lateral spine; endopods of walking legs elongate, slender; eighth thoracic somite with low, rounded median keel.

Abdomen depressed but compact, first 3 somites unarmed; posterior margin of fourth somite armed laterally, smooth mesially; posterior margin of last 2 somites lined with spinules; sixth somite with strong posterolateral spines, lateral portion of dorsal surface with tubercles and a few spinules, near but not on posterior margin; sixth abdominal somite with strong spine in front of articulation of uropod.

Telson broader than long, dorsal surface ornamented with spinules, variable in number and arrangement; median projection produced into 3 rounded lobes, projection flanked laterally by converging row of spinules; dorsal surface with irregular lateral rows of tubercles and spinules; posterior margin, above marginal armature, with pair of spines above each submedian marginal tooth; lateral margins lined with erect spines, fewer on smaller specimens; marginal armature consisting of a row of 8-12 slender submedian denticles, a movable submedian tooth, 4 intermediate denticles, second and fourth slenderest, an intermediate tooth, 1 lateral denticle, and a lateral tooth.

Uropod with short carina, lined with spinules, on inner portion of dorsal surface, carina not extending to distal spine; proximal segment of exopod broad, longer than distal, with 8-9 slender movable spines on outer margin, last extending almost to end of distal segment; inner margin of proximal segment with a few stiff distal setae; endopod slender, curved; basal prolongation produced into 2 flattened, triangular spines, outer larger, inner with proximal spine at articulation of endopod.

*Color.*—Completely faded in all specimens.

*Size.*—Males, TL 20.7-66.8 mm; females, TL 20.2-47.3 mm. Other measurements of male lectotype, TL 38.0: carapace length, 6.8; cornea width, 2.0; rostral plate length, 2.0, width, 2.1; fifth abdominal somite width, 8.1; telson length, 3.5, width, 6.5.

*Discussion.*—The striking armature of the posterior portion of the body will immediately distinguish this species from all others in the genus. The only other stomatopod that occurs offshore in the same geographic area, New England, is *Platysquilla enodis;* that species lacks posterior spinules on the last three abdominal somites and has four pairs of fixed marginal projections on the telson rather than two.

*Remarks.*—Several specimens of *H. armata* were taken in the stomach contents of tilefish. In spite of numerous collections off New England since the 1880's, no specimens of this species have been collected since that time.

The illustration used here is an unpublished original by R. P. Bigelow.

*Ontogeny.*—Verrill (1923), Fish (1925), Coventry (1944), and Deevey (1952) identified larval stomatopods with this species. Gurney (1946) summarized the early records.

*Type.*—U. S. National Museum. The intact 38.0 mm male (USNM 35398) is here selected as a lectotype. A female paralectotype is deposited in the Yale Peabody Museum (YPM 4435).

*Type-locality.*—Off Martha's Vineyard, Massachusetts.

*Distribution.*—Off New England in 96-218 m.

### *Heterosquilla (Heterosquilloides) mccullochae* (Schmitt, 1940)
### Figure 12

*Lysiosquilla mccullochae* Schmitt, 1940, p. 197, text-fig. 23.—Manning, 1963a, p. 321 [listed, discussion].

*Previous records.*—MEXICO: Gulf of California (Schmitt, 1940).

*Material.*—FLORIDA: 1 ♀, 32.2; off Alligator Reef Light, Monroe Co.; 21 July 1961; W. A. Starck, II; USNM 111028.—PUERTO RICO: 1 ♀, 20.5; La Parguera, Lajas; 9 December 1961; J. Randall; UMML 32.2182.

*Diagnosis.*—Rostral plate broader than long; antennal protopod with 2 mesial and 2 ventral papillae; dactylus of raptorial claw with 4 teeth; telson with broad median projection, armed posteriorly with 3 sharp

55

spines; dorsal surface with submedian pair of trispinous projections, each flanked laterally by 2 single spines; 4 intermediate marginal denticles present on each side of telson.

*Description.*—Eye large, cornea strongly bilobed, expanded laterally beyond stalk; ocular scales with apices separate, bases fused; eyes not extending past end of second segment of antennular peduncle.

Antennular peduncle short, half as long as carapace; antennular pro-

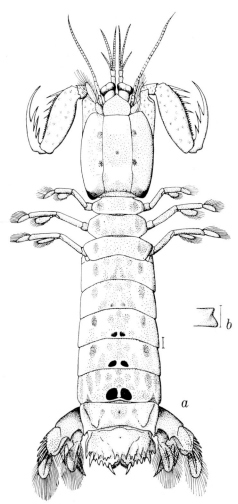

FIGURE 12. *Heterosquilla mccullochae* (Schmitt), female holotype, TL 32.0, San Francisco Id., Gulf of California. *a,* dorsal view; *b,* lateral margin of fifth thoracic somite (from Schmitt, 1940).

cesses produced into broad, anteriorly-directed spines.

Antennal scale short, about one-third as long as carapace; antennal protopod with 2 mesial and 2 ventral papillae.

Rostral plate broad, cordiform, anterolateral angles broadly rounded, anterior margins sloping to obtuse apex; plate completely covers base of eyes.

Raptorial claw stout, dactylus armed with 4 teeth; outer margin of dactylus with prominent basal notch, proximal flanking lobe sharper but smaller than distal lobe; propodus with 4 proximal spines on inner margin, first longest, fourth smallest; dorsal ridge of carpus obscure, terminating in blunt lobe; merus much longer than ischium.

Mandibular palp and 5 epipods present.

Lateral process of fifth thoracic somite a blunt lobe, directed ventrally; lateral processes of sixth and seventh thoracic somites truncate, more rounded on seventh somite; distal segment of endopods of walking legs ovate but elongate, slenderest on last leg; eighth thoracic somite without prominent ventral keel.

Abdomen depressed but compact, unarmed except at posterolateral angles of sixth somite; sixth somite with low, blunt lateral ridges parallel to margin; no spine in front of articulation of uropod.

Telson broader than long, dorsal surface ornamented with 13 spines above marginal armature; posterior margin of median prominence divided into 3 sharp spines, flanked laterally by another trispinous process, and, lateral to this, 2 single spines, outer above intermediate marginal tooth; bifurcate lobe present on either side between submedian group of spines and movable marginal submedian tooth; Florida specimen with additional spine (on 1 side) dorsal to bifurcate lobe; marginal armature consisting of a curved row of 6-10 fixed submedian denticles, outer largest, a movable submedian tooth, 4 intermediate denticles, second and fourth smallest, an intermediate tooth, a lateral denticle and a lateral tooth.

Uropod broad, basal segment with inner dorsal carina terminating in sharp spine; proximal segment of exopod broad, swollen, with 7-8 slender movable spines on outer margin, last extending to or beyond midlength of distal segment; inner margin of proximal segment with distal patch of 8-10 stiff setae; distal segment of exopod slightly longer than proximal; endopod ovate, curved, proximal portion of outer margin not folded over; basal prolongation produced into 2 triangular spines, inner longer.

*Color.*—Background cream, with numerous scattered patches of light-brown chromatophores; carapace with 2 brown bands, posterior darker, and 4 dark spots on gastric grooves; anterolateral plates of abdomen almost black; second to fifth abdominal somites each with pairs of black spots, increasing posteriorly in size and intensity of color.

*Size.*—Females only known, TL 20.5-32.2 mm. Other measurements of female, TL 32.2: carapace length, 6.0; cornea width, 1.6; rostral plate length, 1.4, width, 2.2; fifth abdominal somite width, 7.0; telson length, 3.2, width, 5.4.

*Discussion.*—*H. mccullochae* can be distinguished from all other species by the presence of four teeth on the raptorial claw, 13 dorsal spines on the telson, and the color pattern.

Schmitt (1940) pointed out the relation of this species to *H. latifrons* (de Haan) from Japan. That species differs in having six teeth on the raptorial claw instead of four, an apical spine on the rostal plate, more submedian denticles on the telson, and fewer dorsal spines lateral to the trispinose median projection.

*Remarks.*—The two specimens reported here are the second and third known specimens and the first Atlantic record for this pretty species described by Schmitt from a single specimen taken in the Gulf of California.

*Ontogeny.*—Unknown.

*Type.*—Allan Hancock Foundation.

*Type-locality.*—Off San Francisco Island, Gulf of California.

*Distribution.*—Gulf of California, 55 m (Schmitt, 1940); Florida; Puerto Rico.

## *Heterosquilla (Heterosquilloides) insolita* (Manning, 1963)
### Figure 13

*Lysiosquilla insolita* Manning, 1963, p. 54, text-fig. 1; 1963a, p. 321 [listed].

*Previous records.*—FLORIDA: northern Straits of Florida (Manning, 1963).

*Material.*—FLORIDA: 1 ♂, 47.7; northern Straits of Florida; 25°41′N, 80°02′W; 238-247 m; GERDA Sta. G-29; 21 June 1962; holotype; USNM 109075.—GALAPAGOS IDS.: 1 fragmented ♂, CL 12.7; N of Hood Island; 91-182 m; AHF Sta. 816-38; 23 January 1938; USNM 76392.

*Diagnosis.*—Rostral plate longer than broad; antennal protopod with 1 ventral papilla; dactylus of raptorial claw with 7-8 teeth; sixth abdominal somite with submedian spines on posterior margin; median projection of telson trilobed; 4 intermediate denticles present on each side of telson.

*Description.*—Eye large, cornea strongly bilobed, set obliquely on stalk; ocular scales fused basally, with posterior median projection; eyes extending almost to end of antennular peduncle.

Antennular peduncle short, more than half as long as carapace; antennular processes produced into slender spines extending to base of eyes.

Antennal scale small, slightly more than one-third as long as carapace; antennal peduncle with 1 ventral papilla.

Rostral plate elongate, longer than broad, anterolateral angles obtusely

58

rounded, anterior margins sloping to obtuse apex.

Raptorial claw slender, dactylus armed with 7-8 teeth, penultimate smaller than antepenultimate; notch at base of outer margin of dactylus flanked by acute proximal lobe; second tooth on propodus the smallest; dorsal ridge of carpus terminating in strong spine; ischiomeral articulation terminal, merus over twice as long as ischium.

Mandibular palp and 5 epipods present.

Lateral process of fifth thoracic somite a round lobe directed ventrally; lateral process of sixth somite truncate, with angular anterodorsal projection on either side; lateral process of seventh thoracic somite rounded,

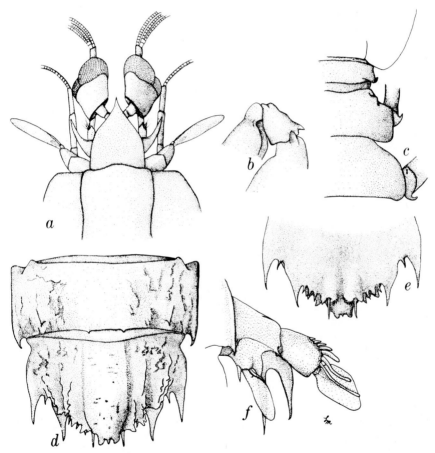

FIGURE 13. *Heterosquilla insolita* (Manning), male holotype, TL 47.7, Straits of Florida. *a*, anterior portion of body; *b*, carpus of claw; *c*, lateral processes of exposed thoracic somites; *d*, last abdominal somite and telson; *e*, telson, ventral view; *f*, uropod, ventral view (setae omitted; from Manning, 1963).

59

anterolateral margin obtuse; walking legs with inner and outer basal spines, outer sharper; eighth thoracic somite with low median ventral prominence.

Abdomen smooth, first 4 somites unarmed, fourth and fifth with posterolateral spines; sixth somite corrugated laterally, with 6 posterior spines, submedians small, intermediates anterior to margin; sixth somite with small ventral spine in front of articulation of uropod.

Telson broader than long, dorsal surface with flat median prominence, apex divided into three rounded lobes; lateral surface of telson with 3 broad rugose areas, increasing in size laterally, each terminating in an upturned tubercle; marginal armature consisting of a curved row of slender submedian denticles, a movable submedian tooth, 4 intermediate denticles, first and third larger, blunter than second and fourth, an intermediate tooth, a lateral denticle, and a lateral tooth.

Basal segment of uropod with dorsal carina terminating in a tooth; proximal segment of exopod with 7 flattened spines, last 2 recurved, last extending well past midlength of distal segment; distal segments of exopod subequal in length; endopod slender, with prominent fixed spine at base of inner margin; inner spine of basal prolongation the longer.

*Color.*—Background pale with faint brown chromatophores; carapace with scattered light-brown chromatophores, arranged in open semicircle near posterior margin; thoracic and abdominal somites with median and posterior diffuse bands of light-brown.

*Size.*—Male holotype, only intact specimen, TL 47.7 mm. Other measurements: carapace length, 8.0; cornea width, 2.7; rostral plate length, 2.7, width, 2.3; fifth abdominal somite width, 9.7; telson length, 5.0, width, 7.5.

*Discussion.*—*H. insolita* is closely related to the Indo-West Pacific *H. insignis* (Kemp), differing from that species in having submedian marginal teeth on the sixth abdominal somite, and in having the rostral plate broader and more sinuate. In all other respects the species seem to very similar.

The fragment from the Galapagos, which consists of the anterior half of a male, agrees with the type in most respects. Unfortunately the posterior part of the abdomen and the telson, which might show distinctive differences, are missing. The Pacific specimen has two papillae on the antennal protopod instead of one, has a larger eye, lacks the posterior ridge on the ocular scales, has one less tooth on the raptorial claw, and has a much sharper projection at the base of the outer margin of the claw. Some of these differences can be explained by the difference in size of the two specimens, but the Pacific specimen could also belong to a separate species.

Another fragmented specimen from off the east coast of Florida, in the collection of the American Museum of Natural History, definitely belongs to a different species. It is a male, CL 6.8, taken in 384–480 m

10-11 miles E of Hillsboro Light, Florida. This specimen has ocular scales that are fused basally, but with apices separate, a broader rostral plate, a more slender claw with 10 teeth, and a truncate posterior margin on the seventh thoracic somite. In other details it resembles the type.

*Remarks.*—The distribution of *H. insolita* and its Indo-West Pacific ally *H. insignis* parallels that of *Bathysquilla microps* and the Indo-West Pacific *B. crassispinosa* (Fukuda). In both cases the species are closely related but distinct.

*Ontogeny.*—Unknown.

*Type.*—U. S. National Museum.

*Type-locality.*—Northern Straits of Florida.

*Distribution.*—Off East Florida, possibly in the Galapagos, in depths between 91 and 247 m.

## Genus *Acanthosquilla* Manning, 1963

*Acanthosquilla* Manning, 1963a, p. 319; 1968, p. 110 [key].

*Definition.*—Size moderate, TL 70 mm or less; body smooth, strongly depressed, loosely articulated; eyes small, cornea subglobular, flattened or faintly bilobed; rostral plate square or rectangular, median spine present, anterolateral spines present or absent; antennal protopod with ventral papilla, mesial papilla present or absent; carapace narrowed anteriorly, without carinae or spines, cervical groove indicated along gastric grooves only; thoracic somites smooth, lateral margins truncate; eighth thoracic somite without median ventral keel or projection; 5 epipods present; mandibular palp usually present; propodus of fifth maxilliped about as broad as long, ventral surface with prominent brush of setae; raptorial claw stout but small, dactylus with 6 or more teeth, outer margin of dactylus with 2 basal projections; raptorial propodus broad, fully pectinate, with 4 movable spines, distal smallest, at base of inner margin; dorsal ridge of carpus undivided, terminating in sharp tooth; ischiomeral articulation terminal, merus much longer than ischium; endopods of walking legs two-segmented, distal segment subricular on first 2 legs, ovate on third; abdomen flattened, without carinae or spines, pleura small, ventral margins sinuous, anterolateral plates of abdomen articulated; sixth abdominal somite with strong posterolateral spines and prominent ventrally-directed process in front of articulation of uropod; telson broader than long, without sharp median carina but with fan-shaped row of 5 or more sharp teeth above posterior armature which consists of a row of slender fixed denticles, a movable submedian tooth, and either 2 or 4 fixed lateral teeth, with 1 or more denticles between each tooth; basal segment of uropod with inner and outer carinae, inner terminating in sharp spine; proximal segment of exopod of uropod with row of slender, curved, movable spines on outer margin, inner margin setose with distal patch of stiff setae on rounded prominence; endopod triangular, proximal portion of outer margin folded over; basal prolongation of uropod produced into 2 slender spines, triangular in cross-section, inner much the longer.

*Type-species.*—*Lysiosquilla multifasciata* Wood-Mason, 1895, by original designation.

*Gender.*—Feminine.

*Number of species.*—Eight, of which two occur in the western Atlantic. The extra-limital species were listed by Manning (1963a).

*Remarks.*—The eight species readily separate into two groups, one with four pairs of primary marginal teeth on the telson, one with two. Within these groups the characters of the species are very similar. In the group containing *A. acanthocarpus, A. multifasciata, A. septemspinosa,* and *A. biminiensis* only seemingly minor features, including the relative size of the penultimate tooth of the raptorial claw, the size of the lobes on the outer margin of the claw, and the size of the intermediate denticles of the telson, characterize the recognized species.

Three of the species are known to be associated with balanoglossid worms, and one, *A. digueti,* lives commensally with a balanoglossid.

Larval stages are definitely known for only three of the species, *A. acanthocarpus, A. multifasciata,* and *A. tigrina* (Alikunhi, 1952). Alikuni was able to show that the larvae of these species could be distinguished from those of two species of *Lysiosquilla.*

The name *acanthocarpus* should date from Claus (1871), not Miers (1880), as listed by me in 1963a. L. B. Holthuis brought to my attention the fact that Claus validated the name first published by White (1847) as a *nomen nudum.*

*Affinities.*—*Acanthosquilla* is similar to *Coronis, Platysquilla,* and *Nannosquilla,* but differs from all in having a series of five or more dorsal teeth on the telson arranged in a fan-shaped row. *Acanthosquilla* and *Nannosquilla* differ from the other two genera by having subcircular distal segments on the inner branches of the first two walking legs. As in all three of the other genera, *Acanthosquilla* has the paired, ventrally-directed processes on the sixth abdominal somite and the uropodal endopod with a folded edge.

## KEY TO WESTERN ATLANTIC SPECIES OF *Acanthosquilla*

1. Telson with 2 pairs of fixed marginal teeth; mandibular palp present
. . . . . . . . . . . . . . . . . . . . . . . . . . . . . . . . . . . . . . . . . . . . . *biminiensis,* p. 63.
1. Telson with 4 pairs of fixed marginal teeth; mandibular palp absent
. . . . . . . . . . . . . . . . . . . . . . . . . . . . . . . . . . . . . . . . . . . . *floridensis,* p. 67.

## *Acanthosquilla biminiensis* (Bigelow, 1893)
### Figures 14, 15

*Lysiosquilla biminiensis* Bigelow, 1893a, p. 102; 1894, p. 504, text-figs. 4-7.
—Kemp, 1913, p. 124 [part].—Schmitt, 1940, p. 180 [discussion].—
Andrews, Bigelow & Morgan, 1945, p. 340.—Chace, 1958, p. 45 [key].—
Manning, 1963a, p. 320 [listed].
not *Lysiosquilla biminiensis* var. *pacificus* Borradaile, 1899, p. 403.—Kemp,
1913, p. 124 [part].—Manning, 1962d, p. 301, text-fig. 1 [= *A. multifasciata* (Wood-Mason, 1895) ].

*Previous records.*—BAHAMAS: Nixies Harbor, Bimini (Bigelow, 1893a,
1894; Andrews, Bigelow & Morgan, 1945).

*Material.*—BAHAMAS: 1 ♂, 46.9; Nixies Harbor, Bimini; R. P. Bigelow;
lectotype; USNM 17999.—1 damaged ♀, CL 8.9; same, paralectotype;
USNM 111035.—FLORIDA: 1 damaged ♀, CL 8.5; W coast of Florida;
sample B-169, Sta. D-20; no other data; USNM.—TEXAS: 1 damaged
spec., CL 5.3; Sabine Pass; 15 June 1957; W. G. Hewatt; USNM 101083.
—CUBA: 1 ♂, 39.3; Matansas Bay; 2-4 m; April 1927; USNM 119134.—
COLOMBIA: 1 brk. ♀, CL 4.9; 1 mi. SW of Cape la Vela; AHF Sta.
A13-39; 8 April 1939; USNM 119162.—BRAZIL: 1 ♂, 61.9; Villa Bella,
Ilha São Sebastião, São Paulo; November 1925; H. Luederwaldt; USNM
111036.

*Diagnosis.*—Rostral plate with single apical spine; cornea subglobular, with
dorsal tubercle; mandibular palp present; dactylus of raptorial claw with
6-7 teeth; 2 large fixed teeth on posterior margin of telson.

*Description.*—Eye very small, cornea subglobular, set obliquely on and
scarcely expanded beyond stalk; cornea with prominent dorsal tubercle;
ocular scales fused; eyes extend almost to end of second segment of
antennular peduncle.

Antennular peduncle short, less than half as long as carapace; antennular processes visible lateral to rostral plate as sharp, anteriorly-directed
spines.

Antennal peduncle with distal segment as long as or much longer than
proximal, extending beyond eyes; proximal segment of peduncle very
setose; antennal scale less than half as long as carapace; antennal protopod
with 1 ventral and 1 mesial papilla.

Length and width of rostral plate subequal; proximal portion of plate
rectangular, convex laterally, anterolateral angles rounded; apical spine
sharp, extending to or beyond cornea; plate completely covering base
of eyes.

Dactylus of raptorial claw with 6-7 teeth, proximal very small, penultimate slightly shorter than antepenultimate; distal lobe on outer margin
of dactylus larger than proximal.

Mandibular palp and 5 epipods present.

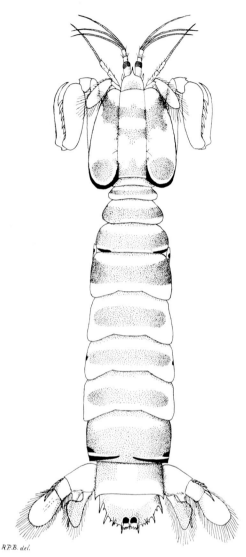

FIGURE 14. *Acanthosquilla biminiensis* (Bigelow), male lectotype, TL 46.9, Bimini, Bahamas, dorsal view (from Bigelow, 1894).

Telson twice as broad as long, with 5 subequal dorsal teeth; 4-5 submedian denticles present, outer largest, inner smallest, slightly higher than outer; 4 fixed denticles present between movable submedian tooth and next fixed tooth, first and third larger than second and fourth.

Uropod with 6 graded, movable spines on outer margin of proximal

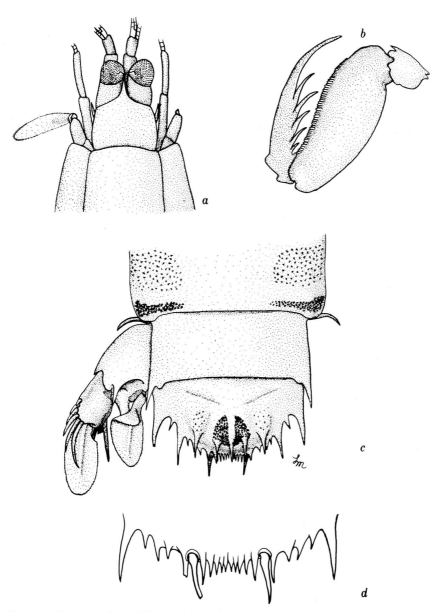

Figure 15. *Acanthosquilla biminiensis* (Bigelow), male, TL 39.3, Matansas Bay, Cuba. *a,* anterior portion of body; *b,* raptorial claw; *c,* last two abdominal somites, telson, and uropod; *d,* marginal armature of telson, ventral view (setae omitted).

segment of exopod, all 6 slender, last few curved, last extending beyond midlength of distal segment; large distal lobe of inner margin of proximal segment armed with 6-10 stiff setae; distal segment of exopod longer than proximal; outer margin of distal segment of exopod more heavily setose than inner.

*Color.*—Carapace and abdomen crossed by diffuse bands of dark chromatophores; posterolateral angles of carapace with dark circle, flanked posteriorly by black semicircle; eighth thoracic somite with pair of short, black posterolateral lines; fifth abdominal somite with black posterolateral lines, interrupted along midline, each with anterior dark spot; submedian dorsal spines of telson each with basal black spot or median and both submedian spines covered by 1 broad spot.

*Size.*—Males, TL 39.3-61.9 mm; no intact females extant. Other measurements of male, TL 61.9: carapace length, 11.5; cornea width, 1.3; rostral plate length, 3.5, width 3.5; fifth abdominal somite width, 12.5; telson length, 5.1, width, 10.0.

*Discussion.*—The presence of only two large fixed marginal teeth on the telson will separate this species from the other American representatives of the genus, *A. floridensis* and *A. digueti*. *A. biminiensis* closely resembles *A. septemspinosa* from the eastern Atlantic, but that species lacks the dorsal tubercle on the cornea, the dark circles on the carapace, and the short dark lines on the eighth thoracic somite. Also, in *A. septemspinosa* the second and fourth intermediate denticles are larger than the first and third, whereas the reverse is true in *A. biminiensis*.

Manning (1962d) showed that *L. biminiensis pacificus* Borradaile is a synonym of *A. multifasciata* (Wood-Mason). In that paper a table is presented in which the distinguishing features of the four species of *Acanthosquilla* which have but two fixed marginal teeth on the telson are summarized.

*Remarks.*—Bigelow (1893) noted that this species was found burrowing in the sand at Bimini.

The most nearly intact specimen of Bigelow's syntypes, the male, is here selected as lectotype; the broken female is a paralectotype.

The large male from Brazil agrees with the other specimens in all but one feature, there is only one large black spot on the telson which extends around the three middle spines on the dorsal surface. In all of the other specimens in which the color is visible there are two black spots on the telson separated by a clear area along the midline.

*Ontogeny.*—Unknown.

*Type.*—U. S. National Museum.

*Type-locality.*—Nixies Harbor, Bimini, Bahamas.

*Distribution.*—Bimini, Bahamas; West Coast of Florida; Texas; Matansas, Cuba; Colombia; Ilha de São Sebastião, Brazil; sublittoral to 24 m.

*Acanthosquilla floridensis* (Manning, 1962)
Figure 16

*Lysiosquilla floridensis* Manning, 1962c, p. 221; 1963a, p. 320 [listed].

*Previous records.*—FLORIDA: Cape Florida (Manning, 1962c).

*Material.*—FLORIDA: 1 ♂, 48.0; shoreline along Cape Florida, Key Biscayne, Dade Co.; 15 May 1961; C. R. Robins and class; holotype; USNM 107875.—BRAZIL: 1 ♂, 45.2; Praia do Araca, São Sebastião, São Paulo; S. Almeida Rodrigues; USP.

*Diagnosis.*—Rostral plate with sharp apex and anterolateral spines; cornea subglobular, without dorsal tubercle; mandibular palp absent; dactylus of raptorial claw with 9-11 teeth; 4 large fixed teeth on posterior margin of telson.

*Description.*—Eye small, cornea subglobular, without dorsal tubercle, set obliquely on and expanded laterally slightly beyond stalk; ocular scales fused; eyes extend past end of second segment of antennular peduncle.

Antennular peduncle short, slightly more than one-half as long as carapace; antennular processes present as sharp, anteriorly directed spines visible lateral to rostral plate.

Antennal peduncle extending to or slightly beyond eyes; antennal scale small, about one-third as long as carapace; antennal protopod with 1 mesial and 1 ventral papilla.

Rostral plate broader than long, with 3 sharp anterior spines, anterolateral spines as long as or longer than median, almost reaching cornea; plate may completely cover base of eyes.

Dactylus of raptorial claw armed with 9-11 teeth, penultimate not noticeably shorter than antepenultimate; outer margin of dactylus with basal notch, flanked on either side by inconspicuous swelling.

Mandibular palp absent; 5 epipods present.

Telson almost twice as broad as long, with 5 subequal dorsal teeth; marginal armature consisting of 3 submedian denticles, inner much the smallest, 1 movable submedian tooth, and 4 fixed lateral teeth, with 1 denticle between each.

Uropods with 6 graded, movable spines on outer margin of proximal segment of exopod, all 6 slender, last few curved, last extending beyond midlength of distal segment; distal segment of exopod longer than proximal; large distal lobe on inner margin of proximal segment with 6 stiff setae; outer margin of distal segment more heavily setose than inner.

*Color.*—Carapace with 3 diffuse bands, posteriormost darkest, and body segments each with 1 diffuse band of dark chromatophores; posterolateral angles of carapace with dark crescent preceded by at most an indistinct dark patch; each thoracic somite with black posterior line; abdominal somites with anterolateral dark patch, median posterior margin darkest; fifth abdominal somite with dark posterior line and 2 dark posterolateral

67

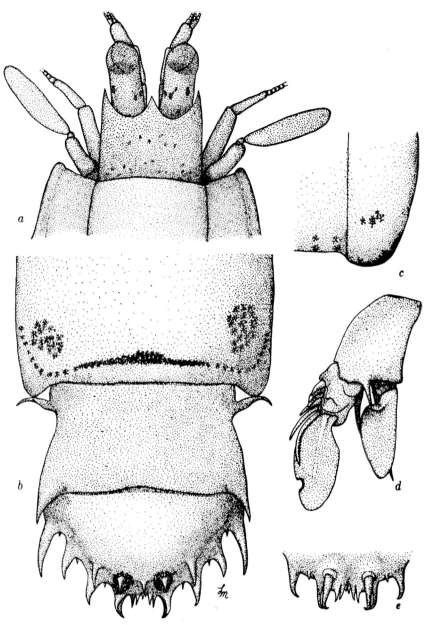

FIGURE 16. *Acanthosquilla floridensis* (Manning), male holotype, TL 48.0, Cape Florida. *a,* anterior portion of body; *b,* last two abdominal somites and telson; *c,* posterolateral angle of carapace; *d,* uropod; *e,* submedian teeth of telson, ventral view (setae omitted).

areas or circles; telson with 1 dorsal pair of black spots, 1 on each submedian dorsal spine.

*Size.*—Males only known, TL 45.2-48.0 mm. Other measurements of male holotype, TL 48.0: carapace length, 7.2; cornea width, 1.0; rostral plate length, 2.5, width, 2.9; fifth abdominal somite width, 7.8; telson length, 3.5, width, 6.0.

*Discussion.*—The presence of four fixed marginal teeth on the telson will distinguish *A. floridensis* from the only other western Atlantic representative of the genus, *A. biminiensis. A. floridensis* closely resembles *A. digueti* (Coutière) from the eastern Pacific, but differs in lacking a mandibular palp, in having more teeth on the raptorial claw (9-11 instead of 6-8), and in lacking the prominent posterolateral black spots on the carapace.

*Remarks.*—Burdon-Jones (1962), in a paper on the biology of *Balanoglossus gigas,* described the habitat in which *A. floridensis* was found by S. A. Rodrigues.

*A. digueti* is the eastern Pacific analogue of *A. floridensis.*

*Ontogeny.*—Unknown.

*Type.*—U. S. National Museum.

*Type-locality.*—Cape Florida, Dade County, Florida.

*Distribution.*—Known only from Florida and Brazil.

### Genus *Nannosquilla* Manning, 1963

*Nannosquilla* Manning, 1963a, p. 318; 1967, p. 111 [key].

*Definition.*—Size small, TL 45 mm or less; body smooth, strongly depressed, loosely articulated; cornea usually subglobular, expanded laterally beyond stalk and set obliquely on it; rostral plate subrectangular, with apical spine; antennal protopod without papillae; carapace short, narrowed anteriorly, carinae and spines absent, cervical groove indicated on lateral plates only; exposed thoracic somites without carinae; eighth thoracic somite without median ventral keel or projection; 4 epipods present; mandibular palp absent; propodus of fifth maxilliped about as broad as long, ventral surface with stiff brush of setae; raptorial claw slender, dactylus with 7 or more teeth, outer margin with basal notch, flanking lobes not prominent; propodus fully pectinate, with 4 movable spines at base, second much the smallest; dorsal ridge of carpus undivided, ending in at most a blunt projection or spine; ischiomeral articulation terminal, merus slender, elongate, longer than carpus; endopods of walking legs two-segmented, distal segment subcircular on first 2 legs, ovate on third; abdomen without longitudinal carinae or spines, articulated anterolateral plates small; sixth abdominal somite with or without posterolateral spines, curved, ventrally directed process present anterior to uropod articulation; telson much broader than long, posterior margin produced into false eave, variable in shape, overhanging true marginal armature; dorsal sculpture of telson variable, surface usually smooth; marginal armature consisting of a row of slender fixed submedian denticles, a movable submedian tooth, and, lateral to this, a variable series of fixed teeth and denticles; uropod broad, flattened, basal segment with inner and outer carinae, inner terminating in spine; proximal segment of exopod of uropod with sharp dorsal carina and row of spatulate spines on outer margin, distal 2 or 3 recurved; inner margin of exopod setose proximally, with distal patch of stiff setae; endopod triangular, with basal portion of outer edge folded over; basal prolongation of uropod produced into 2 spines, length variable, both triangular in cross-section.

*Type-species.*—*Lysiosquilla grayi* Chace, 1958, p. 141, by original designation.

*Gender.*—Feminine.

*Number of species.*—11.

*Remarks.*—Eight of the 11 species now placed in this genus are American, and five occur in the western Atlantic. The three extra-American species agree in most features with the American species, but the presence of many of the generic characters (reduction of epipods, absence of palp and antenna papillae) has not been verified and this will need to be done before the status of these species can be accepted. *N. hystricotelson,* which lacks the mandibular palp, is the only species of the genus in which the telson is completely covered with dorsal spinules. *N. varicosa* is the only species in which the cornea is bilobed; it may prove to belong in another genus.

*Affinities.*—*Nannosquilla* closely resembles *Coronis* and *Platysquilla* in general body form. However, the presence of but four epipods, the absence of the antennal papillae, and the presence of a false eave, under which the true marginal armature of the telson is concealed, will separate *Nannosquilla* from *Coronis* and *Platysquilla.* The marginal armature of the telson is more complex in *Nannosquilla* than *Coronis,* for at least three pairs of lateral teeth are present in the former. Also, the movable spines on the uropod are spatulate in *Nannosquilla,* slender and sharp in *Coronis.* The distal segment of the endopod of the first two walking legs is almost circular in *Nannosquilla,* ovate in *Coronis* and *Platysquilla.*

*Nannosquilla* also resembles *Acanthosquilla* in general body form but the presence of dorsal teeth on the telson will distinguish species of the latter genus.

*Nannosquilla, Acanthosquilla, Coronis,* and *Platysquilla* all have a curved, ventrally-directed process on the sixth abdominal somite in front of the articulation of the uropod, and all share the triangular endopod of the uropod with the folded proximal edge. These features will distinguish members of these genera from both *Lysiosquilla* and *Heterosquilla.*

70

Perhaps the genus most similar to *Nannosquilla* is *Hadrosquilla* Manning, 1966, containing but one species, *H. perpasta* (Hale). *Hadrosquilla* agrees in shape of eyes, rostral plate, and endopod of uropod, it has a false eave on the telson, and it lacks a mandibular palp. It differs as follows: (a) papillae are present on the antennal protopod; (b) five epipods are present; (c) the body is stouter and more compact; (d) there are fewer teeth on the dactylus of the claw, 5-6 rather than 7 or more; (e) the spines of the basal prolongation of the uropod are flattened; (f) there are only 2 intermediate marginal denticles on the telson; (g) the spines on the outer margin of the uropodal exopod are not flattened; and (h) the characteristic long process overhanging the uropod is absent.

Inasmuch as most of the species of *Nannosquilla* are American the following key includes all American species.

### KEY TO AMERICAN SPECIES OF *Nannosquilla*

1. Spines of basal prolongation of uropod subequal in length or outer longer than inner ........................................ 2
1. Inner spine of basal prolongation of uropod longer than outer..... 4
2. False eave of telson with numerous posterior projections .......... ........................... [*N. californiensis* (Manning, 1962)]
2. False eave with rounded median projection, lateral margins not markedly subdivided ........................................ 3
3. Anterolateral angles of rostral plate broadly rounded; telson with 10-14 submedian denticles present either side of midline ................ .................................... [*N. chilensis* (Dahl, 1954)]
3. Rostral plate angled anterolaterally; telson with 5-8 submedian denticles present either side of midline .................. *antillensis*, p. 72.
4. Telson margin with 7 fixed teeth and denticles lateral to each movable submedian tooth ............................... *schmitti*, p. 74.
4. Telson with no more than 3-5 fixed teeth and denticles lateral to each movable submedian tooth ............................... 5
5. Movable submedian teeth of telson situated anterior to outermost submedian denticle, at level of or slightly behind lateral teeth ........... ........................................... *hancocki*. p. 76.
5. Movable submedian teeth of telson adjacent to outermost submedian denticle ................................................ 6
6. Anterolateral angles of rostral plate acute ..................... ........................... [*N. decemspinosa* (Rathbun, 1910)]
6. Anterolateral angles of rostral plate rounded .................. 7
7. Submedian excavations of false eave of telson shallow, scarcely indented; submedian denticles in transverse row ............ *grayi*, p. 78.
7. Submedian excavations of false eave of telson noticeably concave; submedian denticles forming a "W" in posterior view.. *taylori*, n. sp., p. 81.

*Nannosquilla antillensis* (Manning, 1961)
Figure 17
*Lysiosquilla antillensis* Manning, 1961, p. 35, pl. 9; 1963a, p. 319 [listed].

*Previous records.*—VIRGIN ISLANDS: St. John (Manning, 1961).— VENEZUELA: Cubagua Id. (Manning, 1961).

*Material.*—MEXICO: 1 ♀, 14.0; 2 mi WSW Cayo Norte, Banco Chinchorro, Yucatan; coral reef; 5-6 m; 23 June 1961; W. A. Starck, II; USNM 111030.—VIRGIN ISLANDS: 1 ♂, 20.2; 2 ♀, 20.5-20.7; off Yawsi Point, Lameshur Bay, St. John; 9 m; 21 December 1958; J. Randall, L. P. Thomas; paratypes; USNM 106055.—2 ♂, 17.6-20.0; 1 ♀, 20.8; same; paratypes; UMML 32.1174.—VENEZUELA: 1 ♀, 24.5; Cubagua Id.; 4-9 m; AHF Sta. A24-39; 14 April 1939; holotype; AHF 3919.

*Diagnosis.*—Rostral plate with sharp anterolateral angles; raptorial claw with 7-8 teeth; posterior armature of telson consisting of 5-8 fixed submedian denticles, and a curved row of 5-8 fixed teeth and denticles; spines of basal prolongation subequal or outer spine the longer.

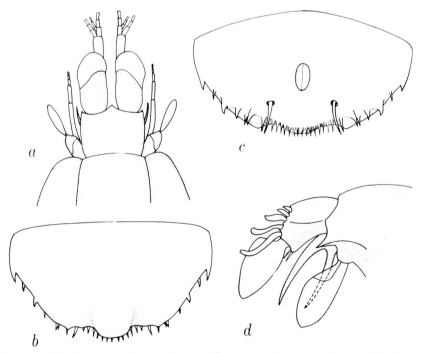

FIGURE 17. *Nannosquilla antillensis* (Manning), female holotype, TL 24.5, Cubagua Id., Venezuela. *a,* anterior portion of body; *b,* telson, dorsal view; *c,* telson, ventral view; *d,* uropod (setae omitted; from Manning, 1961).

*Description.*—Eye small, extending past end of second segment of antennular peduncle; ocular scales fused basally, apices separate.

Antennular peduncle short, about half as long as carapace; antennular processes visible as slender, anteriorly-directed spines on each side of rostral plate.

Antennal scale small, about one-fourth as long as carapace; antennal peduncles extending to cornea.

Rostral plate subquadrate, broader than long; anterolateral angles sharp, acute, extending almost to level of obtuse apex; plate extending to and may completely cover base of eyes.

Dactylus of raptorial claw armed with 7-8 teeth; basal notch on outer margin of dactylus with acute proximal and obtuse distal lobes; dorsal ridge of carpus ending in blunt lobe.

Sixth abdominal somite with posterolateral spines.

Telson much broader than long, smooth dorsally; false eave with broad, rounded median prominence, slightly swollen dorsally, flanked laterally by shallow, sinuous depressions; marginal armature consisting of 5-8 fixed submedian denticles, 1 movable submedian tooth, placed slightly anterior to denticles, and 5-8 fixed teeth and denticles in an irregular, curved row.

Outer margin of penultimate segment of exopod of uropod with 5 spatulate spines, last extending to or beyond midlength of distal segment; distal lobe on inner margin of penultimate segment with 2-4 stiff setae; spines of basal prolongation subequal or outer spine the longer.

*Color.*—Pattern almost completely faded; carapace with dark posterolateral spots and a few other spots on posterior margin; fifth abdominal somite with several scattered dark spots.

*Size.*—Males, TL 17.6-20.2 mm; females, TL 14.0-24.5 mm. Other measurements of female, TL 24.5: carapace length, 4.9; cornea width, 0.7; rostral plate length, 1.2, width, 1.6; telson length, 1.8, width, 3.1.

*Discussion.*—This species can be distinguished from all others in the western Atlantic by the spines of the basal prolongation which are either subequal in length or with the outer spine slightly longer than the inner. *N. antillensis* also has fewer teeth (7-8) on the raptorial claw than any other western Atlantic species.

The eastern Pacific representatives of the group can be distinguished from *N. antillensis* as follows: *N. chilensis* has rounded anterolateral angles on the rostral plate and more (12-17) teeth on the raptorial claw; *N. decemspinosa* has the inner spine of the basal prolongation of the uropod the longer and more teeth (11) on the raptorial claw; *N. californiensis* has the false eave of the telson divided into numerous lobes.

*Ontogeny.*—Unknown.

*Type.*—Allan Hancock Foundation.

*Type-locality.*—Cubagua Island, Venezuela.

*Distribution.*—Yucatan; St. John, Virgin Islands; Cubagua Island, Venezuela; 4-9 m, on coral reefs.

## *Nannosquilla schmitti* (Manning, 1962)
### Figure 18
*Lysiosquilla schmitti* Manning, 1962c, p. 221; 1963a, p. 319 [listed].

*Previous records.*—FLORIDA: Tortugas (Manning, 1962c).

*Material.*—FLORIDA: 1 ♂, 22.4; Harry Harris Park, S Key Largo; Pronoxfish in *Porites* flat; 18 August 1965; D. R. Moore, R. B. Manning; USNM 113319.—1 ♀, 22.7; southernmost rock jetty, Lower Matecumbe Key; 13 August 1964; Davis and Davis; UMML.—1 ♂, 20.2; seining beach, Long Key, Tortugas; dug from sand; 7 August 1930; W. L. Schmitt; holotype; USNM 107874.—1 ♀, 27.4; Tortugas; 5 August 1930; W. L. Schmitt; paratype; USNM 111032.—2 ♂, 21.7-24.7; 5 ♀, 18.5-24.9; Long Key, Tortugas; dynamite in flat of seining beach; 16 August 1930; W. L. Schmitt; paratypes; USNM 111031.—1 mutilated spec.; Tortugas; fish stomach; 13 June 1925; W. L. Schmitt; USNM 111029.—LESSER ANTILLES: 1 ♀, 12.6; off Cocoa Point, Barbuda; dredge around ship's anchorage; 8 m; Smithsonian-Bredin Sta. 99-59; 26 April 1959; USNM 111034.

*Diagnosis.*—Rostral plate with rounded anterolateral angles; raptorial claw with 9-11 teeth; posterior armature of telson consisting of 7-9 fixed submedian denticles, 1 movable submedian tooth, near outermost denticle, and a curved row of 7 fixed teeth and denticles; inner spine of basal prolongation the longer.

*Description.*—Eye small, extending almost to end of antennular peduncle; ocular scales fused.

Antennular peduncle short, about half as long as carapace; antennular processes visible lateral to rostral plate as slender, anteriorly-directed spines.

Antennal scale small, about one-fourth as long as carapace; antennal peduncle extending to cornea.

Rostral plate subrectangular, broader than long, anterolateral angles rounded; concave anterior margins slope to sharp, obtuse apex; plate extends to base of eyes.

Dactylus of raptorial claw armed with 9-11 teeth, outer margin of dactylus with basal notch, proximal lobe sharper but not much larger than distal; dorsal ridge of carpus terminating in blunt spine.

Sixth abdominal somite with acute but not sharp posterolateral angles.

Telson much broader than long, false eave with acute median projection, sharp dorsally but not carinate, flanked laterally by broad, obtuse lobes; posterior armature consisting of 7-9 submedian denticles in a curved row,

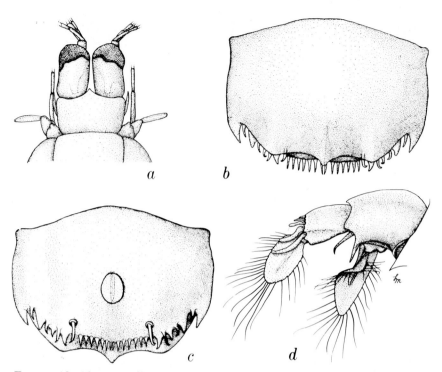

FIGURE 18. *Nannosquilla schmitti* (Manning), male holotype, TL 20.2, Tortugas, Florida. *a,* anterior portion of body; *b,* telson, dorsal view; *c,* telson, ventral view; *d,* uropod (setae of antennal scales omitted).

1 movable submedian tooth adjacent and anterior to outer denticle, and 7 fixed teeth and denticles lateral to movable one, second, third and fifth smaller and sharper than remainder.

Proximal segment of exopod of uropod with 4-5 spatulate spines on outer margin, last extending to or beyond midlength of distal segment; inner margin of proximal segment with 2 stiff distal setae; inner spine of basal prolongation the longer.

*Color.*—Completely faded in all specimens.

*Size.*—Males, TL 20.2-24.7 mm; females, TL 12.6-24.9 mm. Other measurements of female, TL 24.3: carapace length, 3.9; cornea width, 0.9; rostral plate length, 1.1, width, 1.8; fifth abdominal somite width, 4.1; telson length, 2.1, width, 3.8.

*Discussion.*—*N. schmitti* can be separated from all other species in the genus by the configuration of the false eave of the telson, on which the median lobe is crested posteriorly. In other species, the median portion of the false eave, even if raised or projected posteriorly, is broadly rounded or flattened.

*Ontogeny.*—Unknown.

*Type.*—U. S. National Museum.

*Type-locality.*—Long Key, Tortugas, Florida.

*Distribution.*—Known only from Key Largo, Matecumbe Key, localities around Tortugas, Florida, and from Barbuda. The latter locality extends the range well to the south.

*Nannosquilla hancocki* (Manning, 1961)
Figure 19

*Lysiosquilla hancocki* Manning, 1961, p. 32, pl. 8; 1963a, p. 319 [listed].

*Previous records.*—VENEZUELA: Cubagua Id. (Manning, 1961).

*Material.*—VENEZUELA: 1 ♀, 21.0; Cubagua Id.; 4-9 m; AHF Sta. A27-39; 15 April 1939; holotype; AHF 3918.—2 ♂, 16.1-18.3; 2 ♀, 15.3-19.4; 7 mi. N of Margarita Id.; 38-40 m; AHF Sta. A42-39; 21 April 1939; USNM 119154.

*Diagnosis.*—Rostral plate with anterolateral angles rectangular but rounded; raptorial claw with 9-10 teeth; posterior armature of telson consisting of 9-11 submedian denticles, 1 movable submedian tooth, set anterior to denticles, and a curved row of 5 fixed teeth and denticles; inner spine of basal prolongation much the longer.

*Description.*—Eye small, extending past end of second segment of antennular peduncle; ocular scales fused along midline.

Antennular peduncle short, about half as long as carapace; antennular processes visible as anteriorly-directed spines either side of rostral plate.

Antennal scale small, about one-fourth as long as carapace; antennal peduncle not extending beyond eye.

Rostral plate pentagonal, broader than long, angled anterolaterally, anterior margins sloping to obtuse apex; in holotype, plate not extending to base of eyes.

Dactylus of raptorial claw armed with 9-10 teeth; basal notch of dactylus flanked by proximal acute prominence and distal obtuse lobe; dorsal ridge of carpus terminating in blunt tooth.

Sixth abdominal somite with posterolateral spines.

Telson smooth dorsally, much broader than long; false eave with obtusely rounded median projection, flanked on each side by sharp but obtuse lobe separated from median projection by deep concavity; marginal armature consisting of 9-11 fixed submedian denticles in a transverse row, outermost recurved anteriorly, submedian tooth situated anterior to outer-

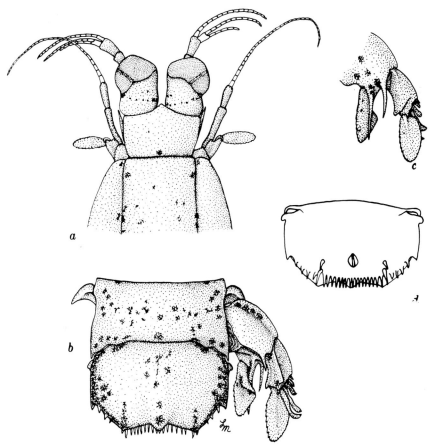

FIGURE 19. *Nannosquilla hancocki* (Manning), female, TL 19.4, N. of Margarita Id., Venezuela. *a,* anterior portion of body; *b,* last abdominal somite, telson, and uropod; *c,* uropod, ventral view; *d,* telson, ventral view (setae omitted).

most denticle, and 5 fixed lateral teeth and denticles, curving anteriorly, of which the second and fourth are smallest.

Outer margin of penultimate segment of uropodal exopod with 5-6 spatulate spines, last extending past midlength of distal segment; distal lobe on inner margin of penultimate segment with 2 slender stiff setae; inner spine of basal prolongation much the longer.

*Color.*—Holotype largely faded to an even cream color, with 1 pair of large pigment spots on carapace; carapace and body with few small chromatophores; telson with faint indication of anterolateral spots.

*Size.*—Males, TL 16.1-18.3 mm; females, TL 15.3-21.0 mm. Other measurements of female holotype, TL 21.0: carapace length( 3.1; cornea width, 0.7; rostral plate length, 0.9, width, 1.5; fifth abdominal somite width, 3.4; telson length, 1.5, width, 2.6.

*Discussion.*—*N. hancocki* is very similar to *N. taylori,* described below, but differs mainly in having the movable submedian teeth of the telson situated anterior to the submedian denticles, at the level of the anus. The angle of the illustration given for the species in 1961 made the teeth appear to be set anterior to the anus. In *N. hancocki* the submedian denticles are transverse, whereas in *N. taylori* they form a "W". Other differences are noted under the discussion of the new species.

*Ontogeny.*—Unknown.

*Type.*—Allan Hancock Foundation.

*Type-locality.*—Cubagua Island, Venezuela.

*Distribution.*—Known only from off Venezuela, in 4-40 m.

*Nannosquilla grayi* (Chace, 1958)
Figure 20

*Lysiosquilla grayi* Chace, 1958, p. 141, pl. 1, figs. 1-5.—Manning, 1961, p. 31 [discussion]; 1963, p. 319 [listed].
*Nannosquilla grayi:* Smith, 1964, p. 127.

*Previous records.*—MASSACHUSETTS: Bass River, Yarmouth (Chace, 1958).—Woods Hole region (Smith, 1964).

*Material.*—MASSACHUSETTS: 2 ♂, 36.3-37.2; Bass River, Yarmouth; burrowing in muddy sand near low water mark; 1 May 1954; M. B. Gray; paratypes; USNM 96976.—1 ♀, 42.1; same; 31 July 1954; paratype; US-NM 96975.—1 ♀, 40.0; same; holotype; USNM 100931.—4 ♂ (one 27.3, rest fragmented); 4 ♀, fragmented; same; 23 May 1955; paratypes; USNM 98151.—1 ♂, 33.7; 1 ♀, 39.6; same; 31 August 1956; paratypes; USNM 100933.—2 ♂, 36.4-37.5; 5 ♀, 36.6-41.5; same; 18 March 1957; paratypes; USNM 100932.—1 ♂, 31.4; same; 1957; USNM 111033.—3 ♂, 32.2-39.8; same; 29 July 1957; SEP 279.—1 ♂, 39.4; same; September 1957; SEP 126.—1 juv. ♀; Squiteague Pond, North Falmouth; September 1957; M. B. Gray; USNM 102281.—1 ♂, 22.3; 1 ♀, 20.5; same; SEP 286. —2 ♂, 36.7-37.3; 2 ♀, 37.2-37.6; Wellfleet; July 1960; S. Gray; USNM 106105.—GEORGIA: 1 ♀, 27.2; off Sapelo Id., Doboy Sound sea buoy 3 mi 274°; 15m; 4 February 1963; Frankenberg and Gray; SRF.

*Diagnosis.*—Rostral plate with rounded or subacute anterolateral angles; raptorial claw with 11-15 teeth; posterior armature of telson consisting of 7-9 submedian denticles, 1 movable submedian tooth, adjacent to outer denticle, and 3-5 fixed teeth and denticles lateral to movable tooth; inner spine of basal prolongation the longer.

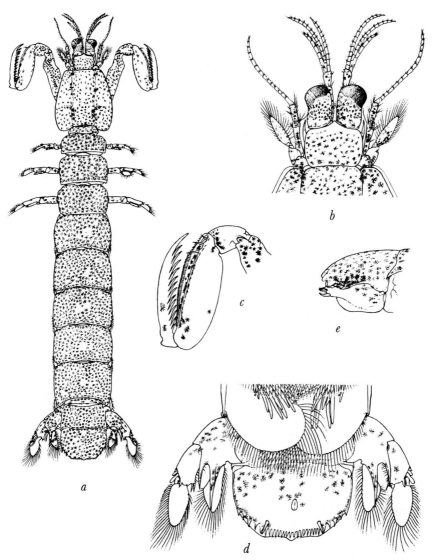

FIGURE 20. *Nannosquilla grayi* (Chace), female holotype, TL 40.0, Bass River, Massachusetts. *a*, dorsal view; *b*, anterior portion of body; *c*, inner view of raptorial claw; *d*, telson and uropods, ventral view; *e*, telson, lateral view (from Chace, 1958).

*Description.*—Eye extending to or beyond end of second segment of antennular peduncle; ocular scales fused.

Antennular peduncle short, about half as long as carapace; antennular processes visible lateral to rostral plate as slender, anteriorly-directed spines.

Antennal scale small, about one-fourth as long as carapace; antennal peduncle short, not extending beyond eye.

Rostral plate broader than long, anterolateral angles truncate or rounded, not extending anteriorly to level of obtuse apex; plate may extend to or beyond base of eyes.

Dactylus of raptorial claw with 11-15 teeth, outer margin of dactylus with basal notch, proximal flanking lobe larger, sharper than distal; dorsal ridge of carpus terminating in obscure lobe.

Sixth abdominal somite without strong posterolateral spines.

False eave of telson with small, obtuse median prominence, flanked laterally by obtuse or almost acute lobe; marginal armature consisting of a curved row of 7-9 submedian denticles, inner highest, 1 movable submedian tooth, adjacent to outermost denticle, and 3-5 fixed lateral teeth and denticles.

Proximal segment of exopod of uropod with 6 spatulate spines, last extending to midlength of distal segment; inner margin of proximal segment with 3-5 stiff distal setae; inner spine of basal prolongation the longer.

*Color.*—Many dark chromatophores scattered over the cream-colored body in no particular pattern.

*Size.*—Males, TL 22.3-39.8 mm; females, TL 20.5-42.1 mm. Other measurements of female, TL 37.6: carapace length, 5.5; cornea width, 1.0; rostal plate length, 1.5, width 2.2; fifth abdominal somite width, 6.0; telson length, 3.2, width, 4.8.

*Discussion.*—N. *grayi* is the largest of the species now placed in this genus. It can be distinguished from other American species by characters given in the key. Of the other western Atlantic species, N. *taylori* is perhaps the most similar. These two species can be distinguished by the configuration of the submedian denticles of the telson, which form a "W" in posterior view in N. *taylori* and a transverse row in N. *grayi*.

*Remarks.*—N. *grayi* and N. *chilensis* are the only known temperate members of the genus; all other species are found in tropical or subtropical waters.

N. *grayi* may well encounter waters of reduced salinity, although this has not been recorded definitely. The species has been collected in association with *Arenicola cristata*. It must be fairly tolerant to varying factors other than temperature, for it has also been taken off Georgia.

*N. decemspinosa* is the eastern Pacific analogue of *N. grayi*, as pointed out by Chace (1958) and Manning (1961).

*Ontogeny.*—Unknown.

*Type.*—U. S. National Museum.

*Type-locality.*—Bass River, Yarmouth, Massachusetts.

*Distribution.*—Massachusetts and Georgia.

## Nannosquilla taylori, new species
Figure 21

*Holotype.*—1 ♀, 28.3; 150 yds. W of Egmont Key, Florida; 1.8 m on fine shell and sand bottom; 16 may 1964; J. L. Taylor; USNM 119155.

*Diagnosis.*—Rostral plate rounded anterolaterally; dactylus of raptorial claw with 11 teeth; posterior armature of telson consisting of 10-11 submedian denticles, 1 movable submedian tooth, adjacent to outer submedian denticle, and 4 fixed lateral teeth and denticles; inner spine of basal prolongation of uropod the longer.

*Description.*—Eye small, not extending past end of antennular peduncle; cornea set obliquely on stalk, expanded laterally; ocular scales fused along midline, apices distinct.

Antennular peduncle short, about half as long as carapace; antennular processes visible as anteriorly-directed spines on either side of rostral plate.

Antennal scale small, about one-fourth as long as carapace; antennal peduncle not extending beyond eye.

Rostral plate pentagonal, broader than long, rounded anterolaterally, concave anterior margins sloping to obtuse apex; lateral margins of plate very divergent; plate completely covering base of eyes in holotype.

Dactylus of raptorial claw with 11 teeth; shallow basal notch on outer margin of dactylus flanked by proximal acute prominence and distal obtuse lobe, both low; dorsal ridge of carpus terminating in blunt tooth.

Sixth abdominal somite acute posterolaterally.

Telson smooth dorsally, broader than long; false eave with obtusely rounded median projection flanked laterally by obtuse lobe separated from median projection by shallow, irregular depression; marginal armature of telson consisting of 11-12 fixed submedian denticles, forming a "W" in posterior view, 1 movable submedian tooth adjacent to outer denticle, and 4 fixed lateral teeth and denticles, curving anteriorly, second and fourth smaller than first and third.

Uropods as in *N. hancocki.*

*Color.*—Background color cream, with numerous dark chromatophores scattered over body, telson and uropods in no particular pattern; lateral margins of telson dark.

*Size.*—Female holotype, only known specimen, TL 28.3 mm. Other measurements: carapace length, 4.5; cornea width, 1.1; rostral plate length, 1.5, width, 1.8; fifth abdominal somite width, 4.8; telson length, 2.6, width, 3.8.

*Discussion.*—*N. taylori* most closely resembles *N. hancocki* but differs in that the median projection of the false eave of the telson is broader, the submedian depressions on the eave are shallower, the submedian denticles are not in a transverse row but form a "W", with outer denticles highest, there are four fixed spines lateral to the movable submedians, rather than five, and that the movable submedian teeth are adjacent to the outermost denticle, not anterior to it. In *N. hancocki,* a line drawn between the apices of the marginal teeth of the telson will cross the movable submedian teeth; in *N. taylori* the submedians would be posterior to such a line.

N. *taylori* also differs in that the cornea noticeably overhangs the stalk and the lateral margins of the rostral plate are very divergent, but these differences may be due to the larger size of the only known specimen.

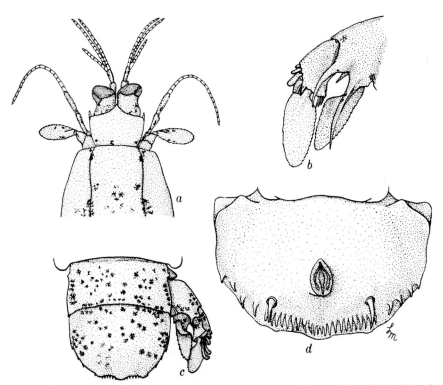

FIGURE 21. *Nannosquilla taylori,* n. sp., female holotype, TL 28.3, W. of Egmont Key, Florida. *a,* anterior portion of body; *b,* uropod, ventral view; *c,* last abdominal somite, telson, and uropod; *d,* telson, ventral view (setae omitted).

*Etymology.*—The species is named for J. L. Taylor, who collected the holotype and made it available for study.

*Ontogeny.*—Unknown.

*Type.*—U. S. National Museum.

*Type-locality.*—Egmont Key, west coast of Florida.

*Distribution.*—Known only from the type-locality.

## Genus *Coronis* Desmarest, 1823

*Coronis* Desmarest, 1823, p. 345 [published September 1823].—Latreille, 1828, p. 474.—Manning, 1963a, p. 318.—Holthuis & Manning, 1964, p. 140.—Manning, 1968, p. 111 [key].
not *Coronis* Huebner, 1823, p. 265 [Lepidoptera; published on or before 21 December 1823].
*Lysiosquilla (Coronis):* Brooks, 1886, p. 48.

*Definition.*—Size moderate, TL 80 mm or less; body smooth, strongly depressed, loosely articulated; cornea subglobular, scarcely expanded beyond stalk and set obliquely on it; rostral plate cordiform, with apical spine; antennal protopod with at least 1 ventral papilla; carapace short, carinae and spines absent, cervical groove indicated on lateral plates only; exposed thoracic somites without carinae; eighth thoracic somite without median ventral keel or projection; 5 epipods present; mandibular palp present; propodus of fifth thoracic appendage about as broad as long, ventral surface with stiff brush of setae; raptorial claw slender, dactylus armed with 13-17 teeth, outer margin of dactylus with small basal notch; propodus fully pectinate, with 4 movable spines, second the smallest, on proximal portion of inner margin; dorsal ridge of carpus undivided, terminating in blunt spine; ischiomeral articulation terminal, merus slender, elongate, slightly longer than carpus; endopods of walking legs two-segmented, broadest on second leg, most elongate on third; each walking leg with 1 pair ventrally-directed posterior spines on basal segment; abdomen strongly flattened, without longitudinal carinae or spines, pleura thin, membranous, anterolateral plates only partially articulated; sixth abdominal somite without posterolateral spines, but with ventrally-directed process on either side in front of articulation of uropod; telson much broader than long, with smooth, unarmed median projection and 1 pair of obtuse submedian projections; posterior armature reduced, consisting of a row of slender, fixed denticles, 1 movable submedian tooth, and 1 sharp lateral tooth; uropod broad, flattened, basal segment with inner and outer carinae, inner terminating in sharp spine; proximal segment of exopod with longitudinal dorsal carina, row of 6-9 slender, sharp movable spines on outer margin; inner margin of proximal segment setose, with 5 stiff terminal setae; endopod triangular, basal portion of outer edge folded over; basal prolongation produced into 2 spines, inner smaller, circular in cross-section, outer much larger, flattened, triangular in cross-section.

*Type-species.*—*Coronis scolopendra* Latreille, 1828, p. 474, by subsequent monotypy.

*Gender.*—Feminine.

*Number of species.*—Two, both of which are discussed below.

*Nomenclature.*—*Coronis* Desmarest has long been considered a junior homonym of *Coronis* Huebner and has been ignored or placed in the synonymy of *Lysiosquilla.* Brooks (1886) used it as a subgenus of *Lysiosquilla.* Manning (1963a) was the first recent author to recognize the genus.

*Remarks.*—This is the only genus of the lysiosquillids which is restricted to the Americas.

The presence of 10 or more submedian denticles on either side of the midline of the telson will distinguish the South American *C. scolopendra* from its North American relative, *C. excavatrix,* which has but 3-6 submedian denticles.

*Affinities.*—*Coronis* is very similar to *Platysquilla,* differing in having the marginal armature of the telson reduced (one fixed marginal tooth or lobe instead of four), and in having the outer spine of the basal prolongation longer than the inner. In addition, *Coronis* has fewer papillae on the antennal protopod and has a mandibular palp. In other features there is close agreement between the two genera.

*Coronis* agrees with *Platysquilla, Acanthosquilla,* and *Nannosquilla* but differs from *Heterosquilla* and *Lysiosquilla* in having broad distal segments on the endopods of the first two walking legs, having a prominent, ventrally-directed process on each side of the sixth abdominal somite in front of the articulation of the uropod, and in having a triangular endopod on the uropod, with the proximal portion of the outer edge folded over.

Some of the species of *Nannosquilla* are similar to those in *Coronis,* but the marginal armature of the telson is far more complex in the former than in either of the species of *Coronis.* Other differences are discussed under *Nannosquilla.*

*Coronis excavatrix* Brooks, 1886

Figures 22-23

*Lysiosquilla:* Brooks, 1885, p. 10 [biology]; 1886b, p. 166 [biology].
*Lysiosquilla (Coronis) excavatrix* Brooks, 1886a, pp. 84-85 [*nomen nudum*]; 1886, p. 48, pl. 10, figs. 8-16; p. 101, pl. 11, figs. 1-3 [ontogeny].
*Lysiosquilla (Erichthus) excavatrix:* Brooks, 1886, p. 101 [larva].
*Lysiosquilla excavatrix:* Brooks, 1892, p. 356.—Hansen, 1895, p. 66.—Wilson, 1900, p. 352 [listed].—Sheldon, 1905, p. 336 [listed].—Kemp, 1913, p. 203 [listed].—Verrill, 1923, p. 193 [discussion].—Lunz, 1935, p. 153, fig. 2 [part].— Schmitt, 1940, p. 180 [discussion].—Anonymous, 1942, p. 5.—Pearse, Humm, & Wharton, 1942, p. 185, figs. 10-13 [ecology].—Gurney, 1946, p. 167 [synonymy of larval stages].—Behre,

1950, p. 19.—Chace, 1954, p. 449 [part]; 1958, p. 145 [key].—Manning, 1959, p. 17.—Dawson, 1963, p. 7 [listed].—Manning, 1963a, p. 318 [listed].—Dawson, 1965, p. 13.
*Lysiosquilla excavitrix:* Hildebrand, 1954, p. 260 [erroneous spelling].
*Lysiosquilla excavata:* Chace, 1958, p. 143 [erroneous spelling].

*Previous records.*—NORTH CAROLINA: Brooks, 1886; Wilson, 1900; Lunz, 1935; Pearse, Humm, & Wharton, 1942; Chace, 1954.—ALABAMA: Lunz, 1935; Chace, 1954.—MISSISSIPPI: Manning, 1959; Dawson, 1965.—LOUISIANA: Anonymous, 1942; Behre, 1950; Chace, 1954.—TEXAS: Hildebrand, 1954.

*Material.*—NORTH CAROLINA: 2 ♀, 63.4-64.1; Ft. Macon; December 1871; Dr. Yarrow; YPM 4414.—1 brk. ♀, CL 8.5; Boque Banks, near Ft. Macon; sand beach; 11 May 1941; K. D. McDougal; USNM 81685.— 1 ♀, 74.5; Bird Shoal, Ft. Macon; on beach; 11 June 1941; A. S. Pearse; USNM 81690.—1 abdomen; Ft. Macon Beach; 20 June 1941; A. S. Pearse; USNM 81686.—1 ♂, ca. 62.0; 2 brk. ♀, CL 9.1-9.3; 1 telson; Bird Shoal, Ft. Macon; 25 June 1941; A. S. Pearse; USNM 81688.— 1 brk. ♂, CL 8.7; outside, Ft. Macon Beach; 5 July 1941; A. S. Pearse; USNM 81687.—1 ♂, 60.5; same; 10 July 1941; A. S. Pearse; USNM 81689.—1 ♂, 69.5; Sheepshead Shoal, Beaufort; 20 June 1941; A. S. Pearse; USNM.—3 ♂, (1 brk.), 55.4-60.2; 3 ♀, 58.0-69.2; Beaufort; syntypes; BMNH 1909.1.20.1-4.—ALABAMA: 1 soft ♂, ca. 63.0; 1 soft ♀, ca. 72.0; Mobile; September 1855; Dr. Nott; MCZ 7904.—MISSISSIPPI: 1 brk. ♀, ca. 29.0; off Petit Bois Id., Mississippi Sound; 4 August 1953; S. L. Wallace; GCRL 161: 206.—TEXAS: 1 ♀, 57.4; near south jetty, Port Aransas; June 1952; D. Darling; USNM 96476.—LOUISIANA: 1 ♂, 44.5; Front Beach, Grand Isle; 19 July 1936; E. Behre; USNM 113315.

In addition to these specimens, there is a male, TL 48.0, in the collection of the U. S. National Museum that bears the label: "Probably Bahamas." This record is doubtful and needs to be verified.

*Diagnosis.*—Eye very small, cornea subglobular, slightly overhanging stalk laterally; ocular scales fused; eyes extend to or slightly beyond end of second segment of antennular peduncle.

Antennular processes visible as sharp, forwardly directed spines lateral to rostral plate; antennular peduncle less than half as long as carapace.

Antennal scale short, about one-half as long as carapace; antennal protopod with 1 mesial and 1 ventral papilla.

Rostral plate as broad as long or slightly broader, greatest width in advance of base; lateral margins broadly rounded; rostral plate extends to cornea, completely covering base of eyes and ocular scales.

Dactylus of raptorial claw armed with 15-17 teeth.

Telson with 3-6 submedian denticles on either side, 8-11 total.

Exopod of uropod with 6-9 movable spines on outer margin, last not extending to midlength of distal segment; distal segment longer than proximal.

*Color.*—Female opaque olive-brown, male transparent gray (Brooks, 1886).

*Size.*—Males, TL 44.5-69.5 mm; females, TL 29.0-74.5 mm. Other measurements of female, TL 74.5: carapace length, 9.4; eye length, 2.2; cornea width, 1.4; rostral plate length, 2.4, width, 3.1; telson length, 5.4, width, 8.1.

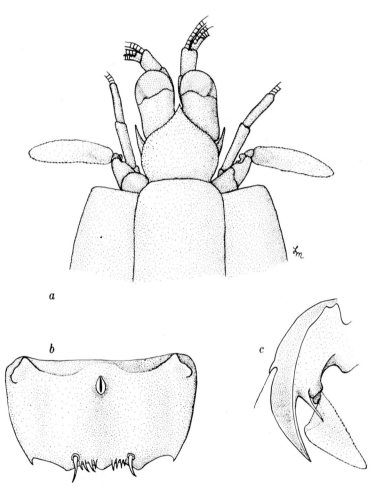

FIGURE 22. *Coronis excavatrix* Brooks, male, TL 60.5, Fort Macon Beach, North Carolina. *a*, anterior portion of body; *b*, telson, ventral view; *c*, uropod, ventral view (setae omitted).

86

*Discussion.*—The small number of submedian denticles on the telson will distinguish this species from *C. scolopendra.* All of the specimens examined had both papillae present on the antennal protopod; only the ventral papilla was present on the only specimen of *C. scolopendra* examined. This character may prove to be reliable in separating the two species, but should be checked in more specimens of *C. scolopendra. C. scolopendra*

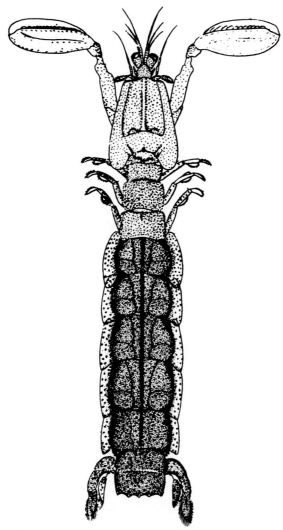

FIGURE 23. *Coronis excavatrix* Brooks, male, Beaufort, North Carolina (from Brooks, 1886).

seems to have fewer (13-15) teeth on the raptorial claw than *C. excavatrix* (15-17) but the overlap reduces the value of this character.

There may also be a difference in the size of the two species. The only extant specimens of *C. scolopendra* have total lengths ranging from 30.0-37.5 mm, whereas *C. excavatrix* can attain a total length of at least 74.5 mm.

*Remarks.*—Brooks (1885, 1886, 1886a) commented on the difficulties of obtaining intact specimens of *C. excavatrix*. The species is very secretive and apparently does not leave its burrow freely. Specimens can be enticed to the mouth of their burrow with a piece of fish; their retreat into the burrow can be cut off then with a trowel. Pearse, Humm, & Wharton (1942) discuss the adaptations of this species to a burrowing existence.

*Ontogeny.*—Brooks (1886) reared this species from a long-spined erichthus. Hansen (1895) pointed out that Brooks possibly had mixed larvae of several species in his account.

*Type.*—British Museum (Natural History).

*Type-locality.*—Beaufort, North Carolina.

*Distribution.*—North Carolina, northern Gulf of Mexico from Alabama to Texas. The specimen from Charlotte Harbor, Florida, reported by Lunz (1935) and Chace (1954) is badly damaged and cannot be identified to species with certainty. The telson armature, even though damaged, indicates that it belongs in *Platysquilla*.

*Coronis scolopendra* Latreille, 1828
Figure 24

> *Coronis scolopendra* Latreille, 1828, p. 474.—Guérin-Méneville, 1829-34, pl. 24, figs. 2-2b.—H. Milne-Edwards, 1837, p. 531.—H. Milne-Edwards, in Cuvier, 1837, p. 161, pl. 55, fig. 3.—Lucas, 1840, p. 212.—Desmarest, in Chenu, 1858, p. 45 [listed].—Manning, 1963a, p. 318 [listed].—Holthuis & Manning, 1964, p. 141 [petition to ICZN to have both generic and specific names placed on Official List].
> *Lysiosquilla scolopendra:* Miers, 1880, p. 9 [discussion].—Kemp, 1913, p. 204 [listed].—Andrade Ramos, 1951, p. 143 [discussion].
> *Lysiosquilla excavatrix:* Lemos de Castro, 1955, p. 39, text-figs. 27-29, pl. 18, fig. 55 [not *L. excavatrix* Brooks, 1886]

*Previous records.*—BRAZIL: Latreille, 1828; Lemos de Castro, 1955.

*Material.*—BRAZIL: 1 ♂, 37.5; São Sebastião; October 1955; USNM 111037.

*Diagnosis.*—Antennal protopod with 1 ventral papilla; dactylus of raptorial claw with 13-15 teeth; telson with 7-11 submedian denticles on either side, 14-21 total.

*Color.*—The only specimen examined has faded. Both Latreille (1828) and Lemos de Castro (1955) noted that their specimens were brown.

88

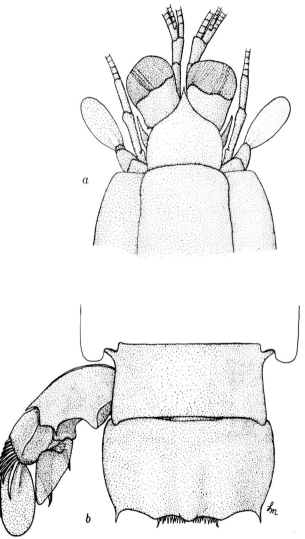

FIGURE 24. *Coronis scolopendra* Latreille, male, TL 37.5, São Sebastião, Brazil. *a,* anterior portion of body; *b,* last abdominal somite, telson, and uropod (setae omitted).

*Size.*—Male, TL 37.5; female, TL 30.0 (Lemos de Castro, 1955). Other measurements of male: carapace length, 7.0; eye length, 1.9; cornea width, 1.3; rostral plate length, 2.1, width, 2.2; telson length, 2.5, width, 5.0.

*Discussion.*—Differences between this species and *C. excavatrix* are summarized in the discussion of *C. excavatrix.*

*Ontogeny.*—Unknown.

*Type.*—Not extant. H. Milne-Edwards (1837) reported that he could not locate the type-specimen.

*Type-locality.*—Brazil.

*Distribution.*—Known only from Brazil, where it has been reported from Santos and São Sebastião.

Genus Platysquilla Manning, 1967
Platysquilla Manning, 1967, p. 238; 1968, p. 111 [key and text].

*Definition.*—Size moderate, TL 75 mm or less; body smooth, depressed, loosely articulated; eyes of moderate size, cornea faintly bilobed, set obliquely on stalk; rostral plate quadrate, with apical spine; antennal protopod with mesial and ventral papillae; carapace narrowed anteriorly, without carinae or spines, cervical groove indicated on lateral plates only; thoracic somites without dorsal carinae, lateral margins truncate; eighth thoracic somite with low, inconspicuous tubercle on midline of ventral surface; at least 4 epipods present; mandibular palp absent; raptorial claw slender, dactylus armed with 9 or more teeth; other margin of dactylus notched basally; propodus fully pectinate, with 4 movable spines at base, first longest, second smallest; carpus with distal, dorsal spine; ischiomeral articulation terminal; merus slender, elongate, longer than ischium; endopods of walking legs two-segmented, distal segment of first two legs ovate, that of last leg slenderest; abdomen depressed, loosely articulated, anterolateral plates with complete suture; sixth somite with or without posterolateral spines, with curved, ventrally-directed process present on each side in front of articulation of uropod; telson broad, with obtuse, triangular median posterior projection; marginal armature consisting of a row of slender submedian denticles, 1 movable submedian tooth, and 4 fixed lateral teeth, with a denticle between each; basal segment of uropod with 2 dorsal carinae, inner terminating in slender spine; proximal segment of exopod with short dorsal carina, outer margin armed with slender, movable spines; endopod triangular, with proximal portion of outer edge folded over; spines of basal prolongation triangular in cross-section, inner spine the longer.

*Type-species.*—*Squilla eusebia* Risso, 1816, by original designation.

*Gender.*—Feminine.

*Number of species.*—Two.

90

*Remarks.*—When I proposed the genus *Heterosquilla,* I noted that further study was needed to clarify the position of some of the species assigned therein. Additional work on *Heterosquilla* for this monograph has shown that two species, *S. eusebia* Risso and *L. enodis* Manning, definitely do not belong in *Heterosquilla* and cannot be assigned to any of the currently recognized genera; they were placed in *Platysquilla* in 1967. Both of these species differ from all of those previously assigned to *Heterosquilla* in having four pairs of fixed marginal teeth on the telson rather than two, in having the curved, ventrally-directed process on each side of the sixth abdominal somite in front of the uropod, and in having the proximal portion of the outer margin of the uropodal endopod folded over.

The differences between the two described species are summarized below under *P. enodis.* In addition to these two species, there may be a third species in the genus from the Gulf of Mexico; see under the *Discussion* for *P. enodis.*

*Affinities.*—*Platysquilla* resembles *Coronis* in most features, differing mainly in lacking the mandibular palp, having four pairs of fixed marginal teeth on the telson, and in having the inner spine of the basal prolongation of the uropod longer than the outer.

*Platysquilla enodis* (Manning, 1962)
Figure 25

*Lysiosquilla armata:* Bigelow, 1894, p. 507 [not *L. armata* Smith].
*Lysiosquilla enodis* Manning, 1962c, p. 220; 1963a, p. 321 [listed]; 1967, p. 238; 1968, p. 112 [listed].

*Previous records.*—MASSACHUSETTS: off Vineyard Sound (Manning, 1962c).—NORTH CAROLINA: Manning, 1962c.

*Material.*—MASSACHUSETTS: 1 ♀, 57.5; off Vineyard Sound; 31-49 m; U. S. Fish Commission Stats. 1247-1251; from the stomach of a flounder; S. I. Smith; 1887; holotype; USNM 12787.—1 fragmented ♂; same; paratype; USNM 111039.—NORTH CAROLINA: 1 brk. ♀; ALBATROSS Sta. 2296; paratype; USNM 8816.

*Diagnosis.*—Rostral plate with apical spine; dactylus of raptorial claw with 9 teeth; sixth abdominal somite lacking paired ventral spines; telson with 4 pairs of fixed marginal projections, inner 2 pair spatulate.

*Description.*—Eye of moderate size, cornea faintly bilobed, set very obliquely on stalk; outer margin of stalk with blunt projection; ocular scales with bases separate, apices appressed; eyes not extending past end of second segment of antennular peduncle.

Antennular peduncle short, more than half as long as carapace; antennular processes produced into sharp, slender, anteriorly-directed spines.

Antennal scale small, one-third as long as carapace; antennal peduncle with 1 mesial and 2 ventral papillae.

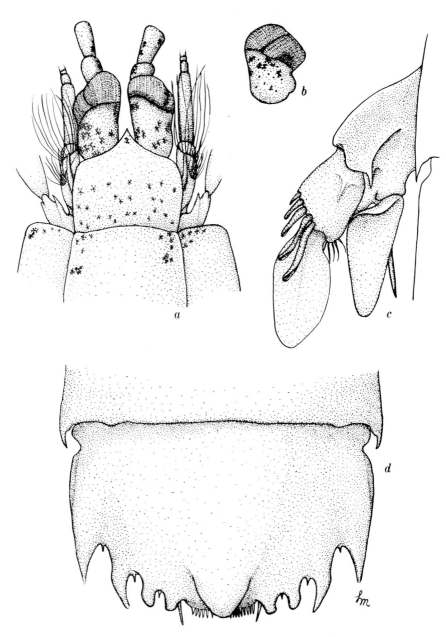

Figure 25. *Platysquilla enodis* (Manning), female holotype, TL 57.5, off Vineyard Sound, Massachusetts. *a,* anterior portion of body; *b,* eye; *c,* uropod; *d,* telson (setae omitted).

Rostral plate quadrate, anterolateral angles rounded, anterior margins sloping to obtuse apex; plate completely covers base of eyes, but apex does not extend to cornea.

Dactylus of raptorial claw with 9 teeth, outer margin of dactylus with basal notch; propodus stout, movable spines damaged in only specimen with claw; dorsal ridge of carpus terminating in strong spine; merus longer than ischium.

Mandibular palp absent; no more than 4 epipods present.

Fifth thoracic somite without noticeable lateral or ventral process; lateral process of sixth somite truncate, more rounded anterolaterally than posterolaterally; lateral process of seventh somite broader and more rounded than that of sixth; outer spine at base of walking legs slenderer than inner; distal segment of endopods of walking legs ovate, longer than broad, that of last leg slenderest.

Sixth abdominal somite with sharp posterolateral spines; spine in front of articulation of uropod slender; ventral surface of sixth somite unarmed.

Telson much broader than long, with blunt, rounded, median triangular projection on dorsal surface; marginal armature consisting of a transverse row of 8-9 slender denticles, a movable submedian tooth, and 4 fixed teeth, inner 2 spatulate, outer sharp, with 1 denticle between each; first spatulate tooth above movable submedian tooth.

Basal segment of uropod with inner and outer carina, inner incomplete proximally, terminating distally in sharp spine; proximal segment of exopod with short dorsal carina, outer margin with 5 spatulate spines, last 2 recurved; inner margin of proximal segment with rounded lobe tufted with setae; distal segment of exopod longer than proximal; endopod triangular, proximal portion of outer margin folded over; spines of basal prolongation triangular in cross-section, inner longer.

*Color.*—Faded in all specimens.

*Size.*—Only intact specimen, female holotype, TL 57.5 mm. Other measurements: carapace length, 9.5; cornea width, 1.8; rostral plate length, 2.7, width, 3.1; fifth abdominal somite width, 10.0; telson length, 5.2, width, 8.3.

*Discussion.*—*P. enodis* differs from *P. eusebia* in having the inner two pairs of marginal teeth on the telson spatulate instead of sharp; also *enodis* lacks the paired ventral spines on the sixth abdominal somite that are characteristic of *eusebia*.

The specimen from off Charlotte Harbor, Florida, in the Gulf of Mexico reported by Lunz (1935) and Chace (1954) as *Lysiosquilla excavatrix* is an undescribed species of *Platysquilla*. It differs from *P. enodis* in lacking an apical spine on the rostral plate and in having the inner two pairs of telson teeth sharp. The specimen, a female TL ca. 14 mm, lacks the raptorial claws and uropods and cannot be adequately characterized.

*Remarks.*—Bigelow (1894) commented on these specimens and pointed out differences between them and *L. armata* Smith (now *Heterosquilla*).

*Ontogeny.*—Unknown.

*Type.*—U. S. National Museum.

*Type-locality.*—Off Vineyard Sound, Massachusetts.

*Distribution.*—Massachusetts; North Carolina, in 31-49 m.

## Family BATHYSQUILLIDAE Manning, 1967

Bathysquillidae Manning, 1967, p. 238; 1968, pp. 109 [key], 113.

*Definition.*—Propodi of last 3 maxillipeds longer than broad, propodus of fourth not markedly broader than that of fifth; telson with median dorsal carina and 4 pairs of marginal teeth, all with movable apices; submedian denticles present, all other denticles absent.

*Type-genus.*—*Bathysquilla* Manning, 1963.

*Number of genera.*—One.

*Ontogeny.*—Unknown. Larvae identifiable with *Bathysquilla* have not been recorded in the literature; an account of a late larva taken by the GALATHEA is in preparation.

*Remarks.*—Manning (1967) commented on the distinctness of the two species of *Bathysquilla* and erected the family Bathysquillidae for the single genus. The presence of movable apices on all marginal teeth of the telson will immediately distinguish members of this family from all other stomatopods.

Ingle (1963) questioned the relationship of one of the species of *Bathysquilla, Lysiosquilla crassispinosa* Fukuda, to *Lysiosquilla s. l.* and pointed out that the presence of a median carina on the telson might indicate some relation to *Pseudosquilla.*

The unique structure of the female genital pores was pointed out by Fukuda (1910) and Gordon (1929). This feature has not been surveyed at the generic level.

## Genus *Bathysquilla* Manning, 1963

Bathysquilla Manning, 1963a, p. 323; 1967, p. 238.

*Definition.*—Body depressed, broad, surface smooth anteriorly, tuberculate posteriorly; eyes with cornea reduced, bilobed or subglobular in shape, set obliquely on stalk; rostral plate subtriangular or trapezoidal, with apical spine; antennal protopod with at least 1 ventral papilla; carapace strongly narrowed anteriorly, without longitudinal carinae or spines; marginal carinae present on posterior portion of lateral plates, not reflected anteriorly; cervical groove not distinct across dorsum of carapace; last 2 or 3 thoracic somites with intermediate carinae; eighth thoracic somite with

94

blunt, noncarinate median prominence on posterior margin of ventral surface; epipods present on first 5 thoracic appendages; mandibular palp present, three-segmented; inferior surface of propodi of last 3 maxillipeds not beaded or ribbed; raptorial claw large, dactylus armed with 10 or more teeth; propodus with a distal spine on inferior margin, a row of small, fixed spines on outer face of upper margin, a row of large fixed spines on inner face of upper margin, and 4 movable spines at base of upper margin; carpus without a prominent dorsal ridge but with 1 or 2 strong dorsal spines; merus stout, grooved inferiorly throughout its length for reception of propodus; ischiomeral articulation terminal; endopod of walking legs slender, two-segmented; abdomen strongly depressed, anterolateral plates not articulated; first 5 abdominal somites with intermediate and marginal carinae, all armed with posterior spines; sixth abdominal somite with 6 posterior spines and a median ventral spine on posterior margin; telson broad, with a median carina and 4 pairs of marginal teeth, all with movable apices; minute submedian denticles present, other denticles absent; uropods with dorsal spine on proximal segment, basal prolongation with inner spine much the longer.

*Type-species.*—*Lysiosquilla microps* Manning, 1961, p. 693, by original designation.

*Gender.*—Feminine.

*Number of species.*—Two.

*Nomenclature.*—This genus was erected for two aberrant species which in the past have been placed in *Lysiosquilla, B. microps* (Manning, 1961) from the western Atlantic and *B. crassispinosa* (Fukuda, 1910) from Japan and South Africa.

*Remarks.*—*Bathysquilla* has the deepest bathymetric range of any living stomatopod. *B. microps* occurs in depths to 952 m, and *B. crassispinosa* is recorded from depths between 242 and 300 m (Komai, 1938; Barnard, 1950).

There is a possibility that the South African and the Japanese specimens referred to *B. crassispinosa* are distinct species. Fukuda (1910) neither mentioned nor showed dorsal carinae lateral to the median carina of the telson in his Japanese material, but such carinae are illustrated by Barnard (1950) from his South African specimen. Further, Fukuda does not show or mention a distal spine on the proximal segment of the exopod of the uropod, whereas Gordon (1929) illustrated this spine in her specimen from South Africa. A direct comparison of specimens from both areas is needed.

*Bathysquilla microps* (Manning, 1961)
Figures 26-28

*Lysiosquilla microps* Manning, 1961a, p. 683, pl. 10, figs. 1-2, pl. 11, figs. 3-4, text-fig. 5; 1963a, p. 323 [listed].

*Bathysquilla microps:* Bullis & Thompson, 1965, p. 13 [listed].—Manning, 1968, fig. 1b.

*Previous records.*—FLORIDA: Manning, 1961a; Bullis & Thompson, 1965 (occurrence at three SILVER BAY Stats., as given below).

*Material.*—BAHAMAS: 1 ♂, 67.7; Santaren Channel; SILVER BAY Sta. 3516; USNM 111108.—FLORIDA: 1 ♀, 44.5; E of Cape Canaveral; SILVER BAY Sta. 445; paratype; UMML 32.1759.—1 ♂, 198.0; SE of Tortugas; SILVER BAY Sta. 1196; holotype; USNM 104109.

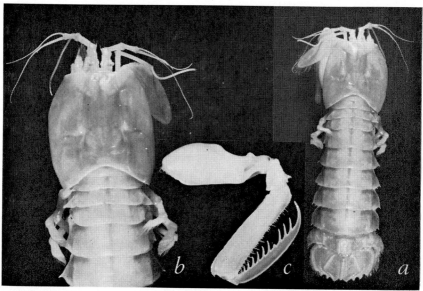

FIGURE 26. *Bathysquilla microps* (Manning), male holotype, TL 198.0, SE of Tortugas, Florida. *a,* dorsal view; *b,* anterior portion of body, enlarged; *c,* raptorial claw (*a, c* from Manning, 1961a).

*Diagnosis.*—Cornea indistinctly bilobed, pigmented portion of cornea reduced to a transverse bar; dactylus of claw with 13 teeth; last 3 thoracic and first 5 abdominal somites with a pair of dorsal carinae in addition to marginals; dorsal surface of telson with 1 pair of oblique submedian carinae.

*Description.*—Eyes very small, pigmented portion restricted to a slender, transverse bar; cornea faintly bilobed, rounded in young but elongate in adult, set obliquely on stalk; ophthalmic somite without ocular scales.

Antennules not as long as carapace; basal segment with outer margin tuberculate.

Antennal scale shorter than carapace in small specimens, longer than carapace in adult; scale 2½ to 3 times as long as broad; antennal protopod with 1 lateral and 1 ventral papilla.

Raptorial claw very large, extending, when folded, from second segment of antennular peduncle to level of cervical groove; outer margin of dactylus a simple curve with a prominent basal notch.

Last 3 thoracic somites with intermediate carinae; lateral processes of thoracic somites spatulate in juveniles, acute in adult, directed laterally; lateral process of seventh somite an acute spine, bifurcate in oldest specimen, spatulate in smallest.

Abdomen with swollen intermediate and marginal carinae on first 5 somites, posterolateral angles all spined; intermediate carinae notched anteriorly, armed with posterior spine or tubercle on second to sixth somite of largest specimen; third abdominal somite unarmed on posterior margin or with up to 3 posterior spinules between intermediate and marginal carinae; fourth somite with 2 or 3 spinules lateral to intermediate carina and with or without 1 spinule mesial to that carina; fifth somite with 2 spinules lateral to intermediate carina and up to 4 mesial to it; largest specimen with a median dorsal tubercle on this somite; submedian carinae of sixth somite almost coalesce anteriorly, all carinae swollen,

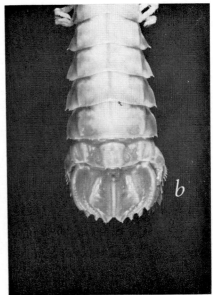

FIGURE 27. *Bathysquilla microps* (Manning), male holotype, TL 198.0, SE of Tortugas, Florida. Posterior portion of body in: *a,* ventral view; *b,* dorsal view (from Manning, 1961a).

spines of intermediates obsolete; no ventral spine on sixth somite in front of articulation of uropod; posterior part of body roughened.

Telson with small apical spine on median carina, obsolete on holotype; submedian carinae on dorsal surface oblique to body axis, with irregular line of tubercles leading to dorsal surface of submedian tooth; an irregular, curved line of tubercles leading anteriorly from intermediate teeth; distal portion of median carina with converging tubercles; 16-23 minute submedian denticles present; ventral surface of telson with a long postanal keel.

Uropods with 7 or 8 graded, movable spines on outer margin of penultimate segment, last extending past midlength of distal segment; proximal segment of uropod with distal, dorsal spine, distal ventral spine not present; proximal segment of endopod, in addition to outer marginal spines, with fixed spines on ventral surface between movable spines, increasing in size distally; ventral distal spine present on this segment, distal dorsal spine absent; basal prolongation with a strong spine at base of endopod, in addition to 2 distal spines.

*Color.*—Pastel orange in life; all pigment faded in preservative.

*Size.*—Males, TL 67.7-198.0 mm; only known female, TL 44.5 mm. Other measurements of male holotype, TL 198.0: carapace length, 38.7; cornea width, 5.6; antennal scale length, 43.5; raptorial propodus length, 50.7; fifth abdominal somite width, 52.0; telson length, 40.0, width, 51.1.

*Discussion.*—*B. microps* is closely related to *B. crassispinosa,* but differs in having more teeth on the raptorial claw (13 instead of 10-11) and in having slender rather than globular eyes. The uropod structure may also be different, for *B. microps* lacks a ventral spine on the basal segment of the uropod and a dorsal spine on the penultimate segment of the exopod, and Gordon (1929) found spines in both of these places in *B. crassispinosa.* Finally, the carpus of *B. microps* bears one dorsal spine, that of *B. crassispinosa* two.

*Remarks.*—In the smaller specimens the eyes are more globular and the abdominal spination is less pronounced than in the adult. The swollen merus of the raptorial claw of the holotype is undoubtedly a sexual character.

The asymmetrical rostal plate of the holotype, with a bilobed anterolateral spine on one side only, is probably aberrant. Both of the smaller specimens have a single apical spine on the plate.

The resemblance of the raptorial claw of *Bathysquilla* to that of *Harpiosquilla* is superficial. The long spines found on the propodus of both species are restricted to the outside of the upper surface in *Harpiosquilla* but are found on both surfaces in *Bathysquilla,* and the carpus of the raptorial claw is unarmed in *Harpiosquilla.* This is a remarkable case of convergence.

Although Barnard (1950) noted that his specimens of *B. crassispinosa*

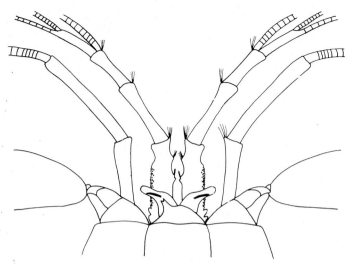

FIGURE 28. *Bathysquilla microps* (Manning), female paratype, TL 44.5, E of Cape Canaveral, Florida, outline of anterior portion of carapace (setae omitted; from Manning, 1961a).

lacked papillae on the antennal protopod, 2 papillae are present in all 3 of the specimens of *B. microps*. Reexamination of the material from South Africa may well show that the papillae are present.

*Ontogeny.*—Unknown.

*Type.*—U. S. National Museum.

*Type-locality.*—Southwest of Tortugas, Florida, in 732 m.

*Distribution.*—Eastern and southern coasts of Florida, and Santaren Channel, Bahamas, in 732-952 m.

## Family SQUILLIDAE Latreille, 1803

Squillares Latreille, 1803, p. 36.
Squillidae White, 1847, p. 83 [correction of Squillares].—Manning, 1968, pp. 105, 109 [key], 113 [revision of genera].

*Definition.*—Propodi of third and fourth thoracic appendages as long as or longer than broad, lacking ventral ribbing; telson with distinct median carina, at most submedian marginal teeth with movable apices; submedian denticles always present; more than 4 intermediate denticles present.

*Type-genus.*—*Squilla* Fabricius, 1787.

*Number of genera.*—Fourteen, of which five occur in the western Atlantic. Manning (1968) presented a key to all of the genera.

*Ontogeny.*—Early larva a pseudozoea with 4 pairs of pleopods; late larva an alima with 4 or more intermediate denticles on the telson.

*Remarks.*—Until recently all students of the stomatopods placed the Recent species and genera in the single family Squillidae. Manning (1967, 1968) reviewed the classification of the family and recognized four families. The family Squillidae was restricted to those species with slender maxillipeds, a sharp median carina on the telson, and four or more intermediate marginal denticles on the telson.

Fourteen genera were assigned to the Squillidae by me; of these, eight were described as new. All known species were listed for each of the genera and a key to all genera was presented. Interrelationships of the genera were discussed briefly. Those papers should be consulted for a more complete account of the family.

Features used in the descriptive accounts are shown in figs. 1 and 29. The following key will serve to distinguish the five American genera.

### KEY TO AMERICAN GENERA OF THE FAMILY SQUILLIDAE

1. Submedian teeth of telson with movable apices ................2
1. Submedian teeth of telson with fixed apices ..................3
2. Dactylus of raptorial claw with 4 teeth; ocular scales rounded; lateral process of fifth thoracic somite not well-developed. . *Meiosquilla,* p. 100.
2. Dactylus of raptorial claw with 6 or more teeth; ocular scales produced into erect dorsal spines; lateral process of fifth thoracic somite a slender, well-developed spine .............. *Pterygosquilla,* p. 122.
3. Lateral process of fifth thoracic somite bilobed ...... *Alima,* p. 127.
3. Lateral process of fifth thoracic somite single .................4
4. No more than 3 epipods present; eyestalks dilated, eyes flask-shaped ........................................ *Cloridopsis,* p. 139.
4. More than 3 epipods present; eyes T-shaped, stalk not dilated ........ ........................................ *Squilla,* p. 146.

Genus *Meiosquilla* Manning, 1968
*Meiosquilla* Manning 1968, pp. 120 [key], 125.

*Definition.*—Body smooth, compact, size moderate or small, TL 100 mm or less; eyes large, cornea bilobed, noticeably broader than stalk; ocular scales separate; carapace smooth, narrowed anteriorly, carinae reduced, at most reflected marginals and posterior portion of each intermediate carina present; cervical groove present, indistinct; posterior median margin evenly concave; posterolateral margins rounded, anterolateral angles usually unarmed; exposed thoracic somites with at most intermediate carinae, submedians absent; lateral process of fifth thoracic somite an inconspicuous diagonal or more prominent flattened lobe, ventral spines usually present; lateral processes of next 2 somites not bilobed, usually

100

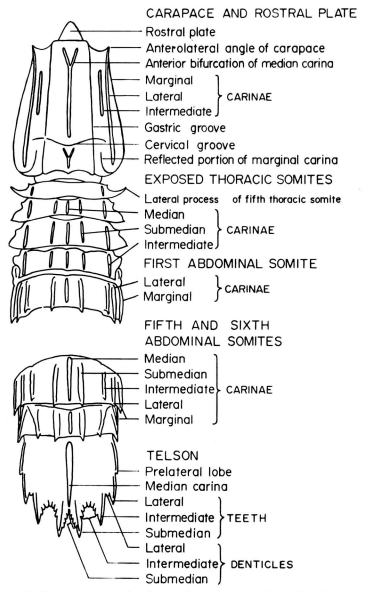

CARAPACE AND ROSTRAL PLATE
Rostral plate
Anterolateral angle of carapace
Anterior bifurcation of median carina
Marginal ⎫
Lateral   ⎬ CARINAE
Intermediate ⎭
Gastric groove
Cervical groove
Reflected portion of marginal carina

EXPOSED THORACIC SOMITES
Lateral process   of fifth thoracic somite
Median ⎫
Submedian ⎬ CARINAE
Intermediate ⎭

FIRST ABDOMINAL SOMITE
Lateral ⎫ CARINAE
Marginal ⎭

FIFTH AND SIXTH
ABDOMINAL SOMITES
Median ⎫
Submedian │
Intermediate ⎬ CARINAE
Lateral │
Marginal ⎭

TELSON
Prelateral lobe
Median carina
Lateral ⎫
Intermediate ⎬ TEETH
Submedian ⎭
Lateral ⎫
Intermediate ⎬ DENTICLES
Submedian ⎭

FIGURE 29. Terms used in the descriptive accounts of the Family Squillidae.

101

rounded, occasionally with posterior spines; ventral keel of eighth thoracic somite well-formed, rounded or sharp; no more than 4 epipods present; mandibular palp absent; dactylus of raptorial claw with 4-5 teeth, outer margin of dactylus sinuate or flattened; propodus stout, upper margin pectinate, with 3 proximal movable teeth, middle smallest; dorsal ridge of carpus undivided; merus stout, inferodistal angle unarmed; ischiomeral articulation terminal; endopods of walking legs linear; abdomen stout, rounded, first 5 somites without submedian carinae, intermediates, laterals, and marginals present; sixth abdominal somite with 3 pairs of carinae; telson broad, flattened, median carina present, supplementary dorsal carinae present or absent; 3 pairs of slender marginal teeth present, submedians with movable apices; prelateral lobes absent; postanal keel usually absent; inner margin of basal prolongation of uropod usually armed with slender, fixed spines; outer margin of longer, inner spine of basal prolongation with variously-shaped lobes.

*Type-species.*—*Squilla quadridens* Bigelow, 1893, by original designation.

*Gender.*—Feminine.

*Number of species.*—10, of which five occur in the western Atlantic.

*Remarks.*—The species assigned herein can be separated into two broad groups on the basis of the presence or absence of spines on the inner margin of the basal prolongation of the uropod. The first group, including the eastern Atlantic *M. desmarestii* (Risso) and *M. pallida* (Giesbrecht) and the eastern Pacific *M. polita* (Bigelow), lacks these spines. The eastern Atlantic species further differ from all of the others in the genus in having five teeth on the dactylus of the raptorial claw. *M. polita* differs from all others in the genus in having the anterolateral angles of the carapace spined. These three species closely resemble the Indo-West Pacific *Squilloides tenuispinis* (Wood-Mason) and *Squilloides leptosquilla* (Brooks) but the latter two species both differ in having a median carina on the carapace, submedian carinae on the abdomen, fixed submedian teeth on the telson, and long inner spines on the basal prolongation of the uropod.

The second group of species in *Meiosquilla* are all very similar in general facies. This group includes the five western Atlantic species discussed in more detail below as well as two additional species from the eastern Pacific, *M. swetti* (Schmitt) and *M. oculinova* (Glassell). These seven species show no close relationship or basic similarity with any of the other genera in the Squillidae. One major distinguishing feature of the second group of species of *Meiosquilla* is the reduction of characters found in most other genera. There are few teeth on the claw, only four epipods, the palp is absent, and the carinae of the carapace and body are reduced from the normal or full complement.

As Manning (1962b) pointed out, the reduction in basic features found in *M. quadridens* is also normally found in postlarvae of known species of *Squilla*. The lack of anterolateral spines on the carapace, reduced carination of carapace and body, and movable submedian teeth on the telson are all characteristic of squillid postlarvae. This brings up the interesting possibility that the species of *Meiosquilla* are neotenic in origin and may have been derived from the postlarvae of another species. Their origin, however it may have come about, is unknown and may remain so until we know more about the development of stomatopod larvae. As noted above, there is no indication of close relationship with any other member of the Squillidae.

KEY TO WESTERN ATLANTIC SPECIES OF *Meiosquilla*

1. Submedian carinae of telson short, not extending anteriorly past base of spine of median carina . . . . . . . . . . . . . . . . . . . . . . . . . . . . . . . . . . . . .2
1. Submedian carinae of telson long, extending anteriorly almost to base of median carina . . . . . . . . . . . . . . . . . . . . . . . . . . . . . . . . . . . . . . . . .3

2. Lateral processes of sixth and seventh thoracic somites spined posteriorly; 3 epipods present . . . . . . . . . . . . . . . . . . . . . *randalli,* p. 103.
2. Lateral processes of sixth and seventh thoracic somites rounded; 4 epipods present . . . . . . . . . . . . . . . . . . . . . . . . . . . *quadridens,* p. 106.

3. Telson lacking dorsal carinae in addition to submedians; 4 epipods present . . . . . . . . . . . . . . . . . . . . . . . . . . . . . . . . . *schmitti,* p. 111.
3. Dorsal surface of telson with several short carinae lateral to submedians; 2 epipods present . . . . . . . . . . . . . . . . . . . . . . . . . . . . . . . . . . . . . . .4

4. Fifth thoracic somite with small ventral spine on each side; eighth thoracic somite with sharp median ventral spine; intermediate and lateral carinae of fifth abdominal somite armed posteriorly . . . . . . . . . . 
. . . . . . . . . . . . . . . . . . . . . . . . . . . . . . . . . . . . . . . . . *tricarinata,* p. 114.
4. Fifth thoracic somite with at most a rounded ventral projection; eighth somite with rounded median ventral projection; intermediate and lateral carinae of fifth abdominal somite unarmed . . . . . . . . . *lebouri,* p. 119.

*Meiosquilla randalli* (Manning, 1962)
Figure 30

*Squilla randalli* Manning, 1962c, p. 218; 1968, p. 125 [listed].

*Previous records.*—VIRGIN ISLANDS: St. John (Manning, 1962c).

*Material.*—FLORIDA: 1 ♀, 34.2; S of Alligator Reef Light, Monroe Co.; 5-6 m; 4 January 1963; R. Schroeder; UMML.—1 ♀, 32.4; same; USNM 119152.—HONDURAS: 1 ♂, 38.7; 1 ♀, 35.6; OREGON Sta. 4930; USNM 119151.—VIRGIN ISLANDS: 1 ♀, 23.8; off Yawzi Point, Lameshur Bay, St. John; 9 m; 21 December 1958; J Randall, L.P. Thomas; holotype; USNM 107873.

*Diagnosis.*—Dactylus of raptorial claw with 4 teeth; 3 epipods present; lateral processes of sixth and seventh thoracic somites spined posteriorly; submedian carinae of telson short; denticles 4-5, 6-8, 1.

*Description.*—Eye large, triangular, cornea strongly bilobed, set almost transversely on stalk; ocular scales truncate; anterior margin of ophthalmic somite produced into small spine. Corneal indices of 5 specimens as follows:

| Sex | TL | CL | CI |
|-----|------|-----|-----|
| ♀ | 23.8 | 5.4 | 382 |
| ♀ | 34.2 | 8.0 | 444 |
| ♀ | 34.2 | 8.3 | 461 |
| ♀ | 35.6 | 8.4 | 420 |
| ♂ | 38.7 | 9.7 | 441 |

Antennular peduncle longer than carapace; antennular processes produced into sharp spines directed anterolaterally.

Rostral plate cordiform, without carinae, length and width subequal or length greater.

Carapace smooth, ornamented only with reflected marginal and short lateral carinae on posterior fourth.

Dactylus of raptorial claw with 4 teeth, strongly notched at base, outer margin a simple curve; dorsal ridge of carpus undivided, ending in a blunt lobe.

Three epipods present.

Thoracic somites with prominent intermediate carinae on last 3 somites; lateral process of fifth somite an oblique, compressed lobe, rounded in anterior view; 2 strong ventral spines present on this somite; lateral processes of sixth and seventh thoracic somites broadly rounded laterally, each produced into a sharp spine posteriorly; ventral keel of eighth somite a sickle-shaped spine, directed posteriorly.

Sixth abdominal somite without ventral spine in front of articulation of uropod; abdominal carinae spined as follows: submedian, 6; intermediate, 5-6; lateral, 1-6; marginal, 1-5.

Median carina of telson faintly notched anteriorly, damaged posteriorly in holotype; submedian carinae short, extending to base of posterior portion of median carina; denticles spiniform, 4-5, 6-8, 1; postanal keel absent.

Penultimate segment of exopod of uropod armed with 7 graded, movable spines, last extending past midlength of ultimate segment; basal prolongation with 6-9 long immovable spines on inner margin.

*Color.*—Gastric grooves and posterior median margin of carapace outlined with dark chromatophores; median portion of thoracic and first to fifth abdominal somites outlined in dark pigment, extent of posterior outline of abdominal somites increasing posteriorly; posterior median area of intermediate spines on first to fifth abdominal somites with patches of dark chromatophores; telson with curved lines of chromatophores; basal prolongation of uropod with dark distal patch; endopod of uropod outlined distally with dark chromatophores.

*Size.*—Only known male, TL 38.7 mm; females, TL 23.8-35.6 mm. Other

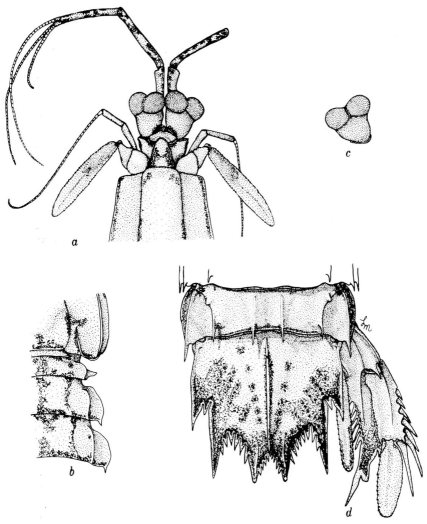

FIGURE 30. *Meiosquilla randalli* (Manning), female, TL 32.4, S of Alligator Reef light, Florida. *a,* anterior portion of body; *b,* lateral processes of fifth to seventh thoracic somites; *c,* eye; *d,* last abdominal somite, telson, and uropod (setae omitted).

105

measurements of female holotype TL 23.8: carapace length, 5.4; cornea width, 1.4; rostral plate length, 0.7, width, 0.7; telson length, 3.7, width, 3.6.

*Discussion.*—The presence of only three epipods and sharp posterior spines on the lateral processes of the sixth and seventh thoracic somites will distinguish this species from its close relatives in the western Atlantic, *M. quadridens, M. schmitti, M. lebouri* and *M. tricarinata,* as well as from the two similar species in the eastern Pacific, *M. swetti* (Schmitt) and *M. oculinova* (Glassell). Among the American species of Squillidae, only the unrelated *Cloridopsis dubia* has three epipods.

*Remarks.*—In the only male examined, the intermediate marginal teeth of the telson extend almost to the bases of the movable apices of the submedian teeth. The bases of the marginal teeth of the telson are swollen in the male, and the teeth appear to be divergent. In the females the longitudinal axes of the teeth are subparallel.

*Ontogeny.*—Unknown.

*Type.*—U. S. National Museum.

*Type-locality.*—Lameshur Bay, St. John, Virgin Islands.

*Distribution.*—Southern Florida, off Honduras, and St. John, Virgin Islands, in depths between 5 and 55 m. The present material extends the range well to the north and west of the type-locality.

## *Meiosquilla quadridens* (Bigelow, 1893)
### Figures 31, 33a

*Squilla quadridens* Bigelow, 1893, p. 101; 1894, p. 511 [description], p. 546, text-figs. 27-28 [larval forms].—Andrews, Bigelow, & Morgan, 1945, p. 340, figs. 3-4 [semi-popular account].—Springer & Bullis, 1956, p. 23.—Holthuis, 1959, p. 189, pl. IX, fig. 6.—Manning, 1959, p. 20 [part]; 1961, p. 14, pl. 3, figs. 1-2.—Bullis & Thompson, 1965, p. 13 [listed].—Manning, 1968, p. 125 [listed].
? *Alima bigelowi* Hansen, 1895, p. 93, pl. 8, figs. 9-10 [larva].—Gurney, 1946, p. 157 [listed].
not *Squilla quadridens:* Holthuis, 1941, p. 32 [= *M. schmitti* (Lemos de Castro) ].
not *Chloridella quadridens:* Coventry, 1944, p. 43 [larvae; most are *Alima hyalina* Leach].

*Previous records.*—FLORIDA: Biscayne Bay (Manning, 1959); off Key Largo (Bigelow, 1893, 1894); Straits of Florida (Bullis & Thompson, 1965).—GULF OF MEXICO: Springer & Bullis, 1956; Manning, 1959; Bullis & Thompson, 1965. — COLOMBIA: Cape la Vela (Manning, 1961).—BRITISH GUIANA: Manning, 1961; Bullis & Thompson, 1965.—SURINAM: Holthuis, 1959; Manning, 1961.

*Material*—NORTH CAROLINA: 1 ♂, 25.9; ALBATROSS Sta. 2596; USNM

106

21481.—BAHAMAS: 1 ♂, 29.3; Little Bahama Bank; SILVER BAY Sta. 3468; USNM 111565.—FLORIDA: 1 ♀, 20.1; off Palm Beach; rocky reef, 55-73 m; 20 April 1950; Thompson and McGinty; USNM 113313.— 1 ♀, 20.8; same; sand and rocky reef, 55-73 m; January 1951; TRITON, Thompson and McGinty; USNM 119236.—1 ♀, 23.5; 1.75 mi. ENE Hillsboro Light, Palm Beach Co.; 38m; AHF Sta. A73-45; 5 June 1945; L. A. Burry; AHF.—1 ♂, 22.0; Bear Cut, Virginia Key, Miami; sublittoral, in *Pinna* shell; 15 February 1958; R. Work; USNM 111567.—1 ♀, 23.0; off Key Largo; 25°05′N, 80°15′W; 102 m; ALBATROSS Sta. 2640; 9 April 1886; holotype; USNM 11547.—1 ♂, 25.1; 1 ♀, 24.7; 4 mi. off Islamorada, Monroe Co.; 73 m; D. deSylva, *et al.;* UMML 32.2075.—1 dam. ♀, 23.5; Straits of Florida; SILVER BAY Sta. 2402; USNM 111568.—1 ♂, 24.8; Tortugas, Monroe Co.; 22 July 1924; W. L. Schmitt; USNM 105353.— 1 ♂, ca. 18.6; same; 46 m; 11 June 1925; W. L. Schmitt; USNM 105354. —1 dam. ♀, CL 4.5; same; from fish stomach; 9 June 1925; W. L. Schmitt; USNM 111566.—1 ♂, 17.0; off SW Florida; ALBATROSS Sta. 2411; USNM 107743.—1 ♂, 14.2; S of Cape St. George, Florida; OREGON Sta. 896; USNM 96403.—COLOMBIA: 1 ♂, 29.8; 2 mi. SW of Cape la Vela; 40 m; AHF Sta. A14-39; 8 April 1939; AHF.—1 damaged ♂; same; USNM 106351.—1 dam. ♀, CL 5.5; same; USNM 113314.— VENEZUELA: 1 ♂, 21.1; 3 ♀, 19.7-27.1; off Venezuela; UMML 32.1209. —BRITISH GUIANA: 1 ♂, 33.4; OREGON Sta. 2249; USNM 103503.— SURINAM: 1 ♀, 32.3; off the mouth of the Surinam River; COQUETTE Sta. 4; RMNH 318.

*Diagnosis.*—Dactylus of raptorial claw with 4 teeth; 4 epipods present; lateral processes of sixth and seventh thoracic somites rounded posteriorly; submedian carinae of telson short; denticles 4-6, 6-10, 1.

*Description.*—Eye large, cornea triangular, set obliquely to the body line; anterior margin of ophthalmic somite with small median spine. Corneal indices for 9 specimens follow:

| Sex | TL | CL | CI |
|-----|------|-----|-----|
| ♂ | 14.2 | 3.3 | 367 |
| ♂ | 17.0 | 3.7 | 336 |
| ♂ | 22.0 | 4.7 | 361 |
| ♀ | 23.0 | 5.5 | 423 |
| ♀ | 24.7 | 5.7 | 407 |
| ♂ | 25.1 | 6.1 | 436 |
| ♀ | 32.0 | 7.3 | 456 |
| ♂ | 29.3 | 7.6 | 447 |
| ♂ | 33.4 | 8.4 | 467 |

Antennular peduncle shorter than or longer than carapace; antennular processes produced into sharp spines directed anterolaterally.

Rostral plate subtriangular or cordiform, without carinae.

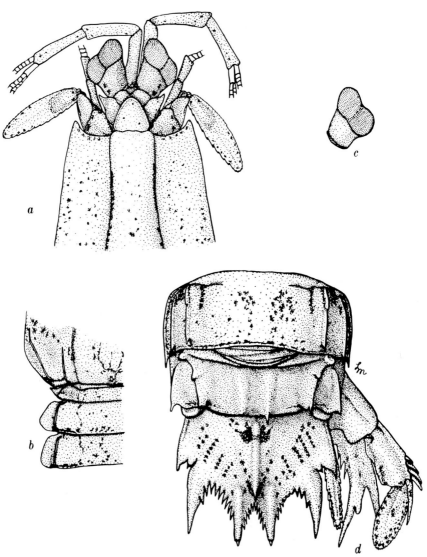

FIGURE 31. *Meiosquilla quadridens* (Bigelow), male, TL 33.4, British Guiana. *a,* anterior portion of body; *b,* lateral processes of fifth to seventh thoracic somites; *c,* eye; *d,* last two abdominal somites, telson, and uropod (setae omitted).

108

Carapace smooth, ornamented at most with lateral and reflected marginal carinae on posterior fourth, laterals often absent.

Dactylus of raptorial claw armed with 4 teeth; outer margin a simple curve, strongly notched at base; dorsal ridge of carpus undivided, ending in a blunt lobe.

Four epipods present, last 2 smaller than first and second.

Intermediate carinae prominent on last 3 thoracic somites; lateral process of fifth thoracic somite an oblique, compressed lobe, varying in shape, in anterior view, from rounded to very acute; sharp ventral spines also present on fifth somite; lateral processes of sixth and seventh thoracic somites slightly flattened anteriorly, broadly rounded posteriorly; eighth thoracic somite with blunt, rounded ventral median keel, curved slightly posteriorly.

Abdominal carinae spined as follows: submedian, 6; intermediate, (5) 6; lateral, (4) 5-6; marginal, (4) 5; sixth abdominal somite without ventral spine anterior to articulation of uropod.

Median carina of telson notched anteriorly, ending posteriorly in sharp spine; submedian carinae short, extending to base of posterior spine of median carina; denticles spiniform, 4-6, 6-10, 1, usually 5, 8, 1; postanal keel absent.

Penultimate segment of exopod of uropod armed with 5-6 graded, movable spines, last extending to or beyond midlength of ultimate segment; basal prolongation with 2-7, usually 3-4 immovable spines on inner margin.

*Color.*—Carapace with broad transverse band of dark color across posterior fourth; rectangular patch of color on dorsum of second abdominal somite; last 3 thoracic and all abdominal somites with posterior line of dark color at intermediate carinae; dark spots at each end of median carina of telson; endopod and distal segment of exopod of uropods lined with dark chromatophores.

*Size.*—Males, TL 14.2-33.4 mm; females, TL 19.7-32.3 mm. Other measurements of male, TL 29.8: carapace length, 6.8; cornea width, 1.6; rostral plate length, 1.0, width, 1.1; telson length, 4.1, width, 5.0.

*Discussion.*—The presence of short submedian carinae on the telson will immediately distinguish this species from the closely related *M. tricarinata* (Holthuis), *M. lebouri* (Gurney), and *M. schmitti* (Lemos de Castro), all of which also occur in the western Atlantic, as well as from the eastern Pacific *M. swetti* (Schmitt) and *M. oculinova* (Glassell). *M. randalli* (Manning), the only similar species with short submedian carinae on the telson, differs in having three epipods and in having posterior spines on the lateral processes of the sixth and seventh thoracic somites. The cornea of *M. quadridens* is set more obliquely on the stalk than in any other related western Atlantic species.

*Alima bigelowi* Hansen, 1895 is a senior homonym of *Squilla bigelowi*

Schmitt, 1940. A replacement name for the latter species will be proposed in a revision of the eastern Pacific stomatopods now in preparation.

*Remarks.*—There is a considerable amount of variation in some of the characters of this species. The rostral plate may be triangular or cordiform, with the latter more common. The projecting lobe of the lateral process of the fifth thoracic somite may be obtusely rounded or may have a sharp, acute apex. In general, in dorsal view, the process appears as a short, acute projection, regardless of its aspect in anterior view. Small specimens usually have a rounded process.

The 22 mm male from Miami and the two small males from Cape le Vela, Colombia have a deeper, more convex abdomen and less conspicuous abdominal carinae than the remainder of the specimens examined. Intergrades are present between these three specimens and those with flattened abdomens strongly ornamented with carinae.

The largest male (TL 33.4), from British Guiana, exhibits sexual dimorphism in that the margins of the telson are noticeably swollen. This and other large specimens also have the anterolateral angles of the carapace projecting more than in specimens with a TL of about 25 mm and less.

The posterior portions of the lateral carinae of the carapace may be present or absent; they are usually better developed in larger specimens.

In 1961 I suggested that *M. polita* (Bigelow) could be considered the Eastern Pacific analogue of *M. quadridens.* Actually, *M. polita* more nearly resembles the Eastern Atlantic *M. desmarestii* (Risso) and *M. pallida* (Giesbrecht). An undescribed species from the Eastern Pacific, to be described in a revision of the Eastern Pacific species, is very similar to *M. quadridens* in general appearance.

The damaged specimen reported from Colombia by Holthuis (1941) as *S. quadridens* proved to be *M. schmitti* (Lemos de Castro). Although the specimen is in very poor shape, the long submedian carinae of the telson are visible.

*Ontogeny.*—Bigelow (1894, p. 546) gave an account of a larval form from Bimini which molted into a postlarva of *M. quadridens;* Bigelow provided illustrations of the late larva and the postlarva. In 1895 Hansen described a larva, *Alima bigelowi,* which he identified with *M. quadridens.* Hansen, however, mentioned none of the diagnostic features of the species and his larval form cannot be identified with certainty. Calman (1917) reported a larva from off Rio de Janeiro as *"Squilla* sp. (near *S. quadridens* Bigelow)" but pointed out differences between his specimen and that described by Bigelow. Calman's larva, which had four epipods, may be the larva of *M. schmitti* (Lemos de Castro), the only other species in the western Atlantic which shares this feature with *M. quadridens.*

*Type.*—U. S. National Museum.

110

*Type-locality.*—Off Key Largo, Florida, in 102 m (56 fms). The depth of 26 fms given by Bigelow (1894) is erroneous.

*Distribution.*—Western Atlantic, where it has been recorded from the Bahamas, North Carolina, southern and western Florida, Colombia, Venezuela, British Guiana, and Surinam. Littoral to 137 m.

## *Meiosquilla schmitti* (Lemos de Castro, 1955)
### Figures 32, 33b

? *Squilla* sp.: Calman, 1917, p. 141 [larva].
*Squilla quadridens:* Holthuis, 1941, p. 32.—Manning, 1959, p. 20 [part, not *S. quadridens* Bigelow].
*Squilla schmitti* Lemos de Castro, 1955, p. 8, text-figs. 5-8, pl. 1, figs. 32-33. Manning, 1961, p. 17, pl. 3, figs. 3-4; 1966, p. 367, text-fig. 4; 1968, p. 125 [listed].

*Previous records.*—FLORIDA: Englewood Beach (Manning, 1959).—COLOMBIA: Cape la Vela (Manning, 1961); Santa Marta (Holthuis, 1941).—BRAZIL: Recife, off Mogiquiçaba, Abrolhos Ids. (Manning, 1966); Rio de Janeiro Bay (Lemos de Castro, 1955).

*Material.*—BAHAMAS: 1 ♂, 21.1; inside Washerwoman Shoal, Cay Sal; 11 March 1869; MCZ.—FLORIDA: 1 ♂, 31.1; SW side Virginia Key, Miami; 2 May 1959; L. Thomas; UMML 32.1208.—1 ♂, 22.6; Lower Matecumbe Key; 18 March 1961; W. A. Starck, II; UMML 32.2067.—1 ♂, 32.5; Broad Creek, Monroe Co.; 21 December 1908; B. Bean; USNM 107744.—1 ♀, 26.5; Englewood Beach, Sarasota Co.; 5 January 1958; M. K. Chadwick; CHML.—1 ♀, 21.6; Bache, W. Florida (?); W. Stimpson; MCZ.—TEXAS: 1 ♂, 16.5; Sabine Pass; June 1956; W. G. Hewatt; USNM 101086.—1 fragmented ♀, CL 3.6; same; USNM 101087.—1 ♀, 20.4; Heald Bank, 30 mi S of High Id.; 94°10′N, 29°10′W; W. G. Hewatt; USNM 101088.—MEXICO: 1 ♂, 27.7; W shore Cayo Norte, Bancho Chinchorro, Yucatan; 22 June 1961; W. A. Starck, II; USNM 111570.—COLOMBIA: 1 damaged ♀; Gairaca, Santa Marta; 30 m; dredge; Yacht CHAZALIE; 29 February 1896; ZMA.—1 ♀, 22.7; 1 mi. SW of Cape la Vela; 10-24 m; AHF Sta. A13-39; 8 April 1939; USNM 119242.—2 ♀, 1 fragmented, other 26.2; 2 mi. SW of Cape la Vela; 40 m; AHF Sta. A14-39; 8 April 1939; AHF.—VENEZUELA: 1 ♀, 26.3; Gulf of Venezuela; 10°17′08′N, 70°45′02′W; 15 November 1958; F. Mago; UCVMB 5014.—1 ♀, 24.5; 4-7 mi. of N of Margarita Id.; 31-38 m; AHF Sta. A42-39; 21 April 1939; USNM 119243.—BRAZIL: 1 ♂, 17.2; off Recife; 08°23.5′S, 34°42′W; 51 m; dredge; CALYPSO Sta. 24; 21 November 1961; USNM 111052.—1 postlarva, ca. 13.0; S of Mogiquiçaba; 16°46′S, 38°53.5′W; 27 m; dredge; CALYPSO Sta. 75; 27 November 1961; MNHNP.—2 ♀, 1 damaged, other ca. 17.5; Abrolhos Ids.; stomach contents of fish; CALYPSO Sta. 87; 29 November 1961; MNHNP.

*Diagnosis.*—Dactylus of raptorial claw with 4 teeth; 4 epipods present;

111

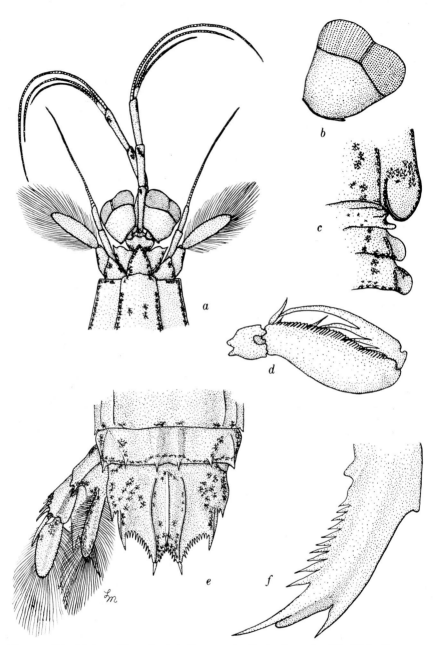

FIGURE 32. *Meiosquilla schmitti* (Lemos de Castro), male, TL 17.2, CALYPSO Sta. 24. *a,* anterior portion of body; *b,* eye; *c,* lateral processes of fifth to seventh thoracic somites; *d,* carpus, propodus, and dactylus of right raptorial claw; *e,* sixth abdominal somite, telson, and uropod; *f,* basal prolongation of uropod, ventral view (from Manning, 1966).

lateral processes of sixth and seventh thoracic somites rounded; submedian carinae of telson long; telson lacking supplementary dorsal carinae; denticles 4-7, 8-12, 1.

*Description.*—Eye large, triangular, cornea bilobed, set slightly obliquely on stalk; anterior margin of ophthalmic somite with small spine; ocular scales truncate, oblique to body line. Corneal indices of 7 specimens follow:

| Sex | TL | CL | CI |
|-----|------|-----|-----|
| ♂ | 16.5 | 3.8 | 345 |
| ♀ | 20.4 | 4.4 | 337 |
| ♂ | 22.6 | 5.0 | 357 |
| ♀ | 26.2 | 5.7 | 380 |
| ♀ | 26.3 | 5.8 | 386 |
| ♂ | 27.7 | 6.9 | 383 |
| ♂ | 31.1 | 6.9 | 363 |

Antennular peduncle shorter than carapace; antennular processes spiniform anterolaterally.

Carapace smooth, without carinae except for reflected marginals and short laterals on posterior fourth, laterals often absent or indistinct.

Dactylus of raptorial claw armed with 4 teeth, outer margin a simple curve with a strong basal notch; dorsal ridge of carpus undivided, ending in a blunt lobe.

Four epipods present.

Thoracic somites with prominent intermediate and lateral carinae on last three somites; lateral process of fifth somite an oblique, compressed lobe, rounded in anterior view; sharp ventral spines present on fifth somite; lateral processes of sixth and seventh thoracic somites broadly rounded; eighth thoracic somite with an acute, blunt ventral keel, curving posteriorly.

Abdominal carinae spined as follows; submedian, 6; intermediate, 5-6; lateral, 5-6; marginal, (4) 5; sixth somite with inconspicuous spine in front of articulation of uropod.

Median carina of telson notched anteriorly, ending posteriorly in sharp spine; submedian carinae long, extending anteriorly almost to base of median carina; denticles spiniform, 4-7, 8-12, 1; postanal keel absent.

Penultimate segment of exopod of uropod with 6-7 graded, movable spines, last extending past midlength of ultimate segment; basal prolongation with 5-11 immovable spines on inner margin.

*Color.*—Gastric grooves and margins of carapace outlined with dark chromatophores; thoracic and abdominal somites with median posterior diffuse band of chromatophores; intermediate abdominal carinae outlined with dark color; posterior margins of lateral plates of abdominal somites with an interrupted band of dark color; second abdominal somite with middorsal patch of chromatophores; submedian carinae of telson outlined in dark color; median carina of telson with a dark patch of color under terminal spine.

*Size.*—Males, TL 16.5-32.5 mm; females, TL 20.4-26.5 mm; postlarva, TL ca. 13.0 mm. Other measurements of female, TL 26.5: carapace length, 5.7; cornea width, 1.5; rostral plate length, 0.9, width, 0.8; telson length, 3.7, width, 4.6.

*Discussion.*—This small species is closely related to the western Atlantic *M. quadridens* (Bigelow) and *M. randalli* (Manning), both of which differ in having short submedian carinae on the telson, among other features. *M. schmitti* also closely resembles two other western Atlantic species, *M. tricarinata* (Holthuis) and *M. lebouri* (Gurney), but it can be distinguished from both of these by the presence of four epipods and the lack of accessory dorsal carinae on the telson.

   *M. schmitti* can be distinguished from the eastern Pacific *M. swetti* (Schmitt) by the presence of long, uninterrupted submedian carinae and the absence of accessory carinae on the telson. *M. oculinova* (Glassell) from the eastern Pacific also resembles *M. schmitti,* but the eastern Pacific species can be distinguished by the geniculate spines or processes on the antennular peduncle and the scalloped anterior border of the eyes.

*Remarks.*—Lemos de Castro (1955) did not mention or figure the short lateral carinae of the carapace. These may be distinct in small specimens but are usually obscure in adults.

   The submedian carinae of the telson are variable in length, usually extending forward to a point between the base of the intermediate tooth and the anterior notch in the median carina.

   As noted under the discussion of *M. quadridens,* the specimen recorded as that species from Colombia by Holthuis (1941) actually has long submedian carinae on the telson and must be assigned to *M. schmitti.*

   *M. schmitti* appears to have a more convex abdomen than specimens of *M. quadridens* of a similar size.

*Ontogeny.*—Larval stages are not known with certainty. The larva reported by Calman (1917) as *Squilla* sp., which has but four epipods, may belong to this species.

*Type.*—Museu Nacional, Rio de Janeiro.

*Type-locality.*—Rio de Janeiro Bay, Brazil.

*Distribution.*—Western Atlantic, from the Bahamas, southern and western Florida, Texas, Yucatan, Venezuela, Colombia, and Brazil; shallow water, sublittoral to 40 m.

*Meiosquilla tricarinata* (Holthuis, 1941)
Figures 33c, 34

*Squilla tricarinata* Holthuis, 1941, p. 32, text-fig. 1.—Manning, 1961, p. 12 [key]; 1966, p. 369, text-fig. 5; 1968, p. 125 [listed].

*Previous records.*—LESSER ANTILLES: Testigos Ids. (Holthuis, 1941).
—BRAZIL: Fernando de Noronha, Mogiquiçaba, Abrolhos Ids. (Manning, 1966).

*Material.*—BAHAMAS: 1 postlarval ♂, 11.9; New Providence; surface, electric light; ALBATROSS; 1886; USNM 21483.—1 postlarval ♂, 12.4; San Salvador; surface; ALBATROSS; 1886; USNM 21482.—FLORIDA: 1 ♀, 34.6; Bear Cut, Key Biscayne, Miami; 28 May 1956; D. DeSylva; UMML 32.1706.—1 ♀, 34.9; Bear Cut, Virginia Key, Miami; 23 March 1961; C. Roper, R. Manning; USNM 119153.—VIRGIN ISLANDS: 1 damaged ♀, ca. 13.5; Yawzi Pt., Lameshur Bay, St. John; from stomach of 279 mm *Epinephelus striatus;* 12 m; 29 December 1958; J. Randall; UMML.—LESSER ANTILLES: 1 ♀, 20.5; Los Testigos Ids.; 11 m, dredged; Yacht CHAZALIE; 20 January 1896; holotype; ZMA.—ARUBA: 1 ♂, 27.1; 8 mi. SW of San Nicolaas Bay; 42-44 m; AHF Sta. A18-39; 10 April 1939; USNM 119244.—BRAZIL: 1 ♂, 33.2; off Fernando de Noronha; 03°49.7'S, 32°26'W; 31 m; CALYPSO Sta. 19; 18 November 1961; USNM 111056.—1 ♂, 16.2; S of Mogiquiçaba; 16°46'S, 38°53.5'W; 27 m; dredge; CALYPSO Sta. 75; 27 November 1961; MNHNP.—1 ♂ abdomen; off Arquipelago dos Abrolhos; 18°00'S, 38°18'W; 48 m; dredge; CALYPSO Sta. 77; 28 November 1961; MNHNP.

*Diagnosis.*—Dactylus of raptorial claw with 4 teeth; 2 epipods present; lateral processes of sixth and seventh thoracic somites rounded, unarmed, fifth somite with 1 pair of ventral spines; eighth thoracic somite with sharp ventral spine; telson with long submedian and shorter accessory carinae; denticles 4-6, 4-9, 1.

*Description.*—Eye large, triangular, cornea with prominent lobes, set almost transversely on stalk; eyes not extending to end of first segment of antennular peduncle; anterior margin of ophthalmic somite with median spine; ocular scales truncate. Corneal indices of 4 specimens follow:

| Sex | TL | CL | CI |
|---|---|---|---|
| ♂ | 16.2 | 3.3 | 330 |
| ♂ | 33.2 | 7.2 | 400 |
| ♀ | 34.6 | 7.2 | 378 |
| ♀ | 34.9 | 7.8 | 433 |

Antennular peduncle shorter than carapace; antennular processes blunt, not spiniform.

Rostral plate cordiform, apex acute but rounded.

Carapace smooth, posterior fourth of lateral plates with reflected marginal carinae, short lateral carinae present or absent.

Dactylus of raptorial claw with 4 teeth; dorsal ridge of carpus undivided, terminating in blunt lobe.

Exposed thoracic somites with unarmed intermediate carinae, lateral

FIGURE 33. Color patterns of species of *Meiosquilla. a, M. quadridens* (Bigelow), female, TL 24.7, off Islamorada, Florida; *b, M. schmitti* (Lemos de Castro), male, TL 27.7, Yucatan; *c, M. tricarinata* (Holthuis), female, TL 34.9, Virginia Key, Miami, Florida.

process of fifth thoracic somite a compressed lobe, not sharp in anterior view; 1 pair of ventral spines present on fifth somite; lateral processes of sixth and seventh thoracic somites rounded, not spined; ventral keel of eighth thoracic somite a slender, sharp spine, directed posteroventrally.

Abdominal carinae spined as follows: submedian, 6; intermediate, 5-6; lateral, 5-6; marginal, (3) 4-5; sixth somite without ventral spine in front of articulation of uropod.

Median carina of telson notched at base, terminating in strong posterior spine, with 2-3 median tubercles under apex of spine; submedian carinae long, extending forward almost to base of median carina; dorsal surface with 3-4 accessory carinae lateral to each submedian, of which 2-3 lie between submedian and intermediate carinae; 1 short carina present lateral to intermediate carina, which may fall short of intermediate tooth; shorter carinae occasionally spined posteriorly; denticles sharp, 4-6, (4) 7-9, 1; postanal keel absent.

Proximal segment of exopod of uropod with 6-7 movable spines, last long, extending past midlength of distal segment; basal prolongation of uropod with 6-7 slender fixed spines on inner margin.

*Color.*—Background cream, with rostral plate and carapace outlined with a fine line of brown chromatophores; grooves and carinae of carapace, thoracic and abdominal somites outlined posteriorly with fine lines of brown chromatophores; median area of lateral plate of carapace with 2 short parallel lines of brown chromatophores; margin and carinae of telson outlined in brown; position of absent submedian carinae on abdomen indicated with lines of brown chromatophores; carinae of first segment of uropod outlined in brown; endopod and distal segments of exopod of uropod lined with brown chromatophores on inner side.

*Size.*—Males, TL 16.2-33.2 mm; females, TL ca. 13.7-34.9 mm; post-larvae, TL 11.9-12.4 mm. Other measurements of female, TL 34.6: carapace length, 7.2; cornea width, 1.9; rostral plate length, 1.2, width, 1.2; telson length, 5.0, width, 6.1.

*Discussion.*—The long submedian carinae and presence of accessory dorsal carinae on the telson and the presence of only two epipods will distinguish this species and *M. lebouri* (Gurney) from the similar east American species, *M. quadridens* (Bigelow), *M. randalli* (Manning), both of which have short submedian carinae, and from *M. schmitti* (Lemos de Castro), which has long submedians but lacks accessory carinae and has four epipods.

The ventral spines of the fifth thoracic somite, slender ventral spine on the eighth thoracic somite, and armed carinae of the fifth abdominal somite will distinguish *M. tricarinata* from *M. lebouri* (Gurney) which it otherwise resembles rather closely. Fresh specimens of *M. tricarinata* can be immediately distinguished from *M. lebouri* by the submedian lines of dark pigment on the abdomen. Other differences between the two species have been summarized in the discussion of *M. lebouri*.

*Remarks.*—Although the short lateral carinae on the lateral plates of the carapace were not mentioned in the original description, they are present in the type and in at least some of the other specimens examined. When present they are low and very difficult to make out.

Both specimens from Miami were collected in shallow water; that from Virginia Key was taken with a pushnet on *Thalassia* flats at night.

The juvenile from Lameshur Bay, St. John, is badly damaged, and most of the diagnostic features cannot be seen; it could well belong to *M. lebouri*.

*Ontogeny.*—Unknown. Although both Gurney (1946a) and Manning

117

(1966) suggested that *Alima lebouri* Gurney was the larva of *M. tricari-nata,* further study has suggested that they are distinct. See also remarks under *M. lebouri.*

The two postlarval stages reported here as *M. tricarinata* are tentatively identified with this species. Until more is known of the larvae and post-

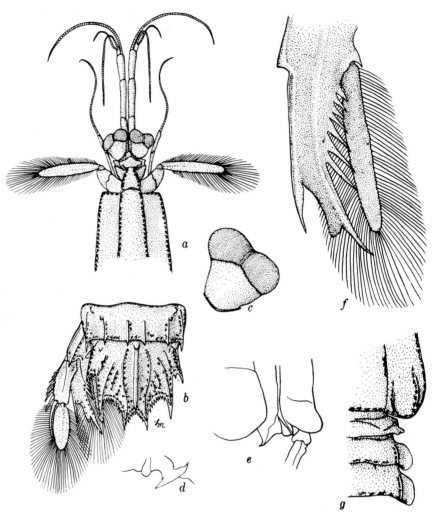

FIGURE 34. *Meiosquilla tricarinata* (Holthuis), male, TL 33.2, CALYPSO Sta. 19. *a,* anterior portion of body; *b,* sixth abdominal somite, telson, and uropod (from Manning, 1966). Female, TL 34.9, Virginia Key, Miami, Florida. *c,* eye; *d,* ventral keel of eighth thoracic somite, lateral view; *e,* lateral process of fifth thoracic somite, lateral view; *f,* basal prolongation of uropod, ventral view; *g,* lateral processes of fifth to seventh thoracic somites, dorsal view.

118

larvae of these species, such postlarvae cannot be identified with certainty.

*Type.*—Zoological Museum, Amsterdam.

*Type-locality.*—Testigos Islands, Lesser Antilles.

*Distribution.*—Western Atlantic, from Miami, Florida, Testigos Ids. and Aruba in the Antilles, and Brazil, from Fernando de Noronha, Mogiquiçaba, and the Abrolhos Ids. Sublittoral to 48 m.

*Meiosquilla lebouri* (Gurney, 1946)
Figure 35

*Alima lebouri* Gurney, 1946, p. 137, text-figs. 1-3; 1946a, p. 734.—Manning, 1968, p. 125 [listed].

*Previous records.*—BERMUDA: Gurney, 1946.

*Material.*—BERMUDA: 1 ♂, 21.7; R. Gurney; no other data; BMNH.—BAHAMAS: 1 ♀, 16.0; Bimini; 25 August 1947; no other data; AMNH.—JAMAICA: 1 ♂, 17.1; 1 ♀, 18.7; no other data; USNM 47355.—LESSER ANTILLES: 1 ♂, 14.2; St. Lucia; Smithsonian-Bredin Sta. 47-56; 22 March 1956; D. V. Nicholson; USNM 107745.—3 ♀, 12.7-20.5; 2 mi SW of Saba; 17°39'N, 63°17'W; 82 m; ARCTURUS Sta. 23; 14 March 1925; AMNH.

*Diagnosis.*—Dactylus of raptorial claw with 4 teeth; 2 epipods present; lateral processes of sixth and seventh thoracic somites rounded laterally, unarmed, fifth somite without ventral spines; eighth thoracic somite with rounded ventral projection; intermediate and lateral carinae of abdomen armed on sixth somite only; telson with long submedian carinae and short accessory carinae; denticles 4-5, 6-10, 1.

*Description.*—Eye large, triangular, lobes of cornea prominent, cornea set almost transversely on stalk; eyes extending to end of first segment of antennular peduncle; anterior margin of ophthalmic somite with small spine; ocular scales truncate. Corneal indices of 6 specimens follow:

| Sex | TL | CL | CI |
| --- | --- | --- | --- |
| ♀ | 12.7 | 2.8 | 311 |
| ♂ | 14.2 | 3.3 | 300 |
| ♂ | 17.1 | 3.5 | 291 |
| ♀ | 18.7 | 3.9 | 300 |
| ♀ | 18.7 | 4.0 | 333 |
| ♀ | 20.5 | 4.8 | 320 |

Antennular peduncle not as long as carapace; antennular processes angled laterally, not spiniform.

Rostral plate cordiform, short, anterior margin rounded.

Carapace smooth, without carinae except for short reflected marginals on posterior fourth.

Dactylus of raptorial claw armed with 4 teeth, outer margin of dactylus evenly curved, strongly notched at base; dorsal ridge of carpus undivided, ending in blunt lobe.

Two epipods present.

Intermediate carinae present on last 3 thoracic somites, unarmed posteriorly and not prominent; lateral process of fifth thoracic somite a blunt, compressed lobe, rounded in anterior view; fifth somite without ventrolateral spines; lateral processes of sixth and seventh thoracic somites rounded laterally; eighth thoracic somite with rounded ventral keel.

Abdomen arched, carinae not prominent; abdominal carinae spined as follows: submedian, 6; intermediate, 6; lateral, 6; marginal, 5; sixth somite without ventral spine in front of articulation of uropod.

Median carina of telson notched anteriorly, terminating posteriorly in strong spine; submedian carinae long, extending anteriorly almost to base of median carina; at most 1 short accessory carina present on each side between submedian carina and carina of intermediate tooth; occasionally

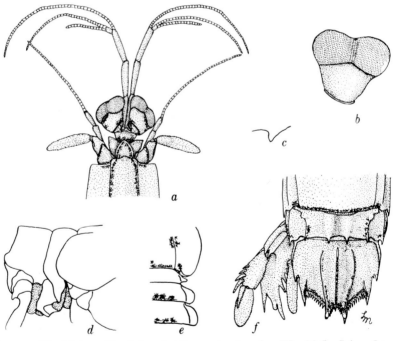

FIGURE 35. *Meiosquilla lebouri* (Gurney), female, TL 20.5, Saba, Lesser Antilles. *a*, anterior portion of body; *b*, eye; *c*, ventral keel of eighth thoracic somite, lateral view; *d*, lateral process of fifth thoracic somite, lateral view; *e*, lateral processes of fifth to seventh thoracic somites; *f*, last abdominal somite, telson, and uropod (setae omitted).

120

1 short carina present lateral to intermediate; denticles 4-5, 6-10, 1; postanal keel absent.

Uropod with 6 movable spines on outer margin of penultimate segment of exopod, last extending past midlength of distal segment; basal prolongation with 4-6 fixed teeth on inner margin.

*Color.*—Rostral plate outlined anteriorly and laterally with dark pigment; gastric grooves and posterior median portion of carapace dark; exposed thoracic somites and abdominal somites with dark posterior line, not extending laterally to pleural margin; pigment bars on abdomen concentrated, not made up of diffuse, stellate chromatophores; telson with median carina, submedian carinae, and margin at intermediate denticles dark; articulation of exopod of uropod dark.

*Size.*—Males, TL 14.2-21.7 mm; females, TL 12.7-20.5 mm. Other measurements of female, TL 18.7: carapace length, 3.9; cornea width, 1.3; rostral plate length, 0.6, width, 0.8; telson length, 2.4, width, 3.2.

*Discussion.*—It is with some hesitation that I assign these small specimens to this species. *M. lebouri* (Gurney), which I synonymized with *M. tricarinata* (Holthuis) in a report on the CALYPSO South American collection (1966), agrees with that species in having but two epipods and long submedian carinae on the telson. Gurney's account, based on postlarvae and juveniles molted from late larvae in captivity, is not particularly detailed and no mention was made of any other diagnostic features. However, the fact that Gurney's species resembles *M. tricarinata* in the two features mentioned above and the presence of a specimen from Bermuda distinct from *M. tricarinata* lends support to the theory that Gurney's species is distinct from *M. tricarinata.* It is possible, however, that *M. lebouri* and *M. tricarinata* are synonymous and the specimens reported herein actually represent an undescribed species. Until Gurney's type can be examined it seems better to use his name rather than introduce a new name here.

*M. lebouri* is very similar to *M. tricarinata,* differing in the following features: (1) there are no ventral spines on the fifth thoracic somite in *lebouri;* one pair of sharp ventral spines is present in *tricarinata;* (2) the ventral keel of the eighth thoracic somite is rounded in *lebouri,* produced into a slender, curved spine in *tricarinata;* (3) intermediate and lateral abdominal carinae are not armed anterior to the sixth somite in *lebouri;* they are armed on the fifth and sixth somites in *tricarinata;* (4) the abdomen and telson appear to be more swollen and convex in *lebouri* than in *tricarinata,* and the abdominal carinae are less prominent in *lebouri;* (5) there are fewer (0-1) accessory carinae between the submedian and intermediate carinae of the telson in *lebouri; tricarinata* has at least two carinae in this area; and (6) *lebouri* lacks the submedian lines of chromatophores on the abdomen that are characteristic of *tricarinata;* also, the marginal pigment of the abdomen is more intense in *lebouri* than in *tricarinata.*

121

Most of the specimens of *M. lebouri* reported here are smaller than specimens of *M. tricarinata* available for study. In the area of size overlap (TL 16-21 mm) all specimens can be assigned to one or the other species on the basis of the characters given above. Although reduced abdominal carination is usually characteristic of immature specimens, the smallest specimens of *M. tricarinata* all have spined intermediate and lateral carinae on the fifth abdominal somite, whereas all specimens of *M. lebouri* lack these spines.

Both *M. tricarinata* and *M. lebouri* can be distinguished from similar American species by the presence of long submedian carinae and accessory carinae on the telson and by the presence of only two epipods.

*Remarks.*—None of the specimens of *M. lebouri* available for study had lateral carinae on the carapace. These carinae may be present or absent in *M. tricarinata*. Also, none of the specimens of *M. lebouri* had any tubercles on the telson under the apex of the crest; one or more of these tubercles is usually present in *M. tricarinata*.

*Ontogeny.*—Gurney (1946) has given an account of the late larvae of this species; he identified his larval forms with those reported from Bermuda by Verrill (1923).

*Type.*—Not traced. It is probably in the British Museum with Gurney's collection of larvae from Bermuda.

*Type-locality.*—Bermuda.

*Distribution.*—Western Atlantic, from Bermuda, Bimini, Jamaica, St. Lucia, and Saba; sublittoral to 82 m. It has not previously been recorded south of Bermuda.

## Genus *Pterygosquilla* Hilgendorf, 1890

*Pterygosquilla* Hilgendorf, 1890, p. 172.—Manning, 1968, pp. 120 [key], 123.

*Definition.*—Body appearing smooth, minutely rugose, convex; size moderate, TL 150 mm or less; eye of moderate size, cornea bilobed, noticeably broader than stalk; ocular scales produced into erect spines; rostral plate subtriangular, without carinae; carapace smooth, narrowed anteriorly, carinae reduced, at most reflected marginals and posterior portions of intermediates present; cervical groove and gastric grooves distinct; posterior median portion of carapace faintly biconcave, median projection inconspicuous; posterolateral margin broadly rounded; anterolateral angles with or without spine; last 3 thoracic somites with low submedian and intermediate carinae; lateral process of fifth thoracic somite a single spine, ventral spines reduced or absent; lateral processes of sixth and seventh thoracic somites rounded laterally, produced into posteriorly-directed spines; ventral keel on eighth thoracic somite prominent, projecting, rounded; 4 epipods present; mandibular palp present; dactylus of raptorial claw with 6-10 teeth, outer margin of dactylus convex or flattened; propodus stout, upper margin pectinate, with 3 movable teeth at base, middle smallest; dorsal ridge of carpus undivided; merus stout, without inferodistal spine; ischiomeral articulation terminal; endopods of walking legs two-segmented; abdomen convex, appearing smooth, normal complement of carinae usually present, submedians occasionally absent from first 5 somites; sixth abdominal somite with 6 carinae; telson broad, dorsal surface with median carina and paired, submedian, anterior obtuse prominences; 3 pairs of marginal teeth present, submedians with movable apices; prelateral lobes absent; ventral surface of telson with or without postanal keel; inner margin of basal prolongation without spines but with erect tubercles; inner spine of basal prolongation of uropod with variously-shaped lobe on outer margin.

*Type-species.*—*Pterygosquilla laticauda* Hilgendorf, 1890, by monotypy. *P. laticauda* is a subjective junior synonym of *P. gracilipes* (Miers).

*Gender.*—Feminine.

*Number of species.*—Two, of which one occurs in the western Atlantic.

*Nomenclature.*—Hansen (1895) synonymized *Pterygosquilla* with *Squilla* where it has remained unused until Manning (1968) recognized it.

*Remarks.*—*Pterygosquilla* is a genus of unknown affinities; it can be recognized immediately by the ocular scales which are produced into erect spines. The two species assigned to the genus occur only in sub-Antarctic waters. One species, *P. gracilipes* (Miers) is west American whereas the other, *P. armata*, occurs off both coasts of Patagonia, off New Zealand, and off South Africa. This is the only genus in the Squillidae which is restricted to cold waters.

*Pterygosquilla armata* (H. Milne-Edwards, 1837)
Figure 36

Restricted synonymy:
*Squilla armata* H. Milne-Edwards, 1837, p. 521.—Bigelow, 1891, p. 94.— A. Milne-Edwards, 1891, p. 53, pl. 7.—Bigelow, 1894, p. 515, text-figs. 9-10.—Manning, 1968, p. 123 [listed].

*Previous records.*—ARGENTINA: S of 44° South Latitude (A. Milne-Edwards, 1891; Bigelow, 1894). It has also been recorded from the eastern Pacific, New Zealand, and South Africa (Schmitt, 1940; Manning, 1966a).

*Material.*—ARGENTINA: 1 ♂, 146.3; Puerto Madryn, Golfo Nuevo; 11 March 1924; Fidel Amadon; USNM 120339.—1 soft ♂, 86.6; 35°30′S,

53°10′W; 109-164 m; Undine, G. Franceschi; 1925; USNM 120338.—
2 ♂, 31.8-54.9; 7 ♀, 51.0-61.7; Gulf of St. George; Albatross Sta. 2769;
USNM 18470.—1 ♂, 102.6; 1 ♀, 89.6; Caleta Cordova, Santa Cruz
Province; 15 April 1954; V. Angelescu; USNM 120341.—1 juv. ♀, ca.
41.0; 44°31′S, 65°56′W; 90 m; Mission Cap Horn; MNHNP.

*Diagnosis.*—Cornea set obliquely on stalk; first 5 abdominal somites with
submedian carinae; telson denticles usually 1, 10-12, 1; ventral surface of
telson with postanal keel.

*Description.*—Eye of moderate size, cornea bilobed, set very obliquely on
stalk; stalk with prominent tubercle on inner, ventral surface; anterior
margin of ophthalmic somite rounded, occasionally tuberculate. Corneal
indices of 8 specimens follow:

| Sex | TL | CL | CI |
|-----|------|------|-----|
| ♂ | 31.8 | 7.4 | 379 |
| ♀ | 55.9 | 12.8 | 432 |
| ♀ | 61.7 | 12.8 | 441 |
| ♀ | 60.7 | 13.0 | 464 |
| ♂ | 86.1 | 17.6 | 451 |
| ♀ | 89.6 | 18.8 | 470 |
| ♂ | 102.6 | 21.8 | 495 |
| ♀ | 146.3 | 32.4 | 540 |

Antennular peduncle shorter than carapace; antennular processes produced into slender spines directed anterolaterally.

Rostral plate subtriangular, as long as broad or slightly longer, sinuous
lateral margins tapering to rounded apex.

Carapace smooth, appearing highly polished; anterolateral angles armed,
spines not extending to base of rostral plate; anterolateral margins straight
or slightly concave.

Raptorial claw stout, dactylus with 7-10, usually 8-9, slender teeth;
outer margin of dactylus rounded or slightly flattened, notched basally;
dorsal ridge of carpus terminating in acute tooth.

Lateral process of fifth thoracic somite a slender spine directed laterally,
apex often bifurcate; obtuse or slightly acute ventral projection present
under each lateral spine; lateral processes of next 2 somites rounded
laterally, with sharp posterior spines; ventral keel of eighth somite usually
a slender projection.

Abdomen smooth, rounded, appearing depressed; submedian carinae
present but low, inconspicuous; abdominal carinae spined as follows:
submedian, 6; intermediate, (1)2-6; lateral, (1)2-6; marginal, 1-5; following spine formula more typical for juveniles: submedian, 6; intermediate,
(2)3-6; lateral, 3-6; marginal, 1-5; anterior carinae unarmed in adult
males; area between intermediate and lateral carinae very swollen in adult

124

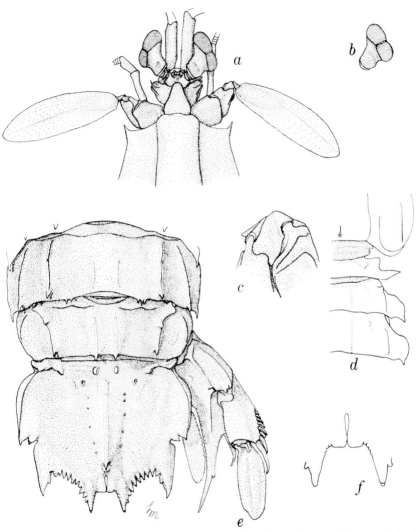

FIGURE 36. *Pterygosquilla armata* (H. Milne-Edwards), male, TL 102.6, Caleta Cordova, Argentina. *a*, anterior portion of body; *b*, eye; *c*, carpus of claw; *d*, lateral processes of fifth to seventh thoracic somites; *e*, last two abdominal somites, telson, and uropod; *f*, submedian teeth of telson, ventral view (setae omitted).

males; fifth abdominal somite with 0-5 accessory spinules on posterior margin between submedian and intermediate carinae, 0-1 spinules present lateral to intermediate carinae; sixth somite with 0-1 accessory spinules present lateral to each submedian spine; sixth somite unarmed in front of articulation of uropod.

Telson broad, with sharp median carina terminating posteriorly in small spine; anterior portion of telson with submedian projections, occasionally tuberculate dorsally; denticles 1, (8)10-12 (16), 1, submedian denticle occasionally with marginal spinules in small specimens; ventral surface of telson with postanal keel.

Uropod broad, exopod with 7-9 graded, movable spines on outer margin of proximal segment; basal prolongation tuberculate on inner margin; lobe on outer margin of inner spine usually inconspicuous in American specimens.

*Color.*—Usually completely faded in preserved material, but in each of the specimens from Caleta Cordova each somite is lined posteriorly in dark pigment.

*Size.*—Males TL 31.8-146.3 mm; females, TL ca. 41.0-89.6 mm. Other measurements of a male, TL 146.3: carapace length, 32.4; cornea width, 6.0; rostral plate length, 5.0, width, 4.8; telson length, 25.9, width, 32.1.

*Discussion.*—The presence of submedian carinae on the abdomen, fewer submedian denticles on the telson, and a prominent postanal keel will distinguish *P. armata* from the only other species in the genus, *P. gracilipes* (Miers), which it otherwise resembles rather closely. Both species occur together off the coast of Chile. *P. gracilipes* also differs in having the cornea set transversely on the stalk and in having many more submedian denticles.

Richardson (1953) was the first to suggest that there might be differences in the populations of *P. armata* from South America, New Zealand, and South Africa. My observations on South American and New Zealand specimens (1966a) supported Richardson but for different reasons. Until specimens from South Africa can be studied in detail, no conclusions can be reached. It is very likely that three subspecies will be recognized in the species.

*Remarks.*—The synonymy and material given above was restricted to western Atlantic records and specimens. Full data on the eastern Pacific forms will be presented in a revision of the eastern Pacific species now in progress. Manning (1966a) gave a complete synonymy for the New Zealand population.

*Ontogeny.*—Larvae attributed to this species have been recorded by Lebour (1954).

126

*Type.*—Musèum National d'Histoire Naturelle, Paris.

*Type-locality.*—Chile.

*Distribution.*—In the western Atlantic, *P. armata* is known only from the coast of Argentina, as far north as 35.5° South Latitude, in depths between 90 and 164 m. It has also been recorded in the eastern Pacific (Schmitt, 1940), off New Zealand (Manning, 1966a), and off South Africa (Barnard, 1950).

## Genus *Alima* Leach, 1817

*Alima* Leach, in Tuckey, 1817, unnumbered plate.—Manning, 1968, pp. 120 [key], 136.

*Definition.*—Body smooth or minutely pitted, flattened; size moderate or small, TL 125 mm or less; eyes flattened, cornea bilobed, noticeably broader than stalk; ocular scales separate, rounded; rostral plate short, triangular or rounded, median carina present or absent; carapace broad, tapering anteriorly, with full complement of carinae, anterior bifurcation of median carina present or absent; cervical groove distinct; posterior median margin biconcave, with slight median projection; carapace rounded posterolaterally, occasionally with obtuse anterior angles; anterolateral angles spined; last 3 thoracic somites with low submedian and intermediate carinae; lateral process of fifth thoracic somite bilobed, anterior lobe a curved spine, posterior lobe usually broadly rounded; lateral processes of next 2 somites rounded or sinuous, not conspicuously bilobed; ventral keel of eighth thoracic somite usually low, inconspicuous; 4 epipods present; mandibular palp present or absent; raptorial claw slender, dactylus with 5-6 teeth, outer margin sinuous; propodus broadening distally, distal spine present or absent; upper margin of propodus pectinate, with 3 movable teeth at base, middle smallest; dorsal ridge of carpus undivided; merus stout, without inferodistal spine, ischiomeral articulation terminal; endopod of walking legs linear or with faint basal suture; abdomen flattened, with 8 or 9 (usually 8) longitudinal carinae on first 5 somites, 6 on last; telson elongate, flattened, with sharp median carina, supplementary dorsal carinae present or absent, and 3 pairs of short marginal teeth, submedians with fixed apices; prelateral lobes present, occasionally inconspicuous; ventral surface of telson usually with postanal keel or tubercle; uropod slender, inner margin of basal prolongation serrate, inner spine of basal prolongation with variously-shaped lobe on outer margin.

*Type-species.*—*Alima hyalina* Leach, 1817, by monotypy.

*Gender.*— Feminine.

*Number of species.*—Four, of which two are restricted to the Indo-West Pacific and two are pantropical except East Pacific.

*Nomenclature.*—*Alima* was first used by Leach (1817) for a larval stomatopod, *A. hyalina*. The term alima has been used as a category name for the type of larva characteristic of *Squilla s.l.* In general, it has been ignored by students of adult Stomatopoda.

*Remarks.*—Manning (1968) listed six nominal species of *Alima*. Examination of material of *S. hieroglyphica* Kemp, 1911, *S. hildebrandi* Schmitt, 1940, and *S. labadiensis* Ingle, 1960 has shown that the latter two species must be synonymized with Kemp's species.

*Alima hieroglyphica* (Kemp) joins *A. hyalina* Leach, *Odontodactylus brevirostris* (Miers), *Pseudosquilla ciliata* (Fabricius), and *P. oculata* (Brullé) as the most widely distributed stomatopods. These five species occur in both the Indo-West Pacific and Atlantic regions; they have not entered the East Pacific area.

## KEY TO WESTERN ATLANTIC SPECIES OF *Alima*

1. Dactylus of claw with 6 teeth; 2 rounded lobes present between spines of basal prolongation of uropod . . . . . . . . . . . . . . . . *hyalina*, p. 128.
1. Dactylus of claw with 5 teeth; 1 rounded lobe present between spines of basal prolongation of uropod . . . . . . . . . . . *hieroglyphica*, p. 135.

### *Alima hyalina* Leach, 1817
Figures 37, 38, 39a

*Alima hyalina* Leach, in Tuckey, 1817, unnumbered fig. on pl. [in Appendix IV to Tuckey; larva]; in Tuckey, 1818, p. 416; 1818, p. 305, fig. 7.— Latreille, 1818, p. 16 [listed], pl. 354, fig. 8.—Desmarest, 1825, p. 253, pl. 44, fig. 1.—Latreille, 1828, p. 475; in Cuvier, 1829, p. 110 [footnote; listed].—H. Milne-Edwards, 1837, p. 507; in Cuvier, 1837, p. 162 [footnote; *A. gracilis* on pl. 57].—H. Milne-Edwards, in Lamarck, 1838, p. 326 [listed]; in Lamarck, 1839, p. 377 [listed].—Lucas, 1840, p. 208.— White, 1847, p. 83 [listed].—Herklots, 1851, p. 26 [listed].—Guérin-Méneville, in Sagra, 1857, p. lxvi.—Desmarest, in Chenu, 1858, p. 44, pl. 4, fig. 2 [listed].—Bate, 1868, pp. 443, 446.—Cunningham, 1870, p. 497.—Claus, 1871, p. 41 [discussion].—Hansen, 1895, p. 92, pl. 8, fig. 8. —Jurich, 1904, p. 382 [p. 25 on separate].—Borradaile, 1907, p. 216.— Foxon, 1932, p. 378 [discussion].—Lebour, 1934, p. 14, text-figs. 3-4.— Gurney, 1946, pp. 141, 158, text-fig. 4.—Manning, 1962b, p. 496, text-figs. 1-2; 1968, p. 136 [listed], fig. 5g, 7e-f.
*Erichthus hyalinus:* Schinz, in Cuvier, 1823, p. 64 [larva ?].
*Squilla (Alima) hyalina:* Voigt, in Cuvier, 1836, p. 195 [larva].
*Alima gracilis* H. Milne-Edwards, 1837, p. 509 [larva]; in Cuvier, 1837, pl. 57, fig. 3 [*A. hyalina* in text].—Claus, 1871, p. 45, pl. 8, fig. 35.—Brooks, 1886, p. 84, pl. 4, figs. 4-5 [6?], pl. 5, fig. 3 [?], pl. 6, figs. 3-5, pl. 8, figs. 4-6.—Gurney, 1946, p. 158 [listed].
*Alima angusta* Dana, 1852, p. 631 [larva]; 1855, Atlas, p. 13, pl. 42, fig. 2. —Gurney, 1946, p. 157 [listed] [*angustata* in Manning, 1962b, discussion of *A. hyalina*].
*Hyalopelta gracilis:* Guérin-Méneville, in Sagra, 1857, p. lxvi [larval].
*Squilla alba* Bigelow, 1893a, p. 103; 1894, p. 539, pl. 22.—Kemp, 1913, p. 200 [listed].—Edmondson, 1921, p. 288.—Andrews, Bigelow & Morgan, 1945, p. 340, fig. 5.—Townsley, 1953, p. 408, text-figs. 6-7.—Ingle, 1958,

p. 49, text-figs. 1-6.—Manning, 1959, p. 18; 1961, p. 13 [key]; 1962b, p. 496, text-figs. 3-4.—Bullis & Thompson, 1965, p. 13 [listed].
?*Alima gracillima* Borradaile, 1907, p. 216, pl. 22, fig. 5.—Gurney, 1946, p. 158 [listed; larva].
*Squilla* (*Alima*-form): Verrill, 1923, p. 202, pl. 53, figs. 1-2, pl. 54, figs. 4A-B [larva].
*Chloridella alba:* Schmitt, 1924, p. 81.
*Squilla* (*Alima*) *gracilis:* Boone, 1930, p. 42 [larva].
not *Squilla alba:* Boone, 1930, p. 35, pl. 5 [= *S. dubia* H. Milne-Edwards].
*Chloridella quadridens:* Coventry, 1944, p. 543 [larvae; most are *A. hyalina*].
*Squilla* sp.: Gurney, 1946, p. 145, text-figs. 8B-D [not 8A].
*Squilla hieroglyphica:* Barnard, 1950, p. 846, text-figs. 2c-e [misidentified, not *S. hieroglyphica* Kemp].
?*Squilla* sp.: Townsley, 1953, p. 428, text-fig. 22d [larva].
*Squilla hyalina:* Manning, 1967b, p. 105 [larva].

*Previous records.*—Adults: BERMUDA: Gurney, 1946.—BAHAMAS: Bigelow, 1893, 1894; Andrews, Bigelow & Morgan, 1945.—FLORIDA: Manning, 1959.—MEXICO: Bullis & Thompson, 1965.—CURAÇAO: Schmitt, 1924.—ST. HELENA: Ingle, 1958.—SOUTH AFRICA: Barnard, 1950.—HAWAII: Edmondson, 1921; Townsley, 1953.

Larvae have been reported from the eastern and central Atlantic, the Indian Ocean, and the western Pacific; records for the larvae are summarized by Gurney (1946). Most records of *A. hyalina* in the above synonymy refer to the larval form.

*Material.*—BAHAMAS: 1 ♀, 41.2; Bimini; R. P. Bigelow; lectotype of *S. alba* Bigelow; USNM 18495.—1 ♀, 36.4; same; paralectotype of *S. alba* Bigelow; USNM 111092.—3 late alima larvae; Lyford Cay, Nassau; found swimming at night; 1 August 1964; T. Pederson; USNM 113318.—1 ♂, 42.9; along E side of anchorage at Highborne Cay, Exuma Chain, Bahamas; 21 August 1963; H. Feddern, *et al.*; USNM 119444.— 1 soft ♀, ca. 28.0; Bahamas; May 1962; ANSP.—FLORIDA: 1 brk. ♀, CL 10.5; off Palm Beach; 55-73 m; dredge, sand and rocky reef; TRITON, Thompson & McGinty; January 1951; USNM 113317.—1 ♀, 20.1, with casts from alima and postlarva; Florida Current, Straits of Florida; 25°23'-42' N, 80°04'-10'W; 37.5 m; GERDA Sta. G-8; 25 May 1962; UMML 32.2215.— 2 postlarvae, 16.0-16.9; Bear Cut, Virginia Key; 11 November 1957; J. Klussmann, W. R. Courtenay, Jr.; USNM 111091.—1 postlarva, 17.1; Virginia Key; 25 February 1960; D. R. Paulson; USNM 111089.—1 ♂, 45.5; 1 postlarva, 18.2; 2 pelagic larvae, 41.0-43.0; Virginia Key, Miami; sublittoral, in *Thalassia* flats; 10 January 1961; D. R. Paulson; UMML 32.2071.—1 ♂, 21.6; Virginia Key, Miami; 4 May 1961; R. B. Manning; USNM 111086.—1 ♀, 21.7; Bear Cut, Key Biscayne; 29 June 1957; R. B. Manning; RMNH 389.—1 ♂, 27.2; 1 ♀, 18.4; Bear Cut, Key Biscayne; 23 November 1957; W. R. Courtenay, Jr., *et al.*; USNM 119136.—1 ♀, 18.3; same; MNRJ.—1 ♀, 40.8; Bache Shoal; 4 May 1960; R. B. Manning; USNM 111087.—2 postlarvae, 16.4-17.5; Margate Fish Shoal; 23 August 1961; UMML Coral Reef Survey; USNM 111090.—1 ♀, 21.6;

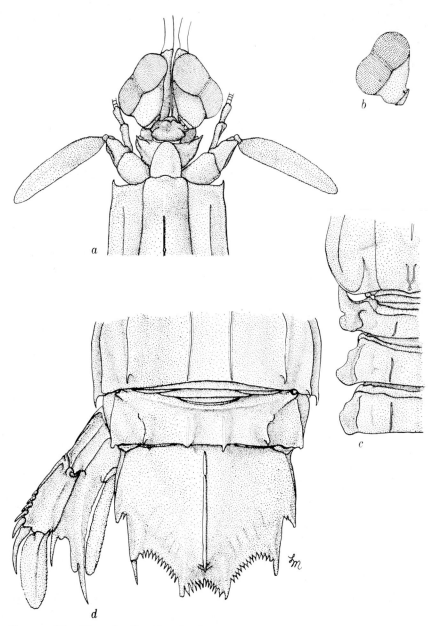

FIGURE 37. *Alima hyalina* Leach, female, TL 18.4, Bear Cut, Key Biscayne, Miami. *a,* anterior portion of body; *b,* eye; *c,* lateral processes of fifth to seventh thoracic somites; *d,* last two abdominal somites, telson, and uropod (setae omitted).

130

Knight's Key; 16 December 1956; USNM 111088.—1 ♀, 28.7; off Alligator Reef Light; 30 April 1961; W. A. Starck, II, *et al.*; UMML 32.2072. —MEXICO: 1 fragment; Orcas Reef, Gulf of Campeche; 18 m; 11 December 1952; Oregon; USNM 94467.— DOMINICAN REPUBLIC: 1 ♂, 20.9; Barahona; J. C. Armstrong; AMNH 8871.—LESSER ANTILLES: 1 ♀, 30.0; E side Cocoa Point, Barbuda; Bredin Sta. 102-59; 27 April 1959; USNM 111569.—CURAÇAO: 1 ♀, 21.9; Spanish Water; 26 April 1920; C. J. van der Horst; USNM 57515.—1 ♀, 27.5; same; ZMA. —1 ♂, 44.0; Santa Barbara Beach, Spanish Water; 13 February 1957; RMNH 436.—SAINT HELENA: 1 brk. ♂; 1 postlarva, 21.6; Sugarloaf; 46 m; 2 February 1930; Th. Mortensen; UZM.—1 ♂ abdomen; Flagstaff Bay; 24 February 1930; Th. Mortensen; UZM.—HAWAII: 1 ♀, 45.7; Waikiki Reef; 1921; C. H. Edmondson; BPBM 517.—1 ♂, 17.4; Kawaihae; at night light; 1 September 1949; K. Ego; UHML.

*Diagnosis.*—Carapace with short anterolateral spines, medium carina lacking anterior bifurcation; dactylus of raptorial claw with 6 teeth; mandibular palp absent; 4 epipods present; both lobes of lateral process of fifth thoracic somite in same plane; 2 rounded lobes between spines of basal prolongation of uropod.

*Description.*—Eye large, triangular, cornea bilobed, set slightly obliquely on stalk; eye broader than long; anterior margin of ophthalmic somite faintly bilobed; ocular scales blunt. Corneal indices of 6 specimens follow:

| Sex | TL | CL | CI |
|---|---|---|---|
| ♀ | 21.9 | 4.7 | 392 |
| ♂ | 21.6 | 5.0 | 357 |
| ♀ | 21.6 | 5.2 | 325 |
| ♂ | 27.2 | 6.5 | 342 |
| ♀ | 28.7 | 6.8 | 324 |
| ♀ | 40.8 | 9.4 | 376 |

Antennular peduncle slightly shorter than carapace; antennular processes blunt, unarmed, appearing sharp in dorsal view.

Rostral plate cordiform, with or without a faint median carina, apex rounded.

Carapace with anterolateral spines which do not extend to level of rostral plate; median carina not bifurcate at either end anterior to cervical groove, open anteriorly behind groove; intermediate carinae short, not converging with laterals; marginal carinae present on posterior portion of carapace, reflected marginals prominent; posterolateral margins lacking anterior angle; posterior margin produced along midline.

Raptorial dactylus armed with 6 teeth; dactylus sinuate, with a prominent lobe at base of outer margin; dorsal ridge of carpus undivided, ending in blunt lobe.

Mandibular palp absent; 4 epipods present.

131

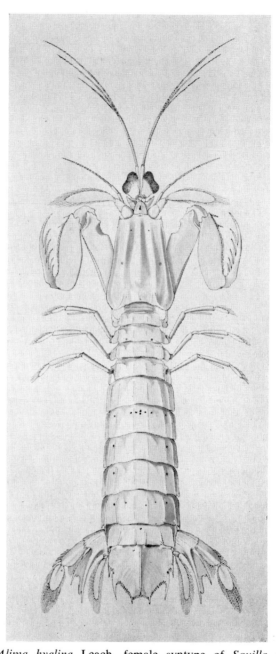

FIGURE 38. *Alima hyalina* Leach, female syntype of *Squilla alba,* Bimini, Bahamas (from Bigelow, 1894).

Exposed thoracic somites with submedian carinae and prominent intermediate carinae; lateral process of fifth thoracic somite bilobed, both lobes in the same plane; anterior lobe a spine directed anteriorly, posterior lobe rounded; no ventral spines on this somite; lateral processes of sixth and seventh somites bilobed, posterior lobe much the larger, angled or rounded in shape; ventral median keel on eighth thoracic somite low, triangular.

Submedian carinae of abdomen subparallel; abdominal carinae spined as follows: submedian, 6; intermediate, 5-6; lateral, (3-4) 5-6; marginal, (1-4) 5; sixth somite with ventrolateral spine in front of articulation of uropod.

Telson with 6 sharp marginal spines, each with short dorsal carina, submedians of postlarva with movable apices; prelateral lobes absent; denticles sharp, 5-7, 11-14, 1; median carina with anterior notch, ending posteriorly in short spine; ventral keel absent.

Penultimate segment of exopod of uropods with 5-6 graded, movable spines, last not extending to or slightly beyond midlength of distal segment; basal prolongation of uropods with 2 rounded lobes between the terminal spines.

*Color.*—Background white; eyestalks yellow-brown, with prominent black dorsal chromatophore; body with many scattered black chromatophores; second abdominal somite with dark, rectangular patch of chromatophores, and telson with 2 median dark patches. Live specimens are also marked with silver and blue.

*Size.*—Males, TL 20.9-45.5 mm; females, TL 18.3-45.7 mm; postlarvae, TL 16.0-21.6 mm. Other measurements of female, TL 40.8: carapace length, 9.4; cornea width, 2.5; rostral plate length, 1.5, width, 1.4; telson length, 6.1, width, 6.3.

Late pelagic larvae may attain a length of 54 mm (Hansen, 1895). The largest larva examined by me was 43 mm long.

*Discussion.*—A. hyalina is very similar to A. hieroglyphica (Kemp), but differs primarily in having six teeth on the raptorial claw and in having two rounded lobes between the spines of the basal prolongation of the uropod.

Manning (1962b) has shown that the well-known stomatopod larva, *Alima hyalina* Leach, is the pelagic larva of *Squilla alba* Bigelow. Leach's name is the oldest available name for the species, which must be known as *Alima hyalina* Leach, 1817.

Ingle (1958) correctly pointed out that Barnard's *Squilla hieroglyphica*, from South Africa, was actually *S. alba* Bigelow. The specimens reported by Boone (1930) as *S. alba* are, as noted by Schmitt (1940), *Cloridopsis dubia* (H. Milne-Edwards).

Leach published his original description of this species in two places, Appendix IV of Tuckey's "Narrative of an expedition to explore the River Zaire . . ." and in "Sur quelques genres nouveaux de Crustaces," both of which were published in 1818. The species is also figured on an unnumber-

FIGURE 39. Color patterns of: *a, Alima hyalina* Leach, male, TL 45.5, Virginia Key, Miami, Florida; *b, Cloridopsis dubia* (H. Milne-Edwards), female, TL 102.8, from Virginia Key, Miami, Florida.

ed plate in the appendix to Tuckey's narrative, and this plate is dated November 1, 1817. Therefore the date of publication of *Alima hyalina* Leach is 1817.

*Remarks.*—Most of the specimens from Florida have been collected on shallow grass *(Thalassia)* beds or on adjacent sand areas; Bigelow took his Bimini specimens from burrows in sand.

   Postlarvae differ from juveniles and adults in several important features, but can always be recognized by the bilobed lateral process of the fifth thoracic somite and the two rounded lobes between the spines of the basal prolongation of the uropod. In the postlarvae, the carinae of the body are poorly developed or absent, the anterolateral spines of the carapace are absent, and the submedian teeth of the telson have movable apices. Man-

134

ning (1962b) has provided descriptions of both juveniles and postlarvae. Although Townsley (1953) noted that the specimen from the Bernice P. Bishop Museum reported by Edmondson (1921) could not be located, the specimen is extant and was loaned to me for study by Dr. Edmondson.

*Ontogeny.*—The late pelagic larvae of this species have long been known as *Alima hyalina* Leach, *A. gracilis* H. Milne-Edwards, and *A. angusta* Dana. Manning (1962b) showed that *A. hyalina* was the larva of the species then known as *Squilla alba* Bigelow; that paper should be consulted for details on the larva and for methods of holding the larvae through the molt to the identifiable postlarval and juvenile stages. Only the later larval stages are known; *A. gracillima* Borradaile may represent a younger stage, and the specimens shown by Townsley (1953, text-fig. 22d) may also belong to this species.

The larvae mentioned by Coventry (1944) as *Choridella quadridens* were examined in the collections of the Academy of Natural Sciences at Philadelphia. All but two of the 36 specimens examined (in four lots, ANSP 5380, 5381, 5385, and 5386) were *A. hyalina;* the two exceptions could be the larvae of *Meiosquilla quadridens* or any of the related species with four teeth on the claw.

Most of the older references to *A. hyalina* repeat one of Leach's original localities, Cape Verde. The "larval" synonymy given here includes all references known to me; as all of the nineteenth century literature is not available to me, some references may have been overlooked.

*Type.*—Not traced, probably British Museum (Natural History). The larger female of Bigelow's syntypes of *S. alba,* USNM 18495, is here selected as a lectotype. The other female, USNM 111092, becomes a paralectotype.

*Type-locality.*—In both his Tuckey account and in the paper published in 1818, Leach gave two localities for his material, Porto Praya [Cape Verde Islands] and 07°37′00″N, 17°34′15″W, off the coast of Africa. The type-locality is here restricted to Porto Praya, Cape Verde Islands.

*Distribution.*—Adults have been recorded from scattered localities in all tropical areas except the eastern Pacific. Specific localities, Western Atlantic: Bermuda, Bahamas, Florida, upper Gulf of Mexico, Dominican Republic, Barbuda, and Curaçao. Adults also have been reported from St. Helena, South Africa, and Hawaii.

*Alima hieroglyphica* (Kemp, 1911)
Figure 40

Squilla hieroglyphica Kemp, 1911, p. 96; 1913, p. 51, pl. 3, figs. 38-41.—
    Komai, 1914, p. 461, pl. 3, fig. 1 [in Japanese].—Kemp, 1915, p. 171.—
    Komai, 1927, p. 313.—Roxas & Estampador, 1930, p. 103.— Alikunhi,
    1944, p. 237, text-fig. 1 [final pelagic larva and post-larval]; 1947, p. 289.

135

—Kurian, 1947, p. 124.—Alikunhi, 1952, p. 264, text-fig. 8 [larva].—
Kurian, 1954, p. 85.—Manning, 1968, p. 136 [listed].
*Squilla hildebrandi* Schmitt, 1940, p. 152, text-fig. 6.—Manning, 1968, p. 136 [listed].
*Squilla labadiensis* Ingle, 1960, p. 566, text-figs. 1-10.—Manning, 1968 p. 136 [listed].
*Squilla (Alima) hieroglyphica:* Alikunhi, 1958, p. 135, text-fig. 14 [larva].

*Previous records.*—INDO-WEST PACIFIC: JAPAN: Nagasaki (Komai, 1914, 1927).—PHILIPPINE IDS.: market in Manila (Kemp, 1915; Roxas & Estampador, 1930).—INDIA: Mahanadi Estuary (Alikunhi, 1958).—Madras (Alikunhi, 1944, 1947, 1952).—Trivandrum (Kurian, 1947, 1954).—WEST AFRICA: GHANA: Ingle, 1960.—WESTERN ATLANTIC: PANAMA: Schmitt, 1940.

*Material.*—PANAMA: 1 ♂, 48.5; Fort Sherman, Canal Zone, Atlantic coast; 3 March 1937; S. F. Hildebrand; holotype of *S. hildebrandi;* USNM 76068.—CUBA: 1 ♂, 51.0; J. Gundlach; UZM.—BRAZIL: 1 ♂, 75.1; Fortaleza, Ceara; Jose Fausto Guimaraes; MNRJ.

EASTERN ATLANTIC: GHANA: 1 ♂, 39.6; off Labadi; Agassiz trawl; 8 m; 5 December 1955; D. T. Gauld; holotype of *S. labadiensis;* BMNH 1958.12.1.12.—1 juv. ♀, 16.4; off Accra; Sta. 30; grab; J. B. Buchanan; paratype of *S. labadiensis;* BMNH 1958.12.1.15.—2 ♀, 50.2-53.2; off Accra; 13 m; Sta. 88; Agassiz trawl; R. S. Bassindale; allotype and paratype of *S. labadiensis;* BMNH 1958.13.1.13-14.—INDIAN OCEAN: INDIA: 1 ♂, 59.7; Trivandrum, Kerala, South India; C. V. Kurian, col.; IM CA 116/1.

*Diagnosis.*—Rostral plate with median carina; no anterior bifurcation present on median carina of carapace, intermediate carinae present; dactylus of raptorial claw with 5 teeth; mandibular palp absent; 4 epipods present; lateral process of fifth thoracic somite bilobed, sharp anterior lobe situated ventral to rounded posterior lobe; telson with faint prelateral lobes, denticles 3-5, 10-12, 1; 1 lobe between spines of basal prolongation of uropod.

*Description.*—Eye large, triangular, length and greatest width subequal; cornea bilobed, set transversely on stalk; ocular scales rounded; anterior margin of ophthalmic somite obtusely angled. Corneal indices of 3 specimens follow:

| Sex | TL | CL | CI |
|-----|------|------|-----|
| ♂ | 51.0 | 11.7 | 454 |
| ♂ | 48.5 | 12.2 | 469 |
| ♂ | 75.1 | 17.6 | 550 |

Antennular peduncle not as long as carapace; apices of antennular processes bluntly rounded.

Rostral plate subtriangular; strong median carina usually present on anterior third, not extending to anterior margin, carina inconspicuous in some specimens.

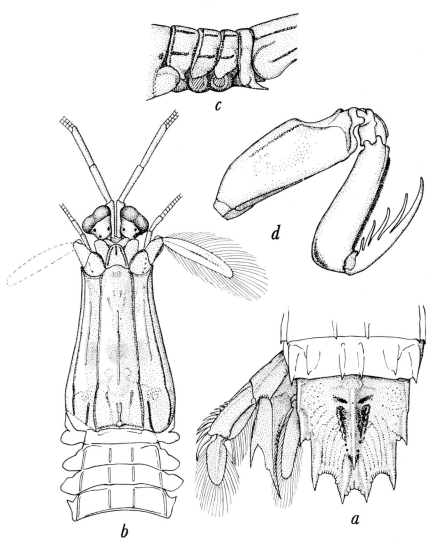

FIGURE 40. *Alima hieroglyphica* (Kemp), male, TL 48.5, Fort Sherman, Panama (holotype of *S. hildebrandi* Schmitt). *a,* last abdominal somite, telson, and uropod; *b,* anterior portion of body; *c,* exposed thoracic somites, lateral view; *d,* raptorial claw (from Schmitt, 1940).

137

Carapace with strong anterolateral spines extending to level of rostral base; median carina bifurcate anteriorly posterior to cervical groove, not bifurcate at either end anterior to cervical groove; intermediate carinae prominent, anteriorly converging with and occasionally meeting laterals; posterolateral margins not angled anteriorly; median posterior margin lobed, with prominent median bead.

Raptorial dactylus armed with 5 teeth, outer margin faintly sinuate, feebly notched at base; dorsal ridge of carpus undivided, ending in a blunt lobe.

Mandibular palp absent; 4 epipods present.

Exposed thoracic somites with submedian carinae, restricted to posterior third on fifth somite; last three somites wth prominent intermediate carinae; lateral process of fifth thoracic somite bilobed, posterior lobe situated dorsal to anterior, bluntly rounded laterally; anterior, ventral lobe produced into a spine directed anterolaterally; fifth thoracic somite without a ventral spine; lateral processes of next two somites rounded laterally, anterior margin faintly sinuate; ventral median keel on eighth thoracic somite low, blunt, inconspicuous.

Second to fourth abdominal somites each with faint median carinule; submedian carinae subparallel on first 4 somites, slightly divergent on fifth somite; sixth somite with at most an inconspicuous lobe in front of articulation of uropod; abdominal carinae spined as follows: submedian, 5-6; intermediate, 4-6; lateral, (3) 4-6; marginal, 1-5.

Telson with 6 sharp marginal spines; prelateral lobes present, inconspicuous; denticles 3-5, 10-12, 1, outer submedian denticle the largest; median carina with a faint anterior notch, ending in a strong posterior spine; ventral surface with a strong postanal keel.

Penultimate segment of exopod of uropod with 8-9 graded, movable spines, last extending to midlength of ultimate segment; basal prolongation of uropod with a prominent lobe, outer margin concave, on outer margin of inner spine.

*Color.*—Eyestalks with three dark chromatophores forming a triangle; gastric grooves and carinae of carapace lined with dark chromatophores; sixth thoracic to fifth abdominal somite lined posteriorly with dark pigment; telson with a triangular dorsal patch of chromatophores, apex of patch under spine of median carina, color densest at base of triangle; a transverse bar of chromatophores lies anterior to triangle.

*Size.*—Males, TL 39.6-75.1 mm; females, TL 16.4-53.2 mm. Other measurements of male, TL 48.5: carapace length, 12.2; cornea width, 2.6; rostral plate length, 1.7, width 2.0; telson length, 9.5, width, 8.3.

*Discussion.*—Comparison of a specimen from Trivandrum with one from Cuba and later examination of the types of *S. labadiensis* leave me no alternative but to conclude that *A. hildebrandi* and *A. labadiensis* must be synonymized with *A. hieroglyphica.* There are no characters that can be

138

used to distinguish the specimens from the three widely separate areas. There is some variation in the carination of the carapace of the West African forms but a parallel variation exists in specimens from the other two areas.

The male holotype of *S. labadiensis* is damaged and the rostral plate does not seem to have an anterior carina; the carina is present in the other specimens reported by Ingle.

Specimens from all three areas also have an identical color pattern which supports the belief that they are conspecific.

The differences between *A. hieroglyphica* and *A. hyalina* are summarized under the discussion of the latter species.

*Remarks.*—Although Schmitt (1940) evidently assumed that the type of *S. hildebrandi* was from the Pacific coast of America, the type-locality, Fort Sherman, is on the Atlantic side.

The specimen from Cuba in the Copenhagen Museum evidently is part of the Gundlach collection, most of which is now housed in the Berlin Museum. The label on the specimen bears the name *S. rubrolineata* Dana. This is not the specimen reported as *S. rubrolineata* by von Martens for he clearly stated (1872: 144) that his specimen had five teeth on one raptorial dactylus and six on the other; the present specimen has five teeth on each dactylus.

Another label in the jar, in K. Stephensen's hand, dated 1938, identifies the specimen as *Squilla strigata* H. J. Hansen. This is a manuscript name, and was never published. As Ingle (1958) pointed out, Hansen also used this name for *Squilla alba* Bigelow.

*Ontogeny.*—The larval forms of *A. hieroglyphica* from the Indian Ocean have been summarized by Alikunhi (1944, 1952, 1958).

*Type.*—The holotype of *S. hieroglyphica* Kemp is in the Indian Museum, Calcutta. The holotype, allotype and paratypes of *S. labadiensis* Ingle are in the British Museum. The holotype of *S. hildebrandi* Schmitt is in the U. S. National Museum.

*Type-locality.*—None given in original description. The type-locality is here restricted to the coast of India.

*Distribution.*—Western Atlantic from Panama, Cuba, and Brazil; Ghana, West Africa; Indo-West Pacific from Japan, the Philippines, and the coasts of India.

## Genus *Cloridopsis* Manning, 1968
*Cloridopsis* Manning, 1968, pp. 120 [key], 128.

*Definition.*—Body smooth or minutely rugose, pits, if present, inconspicuous; size moderate or small, TL 160 mm or less; eye small, stalk dilated or margins subparallel, cornea slightly broader than stalk; ocular scales separate, truncate; rostral plate rounded or subtriangular, with or

139

without median carina; carapace strongly narrowed anteriorly, with normal complement of carinae, median lacking anterior bifurcation, laterals occasionally interrupted; cervical groove distinct; posterior median margin transverse; posterolateral margins evenly rounded, anterolateral angles spined; last 3 thoracic somites with low submedian and intermediate carinae; lateral process of fifth thoracic somite a thick, anteriorly-curved spine, ventral spines of fifth somite present or absent; lateral processes of next 2 somites sinuous or evenly rounded, not conspicuously bilobed; median ventral keel of eighth somite low, rounded; 2-3 epipods present; mandibular palp present or absent; raptorial claw stout, dactylus with 5-6 teeth, outer margin of dactylus sinuous or evenly rounded; propodus broadening distally, upper margin with pectinations and distal obtuse prominence; propodus with 3 movable teeth at base, middle smallest; dorsal ridge of carpus undivided; merus stout, without inferodistal spine; ischiomeral articulation terminal; endopods of walking legs linear; abdomen broad, inflated, with 8 carinae on first 5 somites, 6 on last; telson broad, inflated, with median carina and 6 short marginal teeth, submedians with fixed apices; prelateral lobe present; ventral surface of telson without long postanal keel; uropods stout, basal prolongation short, inner margin tuberculate, with conspicuous rounded lobe on outer margin of inner spine.

*Type-species.*—*Squilla scorpio* Latreille, 1825, by original designation.

*Gender.*—Feminine.

*Number of species.*—Six, as listed by Manning (1968), of which one occurs in the Americas.

*Remarks.*—*Cloridopsis* is characterized by its small eyes, inflated body and telson, reduced number of teeth on the raptorial claw and reduced number of epipods. The carapace carries a normal complement of carinae, with the median carina always present but lacking an anterior bifurcation. The general facies of all six species included herein is very similar.

   *C. dubia* and *C. gibba* differ from the remainder of the species of the genus in having the eyestalk dilated; the eye shape is very similar to that found in *Clorida*. In other features *C. dubia* is very similar to *C. scorpio* and the other Indo-West-Pacific species which I have examined.

   *Squilla gibba* Nobili was tentatively assigned here rather than to *Clorida;* it was grouped by Kemp (1913) with the species now assigned to that genus. Nobili's species differs from all those now in *Clorida* in lacking both dorsal tubercles on the telson and the inner spines on the basal prolongation of the uropod. Although Kemp stated that *S. gibba* seemed to have movable apices on the submedian teeth of the telson, these are not shown in his figure. As in the other Indo-West Pacific species of *Cloridopsis*, *C. gibba* has but two epipods.

   In at least four of the species of *Cloridopsis* sexual maturity in the male is accompanied by a noticeable thickening of the margin of the telson. The male of *C. gibba* exhibits an extreme amount of swelling as shown by

140

Kemp (1913, pl. 11, fig. 1); in that species much of the lateral area of the telson is unusually swollen.

The relationships of *Cloridopsis* are not clear, although the genus resembles *Clorida* in many respects; the former almost seems to be a large *Clorida* which has retained the small eyes but has lost its telson sculpture and movable apices on the submedian teeth of the telson. The similarity to *Clorida* is reflected in the name chosen for the genus.

*Cloridopsis* resembles *Squilloides* in some respects, particularly *S. gilesi* (Kemp) and *S. lata* (Brooks), but similar species in the latter genus completely lack the median carina of the carapace and all species now in *Squilloides* have four epipods. In my opinion the resemblance is superficial.

### *Cloridopsis dubia* (H. Milne-Edwards, 1837)
### Figures 39b, 41

Restricted synonymy:
*Squilla dubia* H. Milne-Edwards, 1837, p. 522.—Gibbes, 1845, p. 70 [listed]; in Tuomey, 1848, p. xvi; 1850, p. 200 [p. 36 on separate].—Herklots, 1861, p. 152 [listed].—Miers, 1880, p. 24.—Rathbun, 1883, pp. 121-130. —Howard, 1883, p. 294.—Thallwitz, 1893, p. 55.—Sharp, 1893, p. 107. —Bigelow, 1894, p. 518 [part].—Moreira, 1901, pp. 2, 70.—Kemp, 1913, p. 200 [listed].—Torralbas, 1917, p. 621, fig. 67 [p. 81 on separate].— Kemp & Chopra, 1921, p. 298.—Lemos de Castro, 1945, p. 29, fig.— Andrade Ramos, 1951, pp. 142, 146, pl. 1, fig. 2.—Lemos de Castro, 1955, p. 13, text- figs. 9-11, pl. 2, fig. 34, pl. 3, fig. 35.—Manning, 1959, p. 18.—Rodriguez, 1959, p. 274.—Manning, 1961, p. 13 [key]; 1967b, p. 105 [discussion]; 1968, p. 128 [listed].
*Squilla rubrolineata* Dana, 1852, p. 618; Atlas, 1855, p. 13, pl. 61, figs. 2a-b.—Smith, 1869, p. 41 [listed].—von Martens, 1872, p. 144.
*Chloridella dubia:* Lunz, 1935, p. 157, text-fig. 5 [part].
*Chlorodiella dubia:* Luederwaldt, 1919, p. 429; 1929, p. 52.—Oliveira, 1940, p. 145 [erroneous spelling of *Chloridella*].
*Squilla alba:* Boone, 1930, p. 35, pl. 5 [misidentification; not *Squilla alba* Bigelow, 1893].

*Previous records.*—AMERICAS: H. Milne-Edwards, 1837.—SOUTH CAROLINA: Charleston (Gibbes, 1850; Bigelow, 1894; Lunz, 1935); no other data (Gibbes, in Tuomey, 1848; Howard, 1883).—GEORGIA: Savannah (Rathbun, 1883; Bigelow, 1894).—FLORIDA: Port Everglades (Manning, 1959); ? Marquesas Keys (Manning, 1967b).—CUBA: Nuevitas Bay (Boone, 1930; Manning, 1967b); no other data (von Martens, 1872; Torralbas, 1917).—DOMINICAN REPUBLIC: Miers, 1880.—TRINIDAD: Sharp, 1893.—BRITISH HONDURAS: Belize (Miers, 1880).—PANAMA: Limon Bay (Boone, 1930; Manning, 1967b). —VENEZUELA: Margarita Island (Rodriguez, 1959).—BRAZIL: Para State, Paraiba State, and Canavierias, Bahia State (all Lemos de Castro, 1955); Rio de Janeiro (Dana, 1852; Thallwitz, 1893; Moreira, 1901; Oliveira, 1940; Lemos de Castro, 1945, 1955); São Sebastião (Luederwaldt, 1919, 1929); Santos and Oguape (Luederwaldt, 1919); São Paulo (Andrade Ramos, 1951); Santa Catarina State (Lemos de Castro, 1955).

It has also been recorded from several localities between El Triunfo, El Salvador to Guayaquil, Ecuador in the eastern Pacific (Schmitt, 1940).

*Material.*—AMERICAS: 1 ♂, ca. 108.0; Amerique; no other data; probably the type; MNHNP.—SOUTH CAROLINA: 1 ♂, 112.5; Charleston; no other data; MNHNP.—1 dam. ♂, 85.0; same; 17 September 1880; USNM 3139.—GEORGIA: 1 ♂, 155.5; Savannah; R. J. Nunn; USNM 2057.—1 ♂, 129.2; same; Dr. G. H. Macon; USNM 2524.—FLORIDA: 1 spec.; Hawks Park (?); February 1915; F. J. Keeley; ANSP 4869.— 1 ♂, 107.1; West Palm Beach; 25 April 1943; Lt. Seabrook; AHF.—1 ♂, 119.0; Lake Worth; in mud; 11 April 1959; L. and S. Thomas; UMML 32.1148.—1 ♀, 116.6; Hillsboro Inlet; 8 July 1945; B. Royal; AHF.— 1 damaged ♂, 111.2; Port Everglades; 24 July 1956; A. Volpe; UMML 32.1176.—1 ♀, 102.8; W side Virginia Key, Miami; mud flats; 20 March 1961; C. R. Robins and class; USNM 119184.—1 ♂, 49.5; Virginia Key, Miami; 27 February 1960; L. Thomas; USNM 111112.—MEXICO: 1 dry, fragmented spec.; Mexico; ANSP 2725.—BRITISH HONDURAS: 1 brk. ♂; Belize; BMNH 54.400.—GUATEMALA: 1 ♀; pier van Puerto Barrios; v. Blydestyn; ZMA.—PANAMA: 1 ♂, 153.6; 1 ♀, 119.3; Fort Randolph, near Colon; March 1924; J. B. Thropshire; MCZ.—1 ♂, 140.9; Sanchez; T. Barbour; MCZ.—CUBA: 1 ♂, 121.0; Gundlach collection; no other data; ZMB 12125a.—2 ♂, 120.0-121.6; same; ZMB 3753.—HAITI: 1 ♀, 93.7; A. Curtiss; 1949, 1950; AMNH.—1 ♂, 105.0; Port-au-Prince; 9 January 1950; A. Curtiss; AMNH.—1 ♂, 96.5; same; February 1950; AMNH.—1 ♂, 82.4; same; no date; AMNH.—1 ♂, 123.0; same; March 1950; AMNH.—2 ♂, 106.5-108.8; same; AMNH.—3 ♂, 87.2-114.5; 5 ♀, 86.7-120.0; same; AMNH.—DOMINICAN REPUBLIC: 1 ♂, 104.8; 1 ♀, 124.6; Samana; 3 May 1922; W. L. Abbott; USNM 55819.—1 ♀, 91.5; Punta Arena, Samana Bay; AMNH 4702.—LESSER ANTILLES: 1 ♂; St. Dominique (Dominica?); Pichon, col.; MNHNP.—1 ♂, ca. 145.5; Fort de France, Martinique; 1900; M. Hédiod; MNHNP.—TRINIDAD: 1 ♂, 110.2; R. Welles, Jr.; USNM 21479.—1 dry ♂, ca. 98.0; Trinidad; Dr. S. Lewis; ANSP 3606.—VENEZUELA: 2 specimens; Laguna Marites, Margarita Island; G. Rodriguez (Specimens in Rodriguez' personal collection).—BRAZIL: 1 ♂, 105.4; Recife; R. von Ihering; USNM 110854.—2 ♂, 98.0-135.2; Bahia (Salvador); 1913; P. Serre; MNHNP.— 1 ♂, 158.2; same; Thayer Expedition; MCZ 7912.—1 ♂, 143.4; 1 ♀, 142.3; Rio de Janeiro; MCZ 7914.—1 ♂, 110.0; same; Jobert; MNHNP. —1 ♀, 126.3; Paranagua; 24°S, 49°W; 22 August 1956; C. H. von Hippel; USNM 99797.—1 ♂, 155.1; Santos; December 1914; H. Luederwaldt; USNM 48305.—1 ♂, 146.0; Cananeia; 4 September 1951; W. Besnard; IOSP 14.

In addition to those listed above, three specimens from three localities were found in one jar in the collection of the Vanderbilt Marine Museum. The specimens (2 ♂, 84.9-95.0; 1 ♀, 83.2) were from Limon Bay, Panama, Nuevitas Bay, Cuba, and SW of Marquesas Keys, Florida. Two of these specimens were mentioned by Boone (1930) and all three by Manning (1967b).

*Diagnosis.*—Eye small, stalk dilated, median carina of carapace without bifurcation; dactylus of raptorial claw with 5-6 teeth; mandibular palp present; 3 epipods present; lateral process of fifth thoracic somite a broad spine, curved forward; lateral processes of next 2 somites rounded; telson denticles 1-4, 3-6, 1.

*Description.*—Eye small, cornea bilobed, set obliquely on stalk; corneal axis of eye smaller than peduncular axis; stalk broad, dilated, distally constricted near cornea; ocular scales separate, broadly rounded; anterior margin of ophthalmic somite obtuse, often with median tubercle or small spine. Corneal indices for 12 specimens follow:

| Sex | TL | CL | CI |
|-----|-----|-----|-----|
| ♂ | ca 85.0 | 17.0 | 713 |
| ♂ | 107.1 | 18.3 | 610 |
| ♀ | 102.8 | 19.7 | 603 |
| ♀ | 116.6 | 23.0 | 657 |
| ♂ | 104.8 | 23.6 | 694 |
| ♂ | 119.0 | 23.8 | 700 |
| ♂ | 121.6 | 25.3 | 791 |
| ♀ | 126.3 | 26.4 | 776 |
| ♂ | 120.0 | 26.7 | 763 |
| ♀ | 124.6 | 27.3 | 803 |
| ♂ | 155.1 | 29.6 | 800 |
| ♂ | 155.5 | 31.5 | 900 |

Antennular peduncle short, about three-fourths as long as carapace; antennular processes produced into sharp spines directed anteriorly.

Rostral plate elongate, triangular, longer than broad, lateral margins tapering to rounded apex; median carina absent.

Median carina of carapace, anterior to cervical groove, not bifurcate at either end; intermediate carinae not extending to anterior margin; anterolateral spines small, not extending to level of rostral plate; posterolateral angles with poorly-marked anterior angle.

Raptorial claw stout, dactylus with 5-6 teeth; outer margin of dactylus sinuate, basal portion swollen; propodus with prominent distal tubercle on upper margin; dorsal ridge of carpus undivided, ending in a blunt lobe.

Mandibular palp usually present; 3 epipods present.

Last 3 thoracic somites with submedian and intermediate carinae, submedians usually divergent; lateral process of fifth thoracic somite directed laterally, anterior margin sinuous, posterior margin convex proximally, concave distally, apex sharp; 1 pair of ventral spines also present on this somite; lateral process of sixth somite a single lobe, anterior margin slightly sinuous, broadly rounded posterolaterally; lateral process of seventh somite with 1 or 2 anterior tubercles, posterior margin broadly rounded; ventral keel on eighth somite rounded, inclined posteriorly.

Submedian carinae of abdominal somites divergent on each somite; fourth and fifth somites with 2 anterior tubercles or swellings between intermediate and lateral carinae, mesial tubercle obscure in small specimens; abdominal carinae spined as follows: submedian, 6; intermediate, (4) 5-6; lateral, (4) 5-6; marginal, (1-2) 3-5; sixth somite with sharp spine

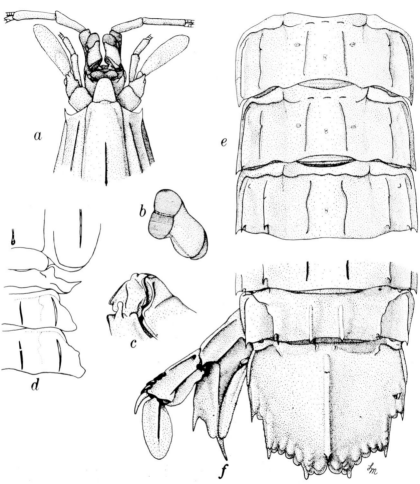

FIGURE 41. *Cloridopsis dubia* (H. Milne-Edwards), female, TL 102.8, Virginia Key, Miami. *a,* anterior portion of body; *b,* eye; *c,* carpus of raptorial claw; *d,* lateral processes of fifth to seventh thoracic somites; *e,* third to fifth abdominal somites, dorsal view; *f,* last abdominal somite, telson, and uropod (setae omitted).

144

in front of articulation of uropod.

Telson swollen, noticeably broader than long, with 3 pairs of short, sharp marginal teeth; prelateral lobes present; median carina swollen, with prominent basal notch and small distal spine; anterior surface with 2 or 3 pairs of dorsal tubercles, at levels of submedian and intermediate carinae of sixth somite; carinae of marginal teeth dorsally tuberculate; denticles swollen, appearing bilobed dorsoventrally, (1) 2-4, (3) 4-5 (6), 1, inner submedian the largest; ventral surface lacking postanal keel.

Uropods with 4-7 (usually 5) graded, short movable spines on outer margin of penultimate segment of exopod, last not extending to midlength of distal segment.

*Color.*—Preserved specimens have no distinctive color pattern. Live specimens are colorfully marked with pastel blue and green, with the abdominal carinae marked with crimson.

*Size.*—Males, TL 49.5-155.5 mm; females, TL 86.7-142.3 mm. Other measurements of male, TL 107.1: carapace length, 18.3; cornea width, 3.0; rostral plate length, 2.9, width, 2.6; telson length, 14.4, width, 16.7.

*Discussion.*—*Cloridopsis dubia* can be separated from the other species in the genus by the long but inflated eyes and the presence of three epipods. It is the only American species of the genus.

There are some differences between specimens from the eastern Pacific and those from the western Atlantic, the most striking of which is the position of the submedian abdominal carinae. In eastern Pacific specimens, the carinae are parallel on each somite; in Atlantic specimens, they are usually noticeably divergent on each somite. The carinae are almost subparallel (but still terminate outside of the submedian carinae of the following somite) in a few Atlantic specimens, so this feature cannot be used to separate Atlantic from Pacific specimens unless some other features can be used with it. The cornea seems to be larger in Pacific specimens than Atlantic, but too few Pacific specimens are available for comparison to draw any conclusions. It seems possible that additional material will indicate two species or subspecies are present.

*Remarks.*—Several features of this species appear to be rather variable. The rostral plate is usually elongate-triangular, with a narrow apex, but may be short, with a broad, blunt apex. The mandibular palp is usually present, but may be present on one side only or totally absent. The proximal tooth on the raptorial claw may be reduced or even absent and the basal swelling of the dactylus is proportionately larger in large specimens. The ornamentation of the telson is very variable; the marginal teeth and their carinae and the denticles may be variously tuberculate or swollen. In one specimen, one of the intermediate denticles is produced into a spine.

There is no well-marked sexual dimorphism, but in large males the carinae, teeth, and denticles of the telson are more swollen than in females.

145

One of the labels accompanying a female from Guayaquil (USNM 14113) notes that the species is "regarded as poisonous and much dreaded." Miers (in Whymper, 1891, p. 124) noted that ". . . . the natives have tried to eat it, but found it poisonous." Also, the lot from the Dominican Republic in the U. S. National Museum has a label stating "sea scorpion— said to be venomous and feared accordingly." These are the only records that I have found of poisonous stomatopods; whether the species is actually poisonous or, as in the case of most stomatopods, capable of inflicting a painful wound, remains to be seen.

*Ontogeny.*—Unknown. Hansen (1895) suggested that *Alima hyalina* Leach might be the pelagic larva of this species, but Manning (1962b) has shown that it is actually the larva of *Squilla alba* Bigelow.

*Type.*—Muséum National d'Histoire Naturelle, Paris.

*Type-locality.*—"les cotes d'Amerique."

*Distribution.*—Western Atlantic, from Charleston, South Carolina, to Santa Catarina State, Brazil, including Georgia, Florida, scattered localities in the West Indies and Caribbean, and several localities in northern and central Brazil. In the eastern Pacific, it is known from several localities between El Triunfo, El Salvador, and Guayaquil, Ecuador (Schmitt, 1940). It is a shallow water species, often found in mud flats.

## Genus *Squilla* Fabricius, 1787

*Squilla* Fabricius, 1787, p. 333.—Hemming, 1945, p. 53.—Manning, 1968, pp. 120 [key], 129 [revision].

*Definition.*—Body smooth or minutely punctate, compact; size moderate to large, TL to over 200 mm, generally 150 mm or less; eyes moderate to large, cornea noticeably bilobed, broader than stalk; ocular scales separate, subtruncate, rounded, or acute; rostral plate variable in shape, median carina present or absent; carapace narrowed anteriorly, carinae well-formed, normal complement present, median carina often with anterior bifurcation; cervical groove distinct; posterior median margin of carapace usually evenly concave, occasionally with obscure obtuse median projection; posterolateral margin of carapace usually with anterior angle; anterolateral angles of carapace armed; last 3 thoracic somites with submedian and intermediate carinae, intermediates occasionally armed; lateral process of fifth thoracic somite a single, variously-shaped spine or lobe, usually an anteriorly-curved spine; ventral spines present on fifth somite; lateral processes of next 2 somites usually bilobed, posterior lobe usually much the larger; ventral keel on eighth thoracic somite prominent, rounded or subtriangular; 4-5 (usually 5) epipods present; mandibular palp usually present; dactylus of raptorial claw with 5-7 teeth, usually 6, outer margin of dactylus sinuate, flattened, or evenly convex; propodus with upper margin pectinate and with 3 movable teeth at base, middle

146

smallest; dorsal ridge of carpus entire or subdivided; merus without infero-distal spine; ischiomeral articulation terminal; endopods of walking legs linear; abdomen compact, carinae well-formed, first five somites with 8 carinae, sixth somite with 6; telson with median carina, supplementary dorsal carinae present or absent; 3 pairs of marginal teeth present, sub-medians with fixed apices; prelateral lobes usually present; ventral surface of telson usually with postanal keel; inner margin of basal prolongation of uropod usually tuberculate, occasionally provided with slender spines; inner spine of basal prolongation with lobe on outer margin.

*Type-species.—Cancer mantis* Linnaeus, 1758, p. 633, by subsequent desig-nation by Latreille, 1810, p. 422. *Squilla* is name no. 619 on the ICZN Official List.

*Gender.—*Feminine.

*Number of species.—*Twenty-seven, as listed by Manning, 1968, of which 16 occur in the western Atlantic.

*Nomenclature.—*In an opinion by the International Commission of Zoolo-gical Nomenclature validating *Squilla* Fabricius, 1787 (Hemming, 1945), the complex early nomenclature of *Squilla* was reviewed. Holthuis & Man-ning (in press) listed synonyms of the genus, and Manning (1968) re-viewed the genus and recognized several new genera.

*Remarks.—*The genus *Squilla* includes the large Atlanto-East Pacific species assigned to *Squilla s.l.* The members of the genus have compara-tively large eyes, well-developed carinae on the carapace and body, and fixed apices on the submedian teeth of the telson. The genus appears to be homogeneous, with the possible exception of *Squilla heptacantha* (Chace). The latter differs from all others in the genus in having seven teeth on the dactylus of the raptorial claw and in having strongly-bilobed lateral processes on the sixth and seventh thoracic somites. In the structure of those processes it resembles *Squilla oratoria* de Haan and its relatives which are now placed in the genus *Oratosquilla.*

   *S. surinamica* Holthuis, a small species known only from off Surinam, has eyes approaching those found in *Cloridopsis,* but in all other details agrees with the other species now placed in *Squilla.*

   The western Atlantic *S. empusa* Say, the eastern Pacific *S. aculeata* Bigelow, the eastern Atlantic *S. calmani* Holthuis, and perhaps the eastern Atlantic *S. mantis* (Linnaeus) form a compact subgroup within the genus. Of some interest is the fact that *S. calmani* and *S. aculeata* seem to be much more closely related to each other than either is to *S. empusa;* in fact, *S. calmani* and *S. aculeata* may prove to be conspecific. Many of the features of *S. neglecta* Gibbes and *S. prasinolineata* Dana also link them to this section of the genus, although both of the latter species have only five teeth on the dactylus of the raptorial claw and differ in other character-istics as well.

147

In 1961 I pointed out that *S. intermedia* Bigelow and *S. edentata* (Lunz) were distinct species, and reported specimens from off northern South America under the latter species. Subsequent study aided by many additional specimens has shown that the South American *S. edentata* is actually two distinct taxa, herein recognized as *Squilla caribaea,* n. sp., and *S. edentata australis,* n. ssp.

*S. intermedia* and its near allies comprise a compact group of six closely related species, of which four, *S. brasiliensis* Calman, *S. caribaea, S. edentata* (with two subspecies), and *S. intermedia,* occur in the western Atlantic. *S. biformis* Bigelow is an eastern Pacific form and an undescribed species, referred to *S. intermedia* by Ingle (1960), occurs in the eastern Atlantic. All of the species in this section of the genus are large, TL to 175 mm, with very large, T-shaped eyes, prominent carinae on the body, and well-marked sexual dimorphism in the adult male. Most are inhabitants of moderate depths; they rarely occur in less than 200 m. The West African form and *S. brasiliensis* have the shallowest bathymetric range.

The characters used to distinguish these species appear to be trivial but also appear to be constant in their occurrence. *S. biformis* differs from the remainder in having a spined postanal keel. In *S. brasiliensis* the submedian carinae of the fourth abdominal somite are armed and the anterolateral margins of the carapace form a straight but sloping line. The West African species, with its small eyes, seems to be most similar to *S. brasiliensis* or perhaps *S. caribaea;* these three species differ from *S. biformis, S. edentata,* and *S. intermedia* in the shape of the lateral process of the fifth thoracic somite. In the "*brasiliensis*" subgroup, the lateral process is gradually curved forward, with its apex directed anterolaterally, whereas in the "*biformis*" subgroup the lateral process is sickle-shaped, with the apex directed anteriorly.

*S. edentata* differs from all of the other species in having the median carina of the carapace composed of two subparallel lines of pits; also the outer submedian denticle is the largest. In *S. caribaea* the submedian denticles of the telson may or may not be subequal, but the median carina of the carapace is a single raised line, as in *S. intermedia.* The latter species can be distinguished by its spined ocular scales.

The development of this group of species is paralleled by the development of an inshore complex consisting of *S. chydaea* Manning, *S. lijdingi* Holthuis, *S. obtusa* Holthuis, and *S. surinamica* Holthuis in the western Atlantic and *S. panamensis* Bigelow and *S. bigelowi* Schmitt in the eastern Pacific. All of these species share the paired dark crescents on the anterior half of the telson.

The interrelationships of the species and species groups within the genus are not clear cut. This can readily be seen by examining Table 7, in which species are listed by several different morphological features, including basic color pattern. The table was originally intended as an aid to quick identification of some of the species. It can help identification but it also

148

shows that species with similar color patterns may have very different morphological features; apparently neither color nor morphology can be used alone as a clue to relationships. For example, *S. obtusa* has a color pattern resembling not only the *chydaea-lijdingi* group of species but those with supplementary dorsal carinae on the telson as well.

Features used in the descriptive accounts given below are illustrated in Figure 29.

TABLE 7

SPECIES OF *Squilla* SHARING DISTINCTIVE MORPHOLOGICAL FEATURES AND

COLOR PATTERN

I. Color Pattern
    A. Median patch of dark pigment present on carapace:

| | |
|---|---|
| *S. hancocki* Schmitt | *S. tiburonensis* Schmitt |
| *S. rugosa* Bigelow | |

    B. Dark vertical bar present distally on merus of claw:

| | |
|---|---|
| *S. deceptrix*, n. sp. | *S. obtusa* Holthuis |
| *S. discors* Manning | *S. rugosa* Bigelow |
| *S. grenadensis*, n. sp. | *S. tiburonensis* Schmitt |
| *S. hancocki* Schmitt | |

    C. Dark median patches present on second and fifth abdominal somites:

| | |
|---|---|
| *S. deceptrix*, n. sp. | *S. prasinolineata* Dana |
| *S. hancocki* Schmitt | *S. rugosa* Bigelow |
| *S. obtusa* Holthuis | *S. tiburonensis* Schmitt |

    D. Anterior dark paired crescents on telson:

| | |
|---|---|
| *S. bigelowi* Schmitt | *S. obtusa* Holthuis |
| *S. chydaea* Manning | *S. panamensis* Bigelow |
| *S. lijdingi* Holthuis | *S. surinamica* Holthuis |

    E. Anterior paired dark rectangles or squares on telson:

| | |
|---|---|
| *S. brasiliensis* Calman | *S. mantis* (Linnaeus) |
| *S. hancocki* Schmitt | *S. tiburonensis* Schmitt |
| *S. heptacantha* (Chace) | |

    F. Anterior and posterior dark patches on telson:

| | |
|---|---|
| *S. deceptrix*, n. sp. | *S. prasinolineata* Dana |

II. Median carina of carapace lacking anterior bifurcation:

| | |
|---|---|
| *S. deceptrix*, n. sp. | *S. parva* Bigelow |
| *S. discors* Manning | *S. prasinolineata* Dana |
| *S. grenadensis*, n. sp. | *S. rugosa* Bigelow |
| *S. hancocki* Schmitt | *S. tiburonensis* Schmitt |
| *S. lijdingi* Holthuis | |

III. Raptorial Claw
    A. Dactylus sinuate:

| | |
|---|---|
| *S. aculeata* Bigelow | *S. neglecta* Gibbes |
| *S. calmani* Holthuis | *S. parva* Bigelow |
| *S. empusa* Say | *S. surinamica* Holthuis |
| *S. mantis* (Linnaeus) | |

    B. Dactylus with 5 teeth:

| | |
|---|---|
| *S. neglecta* Gibbes | *S. prasinolineata* Dana |

    C. Dactylus with 7 teeth:
      *S. heptacantha* (Chace)

IV. Exposed thoracic somites
    A. Intermediate carinae armed:
      *S. discors* Manning

B. Lateral process of fifth thoracic somite not an anteriorly-curved spine:
S. *heptacantha* (Chace)        S. *parva* Bigelow
S. *mantis* (Linnaeus)          S. *prasinolineata* Dana
S. *mantoidea* Bigelow          S. *rugosa* Bigelow
S. *neglecta* Gibbes
C. Lateral process of sixth thoracic somite strongly bilobed:
S. *heptacantha* (Chace)
D. Lateral processes of sixth and seventh thoracic somites spined posteriorly:
S. *grenadensis*, n. sp.        S. *tiburonensis* Schmitt
S. *rugosa* Bigelow
E. Lateral processes of sixth and seventh thoracic somites not bilobed nor armed posteriorly:
S. *prasinolineata* Dana
F. Lateral processes of sixth and seventh thoracic somites with posterior lobe rounded:
S. *neglecta* Gibbes            S. *parva* Bigelow
S. *obtusa* Holthuis            S. *surinamica* Holthuis
S. *panamensis* Bigelow

V. Mandibular palp and epipods
A. Mandibular palp absent:
S. *neglecta* Gibbes
B. Four epipods present:
S. *aculeata* Bigelow           S. *heptacantha* (Chace)
S. *calmani* Holthuis           S. *prasinolineata* Dana
S. *grenadensis*, n. sp.

VI. Telson with dorsal tubercles or carinae:
S. *deceptrix*, n. sp.          S. *hancocki* Schmitt
S. *discors* Manning            S. *rugosa* Bigelow
S. *grenadensis*, n. sp.

VII. Prelateral lobes of telson spined:
S. *bigelowi* Schmitt

VIII. Inner margin of basal prolongation of uropod with spines:
S. *grenadensis*, n. sp.        S. *rugosa* Bigelow


KEY TO WESTERN ATLANTIC SPECIES OF *Squilla*

1. Dorsal surface of telson with numerous longitudinal carinae in addition to carinae of marginal teeth . . . . . . . . . . . . . . . . . . . . . . . . . . . . . .2
1. Dorsal surface of telson lacking supplementary dorsal carinae . . . . .5

2. Inner margin of basal prolongation of uropod armed with slender spines; lateral processes of sixth and seventh thoracic somites produced posteriorly into spines . . . . . . . . . . . . . . . . . . . . . . . . . . . . . .3
2. Inner margin of basal prolongation of uropod at most irregularly tuberculate; lateral processes of sixth and seventh thoracic somites not produced posteriorly into spines . . . . . . . . . . . . . . . . . . . . . . . . . .4

3. Lateral process of fifth thoracic somite an anteriorly-curved spine; 4 epipods present . . . . . . . . . . . . . . . . . . *grenadensis*, n. sp., p. 152.

3. Lateral process of fifth thoracic somite a broad, acute lobe, directed laterally, anterior margin convex; 5 epipods present. .*rugosa*, p. 155.
4. Intermediate carinae of last 3 thoracic somites armed posteriorly
.......................................... *discors*, p. 161.
4. Intermediate carinae of last 3 thoracic somites unarmed..........
................................ *deceptrix*. n. sp., p. 165.
5. Lateral process of sixth thoracic somite strongly bilobed, lobes sub-equal; dactylus of claw with 7 teeth .........*heptacantha*, p. 171.
5. Lateral process of sixth thoracic somite, if bilobed, with posterior lobe much larger; dactylus of claw with no more than 6 teeth .........6
6. Dactylus of claw with 5 teeth; lateral process of fifth thoracic somite a straight spine or spatulate lobe, directed laterally ..............7
6. Dactylus of claw with 6 teeth; lateral process of fifth thoracic somite an anteriorly-curved spine ................................8
7. Median carina of carapace not bifurcate anteriorly; lateral process of fifth thoracic somite a slender spine .........*prasinolineata*, p. 175.
7. Median carina of carapace with anterior bifurcation; lateral process of fifth thoracic somite a spatulate lobe ............*neglecta*, p. 181.
8. Lateral processes of sixth and seventh thoracic somites rounded posterolaterally ........................................9
8. Lateral processes of sixth and seventh thoracic somites acutely pointed posterolaterally ........................................10
9. Rostral plate broad, subquadrate, with short median carina; eye small, cornea only slightly broader than stalk ........ *surinamica*, p. 185.
9. Rostral plate elongate, triangular, without median carina; eye large, T-shaped, cornea much broader than stalk ........ *obtusa*, p. 187.
10. Median carina of carapace with incomplete, poorly-marked anterior bifurcation ........................................11
10. Median carina of carapace with well-defined anterior bifurcation...12
11. Rostral plate broad, subquadrate ...............*lijdingi*, p. 192.
11. Rostral plate elongate, subtriangular ........... *chydaea*, p. 196.
12. On median carina of carapace, distance from dorsal pit to anterior bifurcation less than 1/5 distance from bifurcation to anterior margin
..................................... *empusa*, p. 201.
12. Distance from dorsal pit to anterior bifurcation more than 1/5 distance from bifurcation to anterior margin ........................13
13. Anterior margin of each lateral plate of carapace straight, sloping backward; submedian carinae of fourth abdominal somite armed posteriorly ............................. *brasiliensis*, p. 215.
13. Anterior margin of each lateral plate of carapace concave; submedian car of fourth abdominal somite unarmed ....................14

151

14. Median carina of carapace composed of 2 or more subparallel rows of pits; carina appearing bicarinate under magnification; intermediate carinae of first abdominal somite usually unarmed; telson tumid, median carina convex in lateral view . . . . . . . . . . .*edentata*, p. 220.

14. Median carina of carapace simple or flanked by unconnected pits, not appearing bicarinate; intermediate carinae of first abdominal somite armed posteriorly; median carina of telson sinuous or straight in lateral view . . . . . . . . . . . . . . . . . . . . . . . . . . . . . . . . . . . . . . . . . . . 15

15. Ocular scales sharp laterally; telson broad; in adult males, abdominal width greater than carapace and rostral plate length combined . . . . . . . . . . . . . . . . . . . . . . . . . . . . . . . . . . . . . . . . . . . . . . *intermedia*, p. 228.

15. Ocular scales rounded; telson narrow; in adult male, abdominal width less than carapace length . . . . . . . . . . . . . . . *caribaea*, n. sp., p. 234.

## Squilla grenadensis, new species
### Figure 42

*Holotype.*—1 ♀, 28.9; off Grenada; 311 m; B1AKE Sta. 297; 1878-1879; MCZ 7851.

*Diagnosis.*—Rostral plate without median carina; median carina of carapace lacking anterior bifurcation; dactylus of raptorial claw with 6 teeth; mandibular palp and 4 epipods present; lateral process of fifth thoracic somite an anteriorly-curved spine; intermediate carinae of thoracic somites unarmed; telson with numerous longitudinal carinae, denticles 4, 8-10, 1; basal prolongation of uropod with mesial spines.

*Description.*—Eye large, cornea set almost transversely on stalk; eyes not extending to end of first segment of antennular peduncle; ocular scales rounded laterally; anterior margin of ophthalmic somite rounded; Corneal Index: 309.

Antennular peduncle longer than carapace; antennular processes produced into sharp spines directed anterolaterally.

Antennal scales not present in holotype.

Rostral plate as long as broad, without median carina, upturned lateral margins converging on rounded apex.

Median carina of carapace without anterior bifurcation, portion posterior to cervical groove bifurcate; intermediate carinae not extending to anterior margin; anterolateral spines not extending to base of rostral plate; posterolateral angles of carapace rounded, without anterior angle.

Raptorial claw large, dactylus with 6 teeth, outer margin of dactylus evenly convex; dorsal ridge of carpus undivided.

Mandibular palp present; 4 epipods present.

Lateral process of fifth thoracic somite a sharp spine, curved slightly forward; lateral processes of next 2 somites not bilobed, rounded laterally, each produced into strong posterior spine; lateral margin of process of sixth somite more sinuous than that of seventh; last 3 thoracic somites with

152

prominent submedian and intermediate carinae, fifth somite with 2 pairs of short, curved dorsal carinae; ventral keel of eighth thoracic somite low, inconspicuous.

Abdomen depressed, strongly carinate, submedian carinae subparallel on each somite; second to fifth somites with interrupted median carinule, anteriorly notched submedian and intermediate carinae, and a broad, irregular area, neither carinate nor tuberculate, between submedian and intermediate carinae; abdominal carinae spined as follows: submedian, 6; intermediate, 4-6; lateral, 3-6; marginal, 1-5; fifth somite without accessory

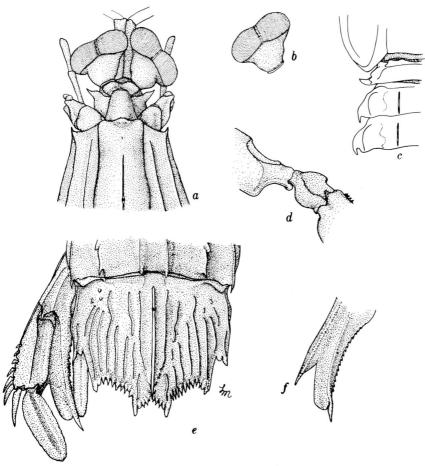

FIGURE 42. *Squilla grenadensis*, n. sp., female holotype, TL 28.9, off Grenada, Lesser Antilles. *a*, anterior portion of body; *b*, eye; *c*, lateral processes of fifth seventh thoracic somites; *d*, carpus of claw; *e*, last abdominal somite, telson, and uropod; *f*, basal prolongation of uropod, ventral view (setae omitted).

153

spinules on posterior margin; sixth somite with strong spine in front of articulation of uropod.

Telson broader than long, prelateral lobes poorly indicated but present; dorsal surface of telson ornamented with numerous longitudinal carinae, including (a) median carina, interrupted anteriorly, terminating posteriorly in strong spine, apex with ventral bituberculate knob; (b) 1 pair of long carinae converging under apex of median carina; (c) 4-5 carinae, some interrupted, between short carina of submedian tooth and long carina of intermediate tooth; (d) 1 anterior carina, in line with submedian teeth; (e) 1 or 2 carinae and several tubercles between carinae of intermediate and lateral teeth; denticles 4, 8-10, 1, outer submedian, outer intermediate, and lateral rounded, remainder sharp; ventral surface with short postanal keel.

Proximal segment of exopod of uropod with 7 movable spines, last short, not extending to midlength of distal segment; basal prolongation with 15 short spinules on mesial margin, and prominent rounded lobe on lateral margin of inner spine.

*Color.*—Largely faded in the holotype; second abdominal somite with dark rectangular dorsal patch, dark lateral patches present on last 3 thoracic and all abdominal somites, most prominent on fifth somite; scattered dark chromatophores present on lateral portions of telson and on uropods; merus of claw with distal vertical bar.

*Size.*—Only known specimen, female holotype, TL 28.9 mm. Other measurements: carapace length, 7.1; cornea width, 2.3; rostral plate length, 2.5, width, 2.5; telson length, 5.1, width, 6.0.

*Discussion.*—This small species closely resembles *S. rugosa, S. discors,* and *S. deceptrix* in lacking an anterior bifurcation on the median carina of carapace, in having a strong raptorial claw with an evenly-curved outer margin, a vertical bar of dark color on the merus of the claw, and numerous longitudinal carinae on the telson. The presence of only four epipods will immediately distinguish it from all of these species. It agrees with *S. rugosa* and differs from *S. discors* and *S. deceptrix* in having the lateral processes of the sixth and seventh thoracic somites produced into posteriorly-directed spines and in having spines on the mesial margin of the basal prolongation of the uropod. The eyes, with the cornea set obliquely on the stalk, the anteriorly-curved lateral process of the fifth thoracic somite, the absence of an anterior lobe on the lateral process of the sixth thoracic somite, the absence of submedian spines on the fifth abdominal somite, and the presence of a large rounded lobe on the inner spine of the basal prolongation of the uropod will also help to separate this species from *S. rugosa,* which it resembles rather closely.

*Remarks.*—At first I identified this specimen, which is probably immature, as a juvenile *S. rugosa.* Direct comparison of the holotype with a juvenile

154

*S. rugosa* of the same size shows that they must be considered as distinct species. All of the differences between *S. grenadensis* and *S. rugosa* are distinct even at that small size. Larger specimens of *S. grenadensis* may be expected to have a better developed lateral process on the fifth thoracic somite and a larger number of spines on the abdominal carinae.

*S. grenadensis* is apparently found in deeper water than *S. rugosa,* which has not been taken in depths exceeding 71 m.

*Ontogeny.*—Unknown.

*Etymology.*—The name is derived from the type-locality, Grenada.

*Type.*—Museum of Comparative Zoology, Harvard.

*Type-locality.*—Grenada.

*Distribution.*—Known only from the type-locality, in 311 m.

*Squilla rugosa* Bigelow, 1893

Figures 43, 44a

*Squilla rugosa* Bigelow, 1893, p. 102.—Bigelow, 1894, p. 541, text-figs. 23-24.—Kemp, 1913, p. 202 [listed].—Kemp & Chopra, 1921, p. 298 [listed].—Boone, 1927, p. 7.—Schmitt, 1940, p. 162 [discussion].—Chace, 1954, p. 449.—Springer & Bullis, 1956, p. 23.—Holthuis, 1959, p. 174, pl. 8, figs. 1-2.—Manning, 1959, p. 20; 1961, p. 18, pl. 4, figs. 1-3.—Bullis & Thompson, 1965, p. 13 [listed].—Manning, 1968, p. 129 [listed].
*Chloridella rugosa:* Rathbun, 1899, p. 628.—Archer, 1948, p. 10.
*Chloridella rugosa* var. *pinensis* Lunz, 1937, p. 12, text-fig. 6.
*Squilla rugosa pinensis:* Manning, 1959, p. 20 [part].

*Previous records.*—BAHAMAS: Little Bahama Bank (Manning, 1959.—FLORIDA: Biscayne Bay, Tortugas shrimp grounds (Manning, 1959); off Charlotte Harbor (Bigelow, 1893, 1894; Chace, 1954); off Cedar Key (Springer & Bullis, 1956).—ALABAMA: Mississippi Sound (Archer, 1948).—MEXICO: Gulf of Campeche (Springer & Bullis, 1956: Manning, 1959).—CUBA: Isle of Pines (Boone, 1927; Lunz, 1937); southern coast (Manning, 1959).—JAMAICA: Port Royal Cays, Port Antonio, Kingston Harbor (Rathbun, 1899).—BRITISH GUIANA: Manning, 1959, 1961.—SURINAM: Holthuis, 1959, Manning, 1959, 1961. Bullis & Thompson (1965) list the occurrence of *S. rugosa* at one SILVER BAY Station and at 12 OREGON Stations.

*Material.*—BAHAMAS: 1 ♂, 92.4; 1 ♀, 90.0; Little Bahama Bank; 13 November 1952; UMML 32.1182.—2 ♀, 94.4-108.8; 20 mi. NW of Mangrove Cay, Little Bahama Bank; 5m; white sand and grass; 11 November 1952; UMML 32.1184.—FLORIDA: 1 ♂, 76.5; off Madeira Beach; from grouper stomach; USNM 113320.—1 ♂, 93.4; Biscayne Bay, Miami; 20 March 1958; D. Tabb; UMML 32.1197.—9 ♂, 66.3-97.8; 2 ♀, 100.4-102.0; same; 20 March 1958; D. Tabb, E. Iversen; UMML 32.1473 (1 ♂, 1 ♀, BMNH; 1 ♂, MCSN).—1 ♀; same 21 April 1961; D. Tabb; UZM.—

155

1 ♂, 107.0; off Dinner Key, Biscayne Bay, Miami; 2 April 1958; D. Tabb; USNM 111561.—3 ♂, 76.1-109.3; 2 ♀, 84.2-125.2; Barnes Sound, Key Largo; 29 December 1959; D. Allen; USNM 111562.—2 ♂, 2 ♀; 45 mi. NW of Key West; 23-25 m; shrimp trawl; 5 April 1953; RMNH 366.— 1 ♀, 100.9; Tortugas shrimp grounds; 26 m; UMML 32.1159.—1 ♀, 86.0; same; July 1956; J. Regan; UMML 32.1150.—1 ♂, 91.0; 2 ♀, 64.0-86.6; same; 27-28 January 1958; V. G. Springer; UMML 32.1200.—1 ♂, 67.6; 1 ♀, 74.0; same; 10-12 March 1958; R. B. Manning; MNRJ.—1 ♀, 85.0; same; 6-9 April 1959; S. Dobkin; USNM 111572.—1 ♂, 108.2; 1 ♀, 110.0; same; 1-3 June 1959; A. Jones; USNM 111559.—2 ♂, 27.1-44.1; same; 20 June 1959; R. Robinson, H. Foulk; USNM 111573.—2 ♀, 44.6-117.4; 30 mi. NE of Loggerhead Key, Dry Tortugas; 34 m; 30 December 1954; G. H. Eubank; USNM 99890.—2 spec., fragmented; Dry Tortugas; from stomachs of *Sciacyion micrurum;* 21 June 1931; W. L. Schmitt; USNM 66012.—1 dam. ♀, 24.2; same; 1-2 August 1931; USNM 119237. —1 telson; same; USNM 78420.—1 ♂, 35.8; N of Key West; Silver Bay Sta. 71; USNM 101620.—1 ♂, 88.0; off Tortugas; 20°50'N, 82°50'W (probably in error; 24°N ?); 40 m; F. Durant; CHML.—1 ♀, 78.5; off Charlotte Harbor; Albatross Sta. 2412; holotype; USNM 9835.—1 ♂, 73.0; off Boca Grande; 24 May 1959; W. Saenz; USNM 119241.—1 ♀, 120.0; W of St. Petersburg; Oregon Sta. 4089; USNM 119238.—2 ♂, 68.8-84.8; same; Oregon Sta. 4090; USNM 119239.—1 ♂, 78.9; 1 ♀, 79.2; W of Cedar Key; Oregon Sta. 907; USNM 96404.—1 ♂, 66.6; Pensacola; 1881; S. Stearns; USNM 4602.—MEXICO: 4 ♂, 74.5-109.1; 5 ♀, 63.0-111.1; Gulf of Campeche, Yucatan; P. Fuentes; MCZ.—2 ♀, 79.1-100.7; Gulf of Campeche; Oregon Sta. 449; USNM 92655.—1 ♂, 98.4; same; Oregon Sta. 713; UMML 32.1180.—HONDURAS: 2 ♂, 57.8-71.2; 1 ♀, 109.1; Oregon Sta. 1938; USNM 111555.—CUBA: 1 ♂, 72.5; Siguanea Bay, Isle of Pines; 22 m; Pawnee, col.; 6 March 1925; holotype of *C. rugosa* var. *pinensis* Lunz; YPM 4404.—1 ♀, 75.0; SE of Rio, Golfo de Batabano; 5 m; September 1959; I. Perez-Farfante; USNM 104248.—COLOMBIA: 1 ♀, 107.8; Oregon Sta. 4866; USNM 119240. —BRITISH GUIANA: 1 ♀, 56.1; Oregon Sta. 2236; USNM 111557.— 1 ♂, 82.2; Oregon Sta. 2244; UMML 32.1882.—2 ♂, 70.4-73.2; 9 ♀, 51.9-95.5; Oregon Sta. 2249; USNM 103502.—2 ♂, 58.3-74.9; 2 ♀, 86.4-87.0; Oregon Sta. 2250; UMML.—SURINAM: 1 ♀, 101.1; Coquette Sta. 280; USNM 103242.—1 ♀, 114.3; Oregon Sta. 2261; UMML 32.1883.—2 ♂, 92.2-97.7; 5 ♀, 88.2-109.8; Oregon Sta. 2262; UMML 32.1881.—5 ♂, 61.4-101.5; Oregon Sta. 2267; USNM 111564. —2 ♂, 83.5-89.0; 2 ♀, 92.7-85.1; same; IM.—1 ♂, 78.9; Oregon Sta. 2276; UMML.

*Diagnosis.*—Rostral plate lacking median carina; median carina of carapace without anterior bifurcation; dactylus of raptorial claw with 6 teeth; mandibular palp and 5 epipods present; lateral process of fifth thoracic somite an acute lobe, not anteriorly curved, anterior margin convex; intermediate carinae of thoracic somites unarmed; fifth and sixth abdominal somites usually with accessory spinules on posterior margin; telson with numerous longitudinal carinae, denticles 4-6, 7-12, 1; basal prolongation of uropod with mesial spines.

*Description.*—Eye large, cornea set obliquely on stalk; corneal axis of eye greater than peduncular; ocular scales rounded, oblique to body line; anterior margin of ophthalmic somite obtuse, projecting. Corneal indices of 10 specimens of various sizes follow:

| Sex | TL | CL | CI |
|---|---|---|---|
| ♂ | 27.1 | 6.3 | 300 |
| ♂ | 58.3 | 13.6 | 378 |
| ♂ | 61.4 | 14.2 | 373 |
| ♂ | 72.5 | 16.5 | 330 |
| ♂ | 78.9 | 17.9 | 389 |
| ♀ | 78.5 | 18.3 | 365 |
| ♀ | 86.0 | 19.5 | 367 |
| ♀ | 91.4 | 20.5 | 386 |
| ♀ | 100.9 | 21.2 | 407 |
| ♀ | 108.8 | 25.0 | 390 |

Antennular peduncle as long as carapace and rostral plate combined; antennular processes produced into acute spines directed obliquely forward.

Median carina of carapace not bifurcate anteriorly; intermediate carinae not extending to anterior margin; anterolateral spines not extending to base of rostral plate; posterolateral margins of carapace flattened laterally, with anterior angle.

Raptorial claw elongate; dactylus with 6 teeth, outer margin evenly curved; dorsal ridge of carpus undivided.

Mandibular palp present; 5 epipods present.

Last 3 thoracic somites with strong submedian and intermediate carinae, with short, oblique carina lying between them on anterior half of each somite; intermediates and short oblique carinae also present on fifth somite; lateral process of fifth somite an acute lobe directed laterally, anterior margin convex, posterior margin concave; lateral process of sixth somite bilobed, anterior lobe small, angled, posterior lobe produced into strong spine directed posteriorly; lateral process of seventh somite not bilobed, produced into strong spine directed posteriorly; ventral keel on eighth somite flattened anteriorly, rounded and inclined posteriorly.

Abdominal carinae sharp, prominent, submedians divergent on each somite; second to fifth somites each with 2 anterior tubercles between intermediate and lateral carinae; abdominal carinae spined as follows: submedian, 5-6; intermediate, (2) 3-6; lateral, 1-6; marginal, 1-5; posterior margin of fifth and sixth somites with 0-4 accessory spinules on each side between submedian and intermediate carinae; sixth somite with prominent ventral spine in front of articulation of each uropod.

157

Telson with slender marginal teeth; prelateral lobes absent; dorsal surface of telson ornamented with numerous longitudinal carinae and tuberculated ridges; an entire submedian carina on either side of median carina converging under its apex; 4 carinae present on each side between carinae

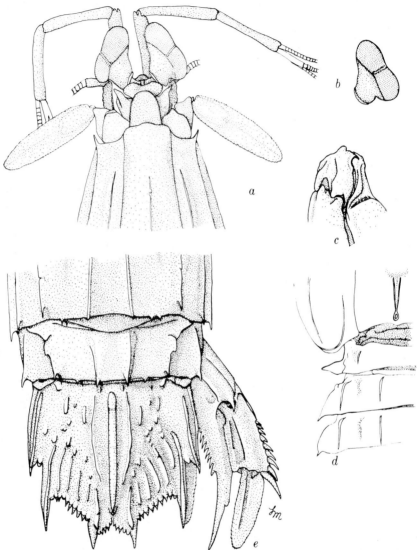

FIGURE 43. *Squilla rugosa* Bigelow, male, TL 73.0, off Boca Grande, Florida. *a,* anterior portion of body; *b,* eye; *c,* carpus of raptorial claw; *d,* lateral processes of fifth to seventh thoracic somites; *e,* last two abdominal somites, telson, and uropod (setae omitted).

of submedian and intermediate teeth; 1 carina present on each side between carinae of intermediate and lateral teeth; intermediate carina occasionally joining a supplementary dorsal carina; several short tuberculated ridges present on anterior half of telson; denticles sharp, 4-6, 7-12, 1; prominent postanal keel present, flanked laterally by 1 tubercle.

Outer margin of penultimate segment of uropodal exopod with 7-8 (13) movable spines, distal not extending to midlength of distal segment; basal prolongation of uropod with 7-13 fixed spines on mesial margin.

*Color.*—Antennal scale with black line on anterodistal fourth; merus of

FIGURE 44. Color patterns of three species of *Squilla: a, S. rugosa* Bigelow, female, TL 84.2, Barnes Sound, Florida; *b, S. discors* Manning, male, TL 58.4, OREGON Sta. 3577; *c, S. deceptrix*, n. sp., male holotype, TL 64.1, OREGON Sta. 3587.

159

raptorial claw with distal vertical slash on lateral face; second and fifth abdominal somites with dorsal black rectangular patch, fifth somite with posterolateral black patches also; telson with median black patch on posterior half; distal portion of endopod and distal segment of exopod of uropod black.

*Size.*—Males, TL 27.1-111.1 mm; females, TL 24.2-125.2 mm. Other measurements of female holotype, TL 78.5: carapace length, 18.3; cornea width, 5.0; rostral plate length, 2.5, width, 2.5; telson length, 14.3, width, 14.5.

*Discussion.*—*S. rugosa* resembles three other western Atlantic species, *S. deceptrix, S. discors,* and *S. grenadensis,* in having numerous longitudinal dorsal carinae on the telson. It may be distinguished from them by the shape of the lateral process of the fifth thoracic somite, which in all those species is an anteriorly curved spine rather than a laterally-directed lobe, by the presence of accessory spinules on the posterior margin of the last two abdominal somites, and by the absence of a prelateral lobe on the telson. *S. grenadensis* may also be distinguished from *S. rugosa* by the presence of only four epipods. Other differences are noted under the discussion of each of the species mentioned above.

S. *rugosa* also resembles *S. hancocki* Schmitt from the eastern Pacific in basic color pattern and in having longitudinal carinae on the telson. However, *S. hancocki,* with its anteriorly-curved lateral process of the fifth thoracic somite, seems to be more closely related to *S. deceptrix* and *S. discors.*

*Remarks.*—The accessory spinules normally present on the last two abdominal somites are variable in number and may be completely absent. The major distinguishing feature of *S. rugosa* var. *pinensis* (Lunz) was apparently the absence of these spinules. Examination of the type of the variety and comparison of it with the holotype of *S. rugosa* shows that this character is not of subspecific importance, as suggested by Holthuis (1959).

S. *rugosa* is the only species in the genus in which the abdominal accessory spinules are present. They are also present in *Pterygosquilla armata,* a subantarctic species.

The specimen from the east coast of Florida reported by me as *S. rugosa pinensis* in 1959 is actually *S. deceptrix.*

*Ontogeny.*—Unknown.

*Type.*—U. S. National Museum.

*Type-locality.*—Off Charlotte Harbor, Florida (26°18′30″N, 83°08′45″W), in 49 m, ALBATROSS Sta. 2412.

*Distribution.*—Western Atlantic, from Little Bahama Bank to Surinam, including southeast and west Florida, Gulf of Campeche, Honduras, Cuba, Jamaica, Colombia and British Guiana. Littoral to 71 m, on sand and shell.

160

## Squilla discors Manning, 1962
### Figures 44b, 45

Squilla discors Manning, 1962c, p. 217 [part, holotype and one paratype only, as listed below].—Bullis & Thompson, 1965, p. 13 [listed].— Manning, 1968, p. 129 [listed].

Previous records.—BAHAMAS: Great Bahama Bank (Manning, 1962c). —VENEZUELA: Manning, 1962c. Other records in Manning (1962c) refer to S. deceptrix, n. sp., which see. Bullis & Thompson (1965) record this species at six U. S. Fish and Wildlife Service stations; of their records, material from SILVER BAY Sta. 2480 is referrable to S. discors. The specimens from other stations are S. deceptrix, n. sp.

Material.—BAHAMAS: 3 ♀, 51.4-63.4; N of Little Bahama Bank; SILVER BAY Sta. 3466; USNM 119180.—1 ♂, 66.5; Little Bahama Bank; 27°22′N, 79°11′W; 222 m; GERDA Sta. 394; 19 September 1964; UMML. —1 ♂, 44.7; Great Bahama Bank; SILVER BAY Sta. 2480; paratype; UMML 32.2164.—1 ♂, 50.0; Santaren Channel; SILVER BAY Sta. 2445; USNM 119181.—NICARAGUA: 10 ♂, 36.7-58.4; 13 ♀, ca. 41.0-58.3; OREGON Sta. 3577; USNM 119183.—VENEZUELA: 1 ♂, 63.3; ALBATROSS Sta. 2120; holotype; USNM 7832.—1 ♂, 54.6; OREGON Sta. 4394; USNM 119182.

Diagnosis.—Rostral plate without median carina; median carina of carapace lacking anterior bifurcation; dactylus of raptorial claw with 6 teeth; mandibular palp and 5 epipods present; lateral process of fifth thoracic somite an anteriorly-curved spine; intermediate carinae of thoracic somites armed posteriorly; telson with numerous dorsal tubercles; denticles (4-5) 6-8, 8-12, 1; basal prolongation of uropod lacking mesial spines.

Description.—Eye large, cornea set very obliquely on stalk; eyes not extending to end of first segment of antennular peduncle; ocular scales obliquely truncate; anterior margin of ophthalmic somite projecting, truncated or faintly emarginate medially. Corneal indices of 11 specimens of various sizes follow:

| Sex | TL | CL | CI |
|---|---|---|---|
| ♂ | 36.7 | 8.3 | 361 |
| ♀ | 43.3 | 10.0 | 385 |
| ♂ | 44.7 | 10.3 | 368 |
| ♀ | 47.8 | 10.8 | 415 |
| ♀ | 50.5 | 11.6 | 387 |
| ♀ | 57.8 | 12.7 | 397 |
| ♂ | 52.7 | 13.1 | 409 |
| ♂ | 58.4 | 13.5 | 397 |
| ♂ | 66.5 | 14.0 | 424 |
| ♂ | 63.3 | 14.5 | 439 |
| ♀ | 63.4 | 14.7 | 397 |

161

Antennular peduncle longer than carapace; antennular processes produced into slender spines directed anterolaterally.

Rostral plate as long as broad, lateral margins tapering to rounded apex; plate in some specimens almost triangular; median carina absent.

Median carina of carapace, anterior to cervical groove, not bifurcate at either end; portion of median carina posterior to cervical groove with anterior bifurcation; intermediate carinae extending to or almost to anterior margin; anterolateral spines short, not extending to base of rostral plate; posterolateral margins with at most a faint anterior angle.

Raptorial claw slender, large, dactylus with 6 teeth; outer margin of dactylus evenly convex, with shallow basal notch; dorsal ridge of carpus irregular but undivided, terminating in an obtuse lobe.

Mandibular palp and 5 epipods present.

Last 3 thoracic somites with submedian and intermediate carinae, intermediates on each somite with posterior spines; at least sixth somite with short anterior carina on each side between submedians and intermediates; fifth somite with 1 pair of short, curved dorsal carinae; lateral process of fifth somite a slender, anteriorly-curved spine, with an irregular basal carina; lateral process of sixth somite bilobed, anterior lobe small, sharp, posterior lobe much larger, triangular, posteriorly-directed apex acute but not spiniform; lateral process of seventh somite not bilobed, obliquely truncate, apex acute but not spiniform; ventral keel of eighth somite rounded, inclined posteriorly.

Abdomen strongly carinate, submedian carinae divergent on each somite, anteriorly notched on second somite; second to fifth abdominal somites with interrupted median carinule, anterior notch on each intermediate carina, and anterior tubercle between intermediate and lateral carinae; abdominal carinae spined as follows: submedian, 5-6; intermediate, 1-6; lateral, 1-6; marginal, 1-5; sixth somite with irregular anterior tubercle between submedian and intermediate carinae, and with small ventral spine in front of articulation of each uropod.

Telson broader than long, appearing very broad due to divergent lateral margins, with sharp marginal teeth; prelateral lobes present; median carina basally notched, terminating posteriorly in small spine; in addition to undivided carinae of marginal teeth, telson ornamented with 6 or more rows of tubercles, of which 2 rows parallel median carina, remainder occur on ridges flanking confluent lines of pits; denticles, (4-5) 6-8, 8-10 (12), 1, outer submedian denticle rounded, remainder sharp; ventral surface with short postanal keel.

Proximal segment of uropodal exopod with 6-7 spines, last short, not extending to midpoint of distal segment; inner margin of basal prolongation tuberculate, not spined.

*Color.*—Median, intermediate, reflected marginal carinae and posterior margin of carapace dark; merus of raptorial claw with distal vertical bar

162

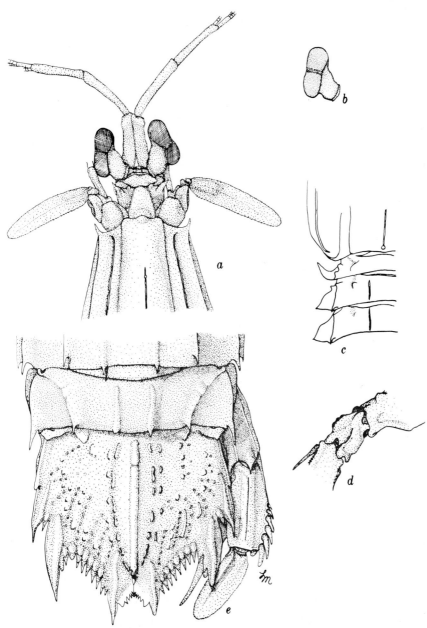

FIGURE 45. *Squilla discors* Manning, male holotype, TL 63.3, Venezuela. *a,* anterior portion of body; *b,* eye; *c,* lateral processes of fifth to seventh thoracic somites; *d,* carpus of claw; *e,* last two abdominal somites, telson, and uropod (setae omitted).

163

on lateral face; submedian abdominal carinae lined with dark pigment; articulated anterolateral abdominal plates dark; second abdominal somite with dark, rectangular median patch; posterolateral angles of abdomen black; endopod of uropod with distal half black; exopod of uropod dark centrally, margin clear.

*Size.*—Males, TL 36.7-66.5 mm; females, TL 41.0-63.4 mm. Other measurements of male holotype, TL 63.3: carapace length, 14.5; cornea width, 3.3; rostral plate length, 2.1, width, 2.1; telson length, 11.4, width, 12.7.

*Discussion.*—*S. discors* is very closely related to *S. deceptrix* with which it has been confused in the past and which is described below. It can be separated from that species by the following characters: (1) the intermediate carinae of the thoracic somites are armed posteriorly in *S. discors,* unarmed in *S. deceptrix;* (2) the lateral processes of the thoracic somites differ slightly in the two species; *S. discors* has a slenderer spine on the fifth somite, a sharper anterior lobe on the process of the sixth somite, and blunter posterolateral apices on the processes of the sixth and seventh somites; and (3) in *S. discors* the submedian abdominal carinae are lined with dark pigment, there is no dark patch on the fifth abdominal somite, and the abdominal somites are not lined posteriorly with dark pigment. The lateral processes of the thoracic somites vary in shape more in *S. deceptrix* than in *S. discors,* but the differences noted above are constant in the material available for study. *S. discors* seems to have more prominent carinae than are present in *S. deceptrix;* both species vary in the prominence and rugosity of the carinae. Also, *S. discors* usually has more dorsal tubercles on the telson than does *S. deceptrix;* in the former species there are usually two more or less complete rows of tubercles parallel to and converging under the median carina.

The spined intermediate carinae of the thoracic segments are the most important characters separating the two species, and will also serve to distinguish *S. discors* from the other species of the genus in the Americas. In some juvenile specimens of *S. discors* the intermediate carinae of the sixth somite are unarmed, but the corresponding carinae of the next two somites are always armed, even in juveniles.

Study of the original series of specimens which led to the description of *S. discors* in 1962 led me to the erroneous conclusion that the presence of spines on the thoracic intermediate carinae was a variable feature, present occasionally in juveniles (as in the case of 1 paratype) and in the largest specimens. It was not until after the description was published that additional material became available which showed the striking color differences (see fig. 44) and stressed the importance of the intermediate carinae. Due to this unfortunate lapse on my part, most of the specimens herein assigned to *S. deceptrix* are also paratypes of *S. discors.*

Both *S. discors* and *S. deceptrix* resemble the only other western Atlantic

164

species of *Squilla* with six teeth on the raptorial claw and numerous longitudinal carinae on the telson, *S. grenadensis* and *S. rugosa.* These latter species, however, both have spines on the mesial margin of the basal prologation of the uropod and have the lateral processes of the sixth and seventh thoracic somites produced into posteriorly-directed spines. *S. grenadensis,* which agrees with *S. discors* and *S. deceptrix* in having a slender, anteriorly-curved lateral spine on the fifth thoracic somite, differs from them in having but four epipods. The lateral process of the fifth thoracic somite in *S. rugosa* does not curve forward.

*S. discors* also resembles *S. hancocki* Schmitt from the eastern Pacific. However, that species has a subquadrate rostal plate, a dark patch on the fifth abdominal somite, and has the submedian abdominal carinae spined on the last somite only. *S. hancocki* seems to be more closely related to *S. deceptrix.*

*Remarks.—S. discors* is a relatively small species; adult males (TL 45-50 mm) exhibit secondary sexual modifications on the abdomen and telson. The median carina of the telson, carinae of the marginal teeth, and marginal denticles are noticeably swollen, as are the carinae of the sixth abdominal somite. The marginal spines of the sixth adbominal somite may be absent. Few males of *S. deceptrix* exhibit this dimorphism and none do at this small size.

The eyes of *S. discors* appear to be smaller than those of *S. deceptrix,* but this observation does not seem to be borne out by the Corneal Indices which overlap broadly.

Both *S. discors* and *S. deceptrix* have been taken together in the southern part of their range off Nicaragua at OREGON Sta. 3577 (155 m) and off Venezuela at OREGON Sta. 4394 (118 m).

*Ontogeny.*—Unknown.

*Type.*—The holotype is in the U. S. National Museum. The single paratype is deposited in the Reference Collection, Institute of Marine Sciences, University of Miami.

*Type-locality.*—Off Venezuela, 11°07′N, 62°14′30″W, in 134 m, ALBATROSS Sta. 2120.

*Distribution.*—Western Atlantic, where it has been recorded from off the Bahamas, Nicaragua, and Venezuela, in depths between 118 and 241 m.

## Squilla deceptrix, new species
Figures 44c, 46

*Squilla rugosa pinensis:* Manning, 1959, p. 20 [part, Florida specimen only; not *S. rugosa* var. *pinensis* (Lunz)].
*Squilla discors* Manning, 1962c, p. 217 [part; most of paratypic series, as discussed below. Paratypes of *S. discors* are marked below with an asterisk (*)].—Bullis & Thompson, 1965, p. 13 [listed; part].

165

*Previous records.*—NORTH CAROLINA, BAHAMAS, E and W coasts of FLORIDA, CARIBBEAN SEA (Manning, 1962c, and Bullis & Thompson, 1965, including OREGON Sta. 36, 920, 944, SILVER BAY Sta. 2480, and COMBAT Sta. 334).

*Holotype.*—1 ♂, 64.1; Panama; 09°18′N, 80°25′W; 137 m; OREGON Sta. 3587; 29 May 1962; USNM 119169.

*Paratype.*—1 ♀, 74.8; data as in holotype; USNM 119170.

*Other material.*—NORTH CAROLINA: 1 ♂, 43.2; 3 ♀, 24.1-40.3; ALBATROSS Sta. 2596; USNM 11260 (*).—1 brk. ♂, CL 9.8; 34°59′N, 75°27.5′ W; 100 m; EASTWARD Sta. 1439; 21 May 1965; DUML.—BAHAMAS: 2 ♂, 39.0-43.7; Santaren Channel; SILVER BAY Sta. 2460; USNM 107936 (*).—1 ♂, ca. 56.0; 4 ♀, ca. 48.0-62.6; Cay Sal Bank; SILVER BAY Sta. 2470; USNM 119178.—FLORIDA: 1 ♀, 57.7; ENE of Daytona Beach; COMBAT Sta. 334; UMML (*).—1 ♂, 58.0; off Melbourne; SILVER BAY Sta. 3279; USNM 119177.—3 ♂, 29.2-51.0; 3 ♀, 36.9-44.4; 4 mi. off Islamorada; 73 m; D. deSylva, *et al.;* 20 August 1961; UMML (*).—3 ♂, 31.9-50.0; 7 ♀, 29.0-53.0; same; USNM 119171 (*). — 1 ♀, ca. 37.0; Straits of Florida; 25°27.5-29.5′N, 79°18-20′W; 309-346 m; GERDA Sta. 274; 30 March 1964; UMML.—1 ♀, 37.7; Straits of Florida; 25°38-47′N, 80°05′W; 69-87 m; GERDA Sta. 282; 1 April 1964; UMML.—2 ♂, 58.4-65.0; S of Tortugas; 24°23-24′N, 82°56-57′W; 86-96 m; GERDA Sta. 572; UMML.—1 ♀, 49.1; same; USNM 119179.—2 ♂, 1 brk., 40.2; 3 ♀, 37.0-56.7; off Charlotte Harbor; ALBATROSS Sta. 2411; USNM 9832 (*).—1 ♀, 52.1; W of Anclote Keys; OREGON Sta. 920; USNM 96400 (*).—1 ♂ abdomen; S of Cape San Blas; OREGON Sta. 36; USNM 91097 (*).—1 ♀, 41.8; NW of Cape San Blas; OREGON Sta. 944; USNM 107742 (*).—HONDURAS: 1 ♀, 73.2; OREGON Sta. 1864; USNM 119175.—NICARAGUA: 1 ♂, 42.0; 1 ♀, 59.6; OREGON Sta. 3577; USNM 119174.—CARIBBEAN SEA: 2 ♀, 28.0-31.6; no other data; UMML 32.2165 (*).—VENEZUELA: 1 ♀, 60.9; OREGON Sta. 4392; USNM 119176.—2 ♀, 37.5-50.4; OREGON Sta. 4393; USNM 119173.—3 ♂, 44.2-58.4; OREGON Sta. 4394; USNM 113038.—TOBAGO: 1 ♀, 57.5; OREGON Sta. 5025; USNM 119172.

*Diagnosis.*—Rostral plate without median carina; median carina of carapace lacking anterior bifurcation; dactylus of raptorial claw with 6 teeth; mandibular palp and 5 epipods present; lateral process of fifth thoracic somite an anteriorly-curved spine; intermediate carinae of thoracic somites unarmed; telson with numerous dorsal tubercles; denticles 5-7, 8-11, 1; basal prolongation of uropod lacking mesial spines.

*Description.*—Eye large, cornea set obliquely on stalk; eyes not extending to end of first segment of antennular peduncle; ocular scales obliquely truncate; anterior margin of ophthalmic somite projecting, medially emarginate. Corneal indices of specimens of various sizes follow:

| Sex | TL | CL | CI |
|-----|------|------|-----|
| ♀ | 24.1 | 5.5 | 344 |
| ♀ | 26.8 | 6.1 | 359 |
| ♀ | 31.6 | 6.9 | 345 |
| ♀ | 37.0 | 8.5 | 369 |
| ♂ | 40.2 | 9.1 | 350 |
| ♀ | 41.8 | 9.5 | 352 |
| ♀ | 52.1 | 12.1 | 378 |
| ♀ | 57.7 | 13.0 | 371 |
| ♂ | 64.1 | 13.8 | 394 |
| ♀ | 59.6 | 14.2 | 406 |
| ♀ | 74.8 | 16.5 | 446 |
| ♀ | 73.2 | 17.1 | 417 |

Antennular peduncle longer than carapace; antennular processes produced into short, sharp spines directed anterolaterally.

Rostral plate slightly broader than long, appearing elongate; shape variable, lateral margins usually sloping toward rounded apex; median carina absent.

Median carina of carapace lacking anterior bifurcation on portion anterior to cervical groove, portion posterior to cervical groove with bifurcation; intermediate carinae usually not extending to anterior margin; posterolateral margins of carapace with faint anterior angle.

Mandibular palp and 5 epipods present.

Dactylus of raptorial claw with 6 teeth, rarely 7, outer margin of dactylus evenly convex, with faint basal notch; dorsal ridge of carpus irregular but undivided, terminating in obtuse lobe.

Last 3 thoracic somites with submedian and intermediate carinae, intermediates unarmed; fifth somite with 1 pair of short, oblique dorsal carinae; lateral process of fifth somite a curved spine, shape and curvature variable, apex directed anteriorly, with irregular basal carina; lateral process of sixth somite bilobed, anterior lobe small, variable in shape, usually rounded, posterior lobe much larger, triangular, apex usually sharp, rarely spiniform; lateral process of seventh somite not bilobed, obliquely truncate, anterior margin a rounded right angle, posterolateral apex usually sharp, acute, occasionally spiniform; lateral margin of process of seventh somite usually straight, occasionally convex; ventral keel of eighth thoracic somite usually rounded, inclined posteriorly, apex acute but not sharp in some specimens.

Abdomen depressed, with prominent carinae, submedians divergent on each somite; submedian carinae of second somite notched anteriorly; second to fifth somites with interrupted median tubercle, anteriorly-notched

167

intermediate carinae, and anterior tubercle between intermediate and lateral carinae; abdominal carinae spined as follows: submedian, 5-6; intermediate, (1-2) 3-6 (usually 1-6 in adults); lateral, 1-6; marginal, 1-5; sixth somite with 1 or more anterior tubercles lateral to each submedian carina, small ventral spine present in front of articulation of each uropod.

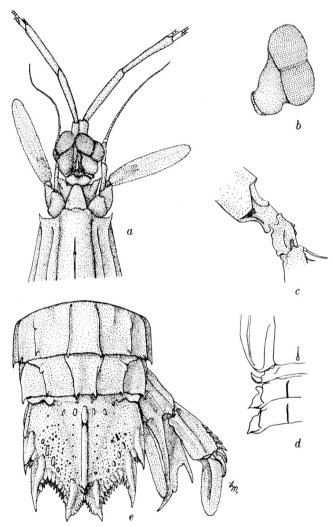

FIGURE 46. *Squilla deceptrix,* n. sp., male, TL 50.0, off Islamorada, Florida. *a,* anterior portion of body; *b,* eye; *c,* carpus of claw; *d,* lateral processes of fifth to seventh thoracic somites; *e,* last two abdominal somites, telson, and uropod (setae omitted).

Telson usually broader than long, with sharp, slender marginal teeth, prelateral lobes present; median carina notched anteriorly, terminating posteriorly in slender spine; dorsal surface, in addition to uninterrupted carinae of marginal teeth, with up to 6 lines of tubercles, of which 1 or 2 parallel the median carina, converging under its posterior apex; remainder of tubercles situated on surface between radiating rows of pits; in some specimens, tubercles restricted to posterior portion of telson; denticles 5-7 (8), 8-11, 1, outer submedian denticle large, rounded, remainder sharp; ventral surface with or without postanal tubercle or short carina.

Proximal segment of uropodal exopod with 6-8 movable spines, last short, not extending to midlength of distal segment; inner margin of basal prolongation tuberculate, not spinulose; lobe on outer spine of basal prolongation variable in size and shape.

*Color.*—Intermediate and reflected marginal carinae of carapace dark; merus of claw with dark dorsal patch, dark longitudinal ventral line, and black distal bar on lateral face; thoracic somites with dark posterior line, faint dark line between submedian and intermediate carinae, and dark patch lateral to each intermediate carina; abdominal somites with dark line extending between intermediate carinae; articulated anterolateral plates of abdomen dark; second and fifth abdominal somites with dark, rectangular dorsal patch, more prominent on second than on fifth and with lightly-pigmented area between intermediate and lateral carinae; posterolateral angles of abdomen dark; telson with anterior and posterior dark patches on median carina, bases of marginal teeth dark, and irregular dark patch on each lateral surface; proximal segment of uropodal exopod with dark distal line, distal segment black centrally, margin clear; endopod with distal half black.

According to notes made on specimens from Silver Bay Sta. 2460, the above color pattern is evident in fresh material. Upon preservation most of the color fades, except for the bar on the merus, the rectangle on the second abdominal somite, the black posterolateral angles of the abdomen, and the pigment on the uropods. The types, however, have lost little of their original pigment pattern.

*Size.*—Males, TL 29.2-65.0 mm; females, TL 24.1-74.8 mm. Other measurements of female, TL 57.7: carapace length, 13.0; cornea width, 3.5; rostral plate length, 2.1, width, 2.1; telson length, 10.9, width, 11.2.

*Discussion.*—*S. deceptrix* closely resembles *S. discors* but differs from it in lacking spines on the intermediate carinae of the thoracic somites and in lacking dark pigment along the submedian carinae of the abdomen. Other differences are discussed under *S. discors.*

*S. deceptrix* differs from *S. rugosa* and *S. grenadensis* in lacking spines on the inner margin of the basal prolongation of the uropod and in lacking posterior spines on the lateral processes of the sixth and seventh thoracic somites. *S. grenadensis* shares with *S. deceptrix* and *S. discors* the anteriorly-curved lateral process on the fifth somite but differs from both species in having but four epipods.

*S. hancocki* Schmitt, from the eastern Pacific, is very similar to *S. deceptrix,* but differs in having a shorter rostral plate, a slenderer lateral process on the fifth thoracic somite, and fewer abdominal spines. In *S. hancocki* the submedian carinae of the fifth abdominal somite are unarmed. *S. hancocki* also differs from both *S. discors* and *S. deceptrix* in having an oval, pigmented area on the midline of the carapace anterior to the cervical groove. *S. hancocki* may be considered as the eastern Pacific analogue of *S. deceptrix.*

As noted under the discussion of *S. discors,* most of the paratypes of that species are herein assigned to *S. deceptrix.* Those lots are indicated in the "material" section with an asterisk (*).

*Remarks.*—There may be more than one species present in the material herein assigned to *S. deceptrix* for there seems to be more variation in morphological features than is usually encountered in a species of *Squilla.* One specimen, from OREGON Sta. 920 in the Gulf of Mexico, has seven teeth on the raptorial claw and completely lacks a postanal keel; in other respects it is indistinguishable from the remainder of the specimens. In other specimens, the shape of the lateral processes of the thoracic somites varies considerably, as follows: (1) the lateral spine of the fifth thoracic somite may be slender or broad and may be directed almost laterally rather than anterolaterally; (2) the anterior lobe of the lateral process of the sixth somite is usually rounded but it is often variable in outline; (3) the posterior lobe of the lateral process of the sixth somite is acute in most specimens, rounded in those from North Carolina; (4) the numbers and arrangement of dorsal tubercles on the telson vary from a full complement arranged over the entire surface of the telson to a few scattered tubercles arranged anterior to the denticles; (5) the postanal keel may be absent or present as a tubercle, row of tubercles, or a keel; (6) the lobe on the inner spine of the basal prolongation of the uropod is variable in size; and (7) only the male abdomen from the Gulf of Mexico (USNM 91097) exhibits any sexual dimorphism in telson shape. The posterior margin of the telson is noticeably swollen, particularly between the submedian and intermediate marginal teeth. The specimen, if intact, would have a total length of ca. 60 mm; none of the other males at that size exhibit any sign of dimorphism.

*S. deceptrix* has a wider bathymetric range than *S. discors;* the former has been taken between 49 and 346 m, whereas the latter has been recorded from 118 to 241 m. The ranges of the two species are similar, but the species have been taken together at only two stations in the Caribbean, OREGON Stats. 3577 and 4394.

One of the specimens from USNM 9832 was sent to the Indian Museum

170

on exchange in 1911. It was originally labelled *S. rugosa*.

Manning (1967b) has noted that among the specimens reported by Boone (1930) are two males, TL 39.5-42.4 m, of *Squilla discors*. Lots from Bimini and the Florida Keys were found together, and there is no way to determine from which locality the two specimens originated. They are now tentatively reassigned to *S. deceptrix* pending their reexamination for the spined intermediate thoracic carinae

*Ontogeny.*—Unknown.

*Etymology.*—The name is derived from Latin, "deceptrix," meaning deceiver.

*Type.*—U. S. National Museum.

*Type-locality.*—Off Panama, 09°18′N, 80°25′W, 137 m, OREGON Sta. 3587.

*Distribution.*—Western Atlantic from North Carolina, the Bahamas, Florida (between Daytona Beach and Choctawhatchee Bay), Honduras, Nicaragua, Panama, Venezuela, and Tobago, in depths between 49 and 346 m.

## *Squilla heptacantha* (Chace, 1939)
### Figures 47, 48a

*Chloridella heptacantha* Chace, 1939, p. 52.—Manning, 1968, p. 129 [listed].
*Squilla heptacantha:* Springer & Bullis, 1956, p. 22.—Manning, 1959, p. 19.
—Bullis & Thompson, 1965, p. 13 [listed].

*Previous records.*—FLORIDA: Off eastern coast (Manning, 1959).— CUBA: northern coast (Chace, 1939; Springer & Bullis, 1956). Bullis & Thompson (1965) record this species at nine OREGON and four COMBAT stations.

*Material.*—BAHAMAS: 1 ♂, 53.5; 1 ♀, 68.2; off Little Bahama Bank; 27°25′N, 78°37.5-41.0′W; 291-309 m; GERDA Sta. 251; 5 February 1964; UMML.—1 ♂, 60.0; COMBAT Sta. 236; RMNH 384.—1 ♂, 59.2; COMBAT Sta. 238; UMML 32.1168.—1 ♂, 82.2; Santaren Channel; SILVER BAY Sta. 2468; USNM 119199.—FLORIDA: 1 brk. ♀, CL 13.0; SW of Sombrero Light; 164-182 m; 6 June 1950; TRITON, Thompson, McGinty; USNM 119196.—STRAITS OF FLORIDA: 1 ♂, 48.5; COMBAT Sta. 445; USNM 101726.—1 ♂, 53.5; off Great Bahama Bank; 25°15′N, 79°14-15′W; 384 m; GERDA Sta. 236; 30 January 1964; USNM 119200.—1 ♂, 48.0; same; 25°17-20′N, 79°14-15′W; 395-404 m; GERDA Sta. 238; 30 January 1964; USNM 119201.—1 postlarva, ca. 17.5; W of Cay Sal Bank; 24°31′N, 81°11-15′W; 137 m; GERDA Sta. 376; 17 September 1964; USNM 119202. —HONDURAS: 1 ♀, 94.7; OREGON Sta. 1879; USNM 111102.— PANAMA: 2 ♂, 92.4-93.4; OREGON Sta. 3585; UMML.—4 ♀, 76.4-99.5; OREGON Sta. 3588; USNM 119196.—1 ♀, 83.2; OREGON Sta. 3590; USNM 111099.—CUBA: 2 dry ♂; 51.9-76.2; OREGON Sta. 1340; USNM 98675.

171

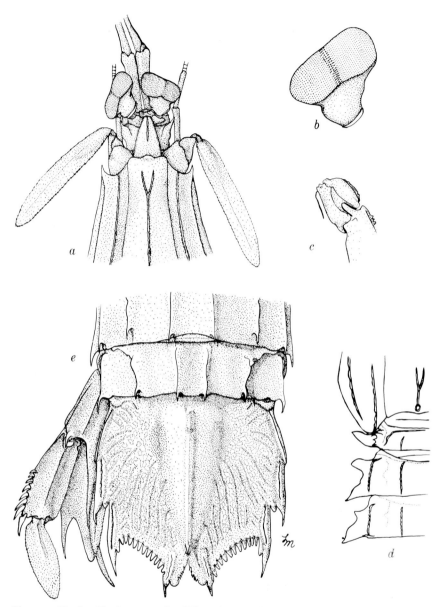

FIGURE 47. *Squilla heptacantha* (Chace), female, TL 78.1, OREGON Sta. 2606. *a*, anterior portion of body; *b*, eye; *c*, carpus of claw; *d*, lateral processes of fifth to seventh thoracic somites; *e*, last two abdominal somites, telson, and uropod (setae omitted).

—1 ♂, 56.9; Bahia de Cochinos, Santa Clara Province; 22°07′N, 81°08′ W; 329-348 m; ATLANTIS Sta. 2962-B; 24 February 1938; holotype; MCZ 10245.—1 ♂, 36.7; Old Bahama Channel, due N of Punta Alegre, Camaguey Province; 22°45′N, 78°45′W; 275-239 m; ATLANTIS Sta. 2982-E; 11 March 1938; paratype; USNM 113641.—DOMINICAN REPUBLIC: 1 ♂, 70.6; 1 ♀, 68.5; SILVER BAY Sta. 5161; USNM 119198.—PUERTO RICO: 1 ♀, 60.2; OREGON Sta. 2664; USNM 111098.—1 ♀, 80.6; OREGON Sta. 2658; USNM 111103.—VIRGIN ISLANDS: 1 ♀, 78.1; OREGON Sta. 2606; USNM 11917.—1 ♀, 60.5; OREGON Sta. 2648; USNM 105355.—1 ♀, 66.0; OREGON Sta. 2649; USNM 111101.

*Diagnosis.*—Rostral plate elongate, triangular, with median carina; median carina of carapace with anterior bifurcation; dactylus of raptorial claw with 7 teeth; mandibular palp present; 4 epipods present; lateral processes of sixth and seventh thoracic somites strongly bilobed; lateral abdominal carinae bicarinate.

*Description.*—Eye of moderate size, cornea set obliquely on stalk; ocular scales tapering laterally, rounded; anterior margin of ophthalmic somite rounded. Corneal indices of 9 specimens of various sizes follow:

| Sex | TL | CL | CI |
|---|---|---|---|
| ♂ | 59.2 | 14.1 | 470 |
| ♀ | 60.5 | 14.5 | 467 |
| ♀ | 66.0 | 15.5 | 517 |
| ♀ | 78.1 | 17.9 | 511 |
| ♀ | 83.2 | 19.5 | 542 |
| ♀ | 94.7 | 21.8 | 589 |
| ♀ | 96.4 | 22.1 | 539 |
| ♂ | 93.4 | 22.2 | 584 |
| ♀ | 99.5 | 23.6 | 576 |

Antennular peduncle slender, longer than carapace and rostral plate combined; antennular processes produced into strong spines directed anteriorly.

Rostral plate elongate-triangular, slender, longer than broad; apex rounded; well-marked median carina present.

Median carina of carapace, anterior to cervical groove, bifurcate anteriorly only; distance from dorsal pit to anterior bifurcation greater than distance from bifurcation to anterior margin; intermediate carinae extending to or almost to anterior margin; anterolateral spines not extending to level of rostral plate; posterolateral margins angled anteriorly.

Raptorial claw large, elongate, propodus longer than carapace; dactylus with 7 teeth, outer margin of dactylus evenly curved; dorsal ridge of carpus multituberculate.

Mandibular palp and 4 epipods present.

Last 4 thoracic somites with submedian and intermediate carinae; fifth

173

 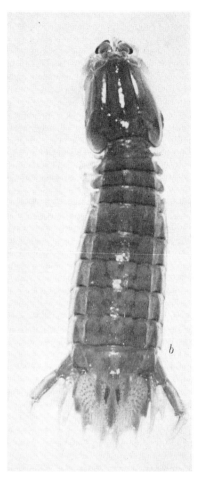

FIGURE 48. Color patterns of: *a, S. heptacantha* (Chace), female, TL 80.6, OREGON Sta. 2658; *b, S. surinamica* Holthuis, male paratype, TL 40.5, Surinam.

somite with irregular, oblique carina between submedians and intermediates; lateral process of fifth somite a broad lobe, anterior margin convex, posterior concave, apex acute, directed anterolaterally; lateral process of sixth somite strongly bilobed, rounded or truncate, anterior lobe almost as large as posterior; lateral process of seventh somite bilobed, anterior lobe small, obtuse, posterior lobe larger, triangular, directed posterolaterally; ventral keel on eighth somite low, triangular, inclined posteriorly.

Submedian abdominal carinae divergent on each somite; second to fifth somites with interrupted median carinule; second, third, and fourth somites with anterior tubercle between prominent intermediate and lateral carinae;

174

lateral carinae longitudinally grooved; abdominal carinae spined as follows: submedian, 5-6; intermediate, (1) 2-6; lateral, 1-6; marginal, 1-6; sixth somite with ventral spine in front of articulation of uropod.

Telson slightly broader than or as broad as long, slightly inflated; pre-lateral lobes present; median carina sinuous in lateral view; denticles 5-9, 8-13, 1, outer submedian denticle much the largest; ventral surface with short but distinct postanal keel.

Uropod slender, with 7-9 graded, movable spines on outer margin of penultimate segment of exopod, last not extending to midlength of distal segment; lobe on outer margin of inner spine of basal prolongation very large, rounded.

*Color.*—Carinae of carapace and posterior margin of abdominal somites lined with dark pigment; telson with a large pair of anterior black spots, each divided by line of pits flanking median carina; distal segment of exopod and distal portion of endopod of uropod dark. Usually all pattern is faded, except for prominent telson spots.

*Size.*—Males, TL 36.7-93.4 mm; females, TL 60.2-99.5 mm; postlarva TL 17.5 mm. Other measurements of female, TL 78.1: carapace length, 17.9; cornea width, 3.5; rostral plate length, 2.8, width, 2.1; telson length, 14.7, width, 14.9.

*Discussion.*—This species is easily recognized by the strongly bilobed lateral processes of the sixth thoracic somite, the large raptorial claw armed with seven teeth, the bicarinate lateral carinae of the abdomen, and the broad, oval dark patches at the base of the telson.

*Remarks.*—In the two largest males examined, TL 92.4-93.4 mm, the longitudinal carinae of the telson are noticeably swollen, and the posterior margin shows some slight swelling. There was no evidence of swelling in any of the females examined.

*Ontogeny.*—Unknown.

*Type.*—Museum of Comparative Zoology, Harvard. A paratype is also in the U. S. National Museum.

*Type-locality.*—Bahia de Cochinos, Santa Clara Province, Cuba.

*Distribution.*—From Little Bahama Bank and the eastern coast of Florida to the southern Caribbean Sea, off Panama, including off the coasts of Cuba, Dominican Republic, Puerto Rico, Virgin Islands and Honduras, in depths between 183 and 439 m.

*Squilla prasinolineata* Dana, 1852
Figures 49, 50a

*Squilla Dufresnii* White, 1847, p. 83 [*nomen nudum*].—Miers, 1880, p. 18, pl. 2, figs. 8-9.

*Squilla prasinolineata* Dana, 1852, p. 620; Atlas, 1855, p. 13 [listed], pl. 61, figs. 3a-c.—Smith, 1869, p. 41 [listed].—Ives, 1891, p. 184.—Sharp, 1893, p. 108.—Rathbun, 1899, p. 628.—Moreira, 1901, pp. 5, 71; 1903b, p. 120 [p. 2 on separate].—Kemp, 1913, p. 201.—Kemp and Chopra 1921, p. 298 [listed].—Oliveira, 1940, p. 145.—Andrade Ramos, 1951, p. 142 [listed].—Lemos de Castro, 1955, p. 15, text-figs. 12-15, pl. 3, fig. 35, pl. 13, fig. 46.—Manning, 1959, p. 20; 1961, p. 13 [key].—Tabb & Manning, 1962, p. 594.—Manning, 1967b, p. 105; 1968, p. 129 [listed].
not *Squilla prasinolineata:* Miers, 1880, p. 19.—Bigelow, 1894, p. 520 [= *S. neglecta* Gibbes].
*Squilla mantis:* Boone, 1930, p. 32, pl. 4 [misidentified; not *S. mantis* (Linnaeus)].
*Squilla empusa:* Glassell, 1934, p. 454 [erroneous correction of *S. mantis:* Boone; not *S. empusa* Say].

*Previous records.* —FLORIDA: Miami (Manning, 1959); northern Florida Bay (Manning, 1959; Tabb & Manning, 1962).—MEXICO: Yucatan (Ives, 1891; Sharp, 1893; Chace, 1954).—JAMAICA: Port Royal Cays, Kingston Harbor, and Port Antonio (Rathbun, 1899).—CUBA: Kemp, 1913; Boone, 1930; Manning, 1959.—ST. THOMAS: Manning, 1959; 1967b.—BRAZIL: Fortaleza (Lemos de Castro, 1955); Pernambuco (Recife) (Moreira, 1901); Rio de Janeiro (Dana, 1852; Moreira, 1901; Oliveira, 1940; Lemos de Castro, 1955); Paranagua Bay (Lemos de Castro, 1955); Sao Francisco de Sul (Moreira, 1903b); Santa Catarina State (Lemos de Castro, 1955).

*Material.*—NO LOCALITY: 1 ♂, 79.6; 1 ♀, 79.1; syntypes of *S. dufresnii* White; BMNH 1939.5.8.21-22.—FLORIDA: 1 ♂, 43.2; 2 ♀, 43.5-60.7; North Bay Island, Biscayne Bay; 5 October 1948; UMML 32.1151.—3 ♀, 92.6-100.0; Biscayne Bay; E. Iversen; 10 July 1961; USNM 111129.— 1 ♀, ca. 29.0; Newfound Harbor Key, Monroe Co.?; 12 December 1906; B. A. Bean; USNM 111123.—1 ♂, 57.5; Conchie Channel, off Flamingo; 24 April 1958; D. Tabb, D. Dubrow; UMML 32.1154.—1 ♀, 39.4; same; RMNH 390.—4 ♀, 45.0-85.2; Sandy Key Basin, off Flamingo; 29-30 October 1958; E. Iversen, R. B. Manning; USNM 119235.—1 ♀, 38.8; off Cape Sable; FISH HAWK Sta. 7367; USNM 111571.—1 ♀, 34.1; off East Cape Sable; 10 November 1958; D. Tabb, R. B. Manning; USNM 111124.—1 ♂, 66.1; Joe Kemp Channel, Florida Bay; D. Tabb, R. B. Manning; 15 April 1959; USNM 111128.—1 ♂, ca. 24.0; Gulf of Mexico, off Shark River; 2 October 1959; A. Jones; USNM 111125.—MEXICO: 1 ♀, 53.5; Silam, Yucatan; 1890; T. E. Ives, Mexican Expedition; ANSP 48.—CUBA: 1 ♀, 80.6; Porto Padre; March 1928; VMM.—1 ♂, 21.9; 3 ♀, 21.3-26.8; Esperanza; shallow water; 11 May 1914; Thomas Barrera Expedition; USNM 48518.—1 ♀, 76.3; Cuba; IM.—1 ♂, 67.3; 1 ♀, 89.6; 1 damaged juvenile; Cuba; Gundlach Collection; ZMB 8295.—1 ♂, 94.0; Cuba; Gundlach Collection; ZMB 12125c.—3 ♂, 60.2-80.2; 2 ♀, 67.4-ca. 88.0; Cuba; Gundlach Collection; ZMB 12125b.—VIRGIN ISLANDS: 2 ♀, 78.5-80.8; Coral Harbor, St. John; 21 March 1961; J. Randall, R.

Schroeder; UMML 32.2074.—1 ♂, 40.8; St. Thomas; shore; 17-24 January 1884; ALBATROSS; USNM 21478.—BRAZIL: 1 ♂, 63.1; Brazil; Consul Sckerblom; UZM.—1 ♀, 118.9; same; UZM.—1 ♂, 58.7; 2 ♀, 54.2-58.8; off Para State; OREGON Sta. 4215; USNM 111126.—1 damaged ♀, 81.4; Rio de Janeiro; Thayer Expedition; MCZ 7883.—1 ♂, 101.4; Rio de Janeiro; MCZ 7685.—1 ♂, 76.2; same; MCZ 7853.—1 ♂, 100.0; Guanabara Bay, Rio de Janeiro; USNM 111127.—1 ♀, 118.5; Praia de Fort, Florianopolis, Santa Cruz; 5 March 1958; Dagoberto; CPP.—1 ♀, 93.8; Praia do Recife, Florianopolis, Santa Catarina; 26 July 1959; E. Tremel; CPP.—1 ♀, 99.2; same; USNM 119234.—1 ♀, 66.0; Muller; MNHNP.

*Diagnosis.*—Rostral plate with median carina; median carina of carapace lacking anterior bifurcation; dactylus of raptorial claw with 5 teeth; mandibular palp and 4 epipods present; lateral process of fifth thoracic somite a blunt spine directed laterally; telson denticles 2-4, 4-8, 1.

*Description.*—Eye large, cornea set obliquely on stalk; ocular scales truncate, inclined laterally, outer margins rounded; anterior margin of ophthalmic somite rounded. Corneal indices of specimens of various sizes follow:

| Sex | TL | CL | CI |
|---|---|---|---|
| ♂ | 24.0 | 5.6 | 280 |
| ♂ | 40.8 | 8.7 | 311 |
| ♀ | 45.0 | 10.1 | 348 |
| ♀ | 45.2 | 10.3 | 355 |
| ♀ | 59.0 | 13.1 | 374 |
| ♂ | 58.7 | 13.7 | 415 |
| ♀ | 58.8 | 14.0 | 400 |
| ♂ | 66.1 | 15.4 | 405 |
| ♂ | 80.2 | 18.7 | 407 |
| ♀ | 85.2 | 18.9 | 429 |
| ♂ | 94.0 | 20.1 | 402 |
| ♀ | 92.6 | 21.3 | 435 |
| ♀ | 100.0 | 23.2 | 455 |

Antennular peduncle about as long as carapace; antennular processes appearing sharp in dorsal view, actually produced into blunt projections directed anterolaterally.

Rostral plate trapezoidal, upraised lateral margins tapering to truncate apex; median carina present on anterior third.

Anterior portion of median carina of carapace with posterior bifurcation, anterior bifurcation completely absent; intermediate carinae extending to or near anterior margin; anterolateral spines strong, not extending to base of rostral plate; posterior margin with median projection; posterolateral margins not angled anteriorly.

Dactylus of raptorial claw armed with 5 teeth, outer margin a simple curve; dorsal ridge of carpus undivided, ending in blunt lobe.

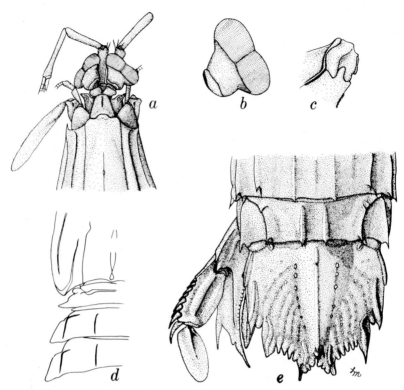

FIGURE 49. *Squilla prasinolineata* Dana, female, TL 85.2, off Flamingo, Florida. *a,* anterior portion of body; *b,* eye; *c,* carpus of claw; *d,* lateral processes of fifth to seventh thoracic somites; *e,* last two abdominal somites, telson, and uropod (setae omitted).

Mandibular palp and 4 epipods present.

Last 4 thoracic somites with submedian and intermediate carinae, submedians of fifth somite short if present; lateral process of fifth thoracic somite a short, blunt, triangular projection; lateral process of sixth somite triangular, not bilobed, apex blunt; lateral process of seventh somite undivided, more truncate than that of sixth; ventral keel on eighth thoracic somite triangular, inclined posteriorly, apex rounded.

Second to fifth abdominal somites each with notched median carinule, notched intermediate carinae, and anterior tubercle between intermediate and lateral carinae; submedian abdominal carinae subparallel or slightly divergent on each somite; abdominal carinae spined as follows: submedian, 5-6; intermediate, (3-4) 5-6; lateral, (3) 4-6; marginal, 1-5; sixth somite with ventral spine in front of articulation of uropod.

Telson as broad as or slightly broader than long, with sharp marginal

178

FIGURE 50. Color patterns of: *a, S. prasinolineata* Dana, female, TL 92.6, Biscayne Bay, Florida; *b, S. neglecta* Gibbes, male, TL 96.2, off Boca Grande, Florida.

teeth; prelateral lobes present; carinae of intermediate teeth extending anterior to apex of prelateral lobe; denticles 2-4, 4-8, 1, submedians truncate, intermediates triangular; ventral surface with inconspicuous post-anal tubercle.

Outer margin of penultimate segment of uropodal exopod with 8-9 movable spines, last not extending to midlength of distal segment.

*Color.*—Carinae and grooves of carapace lined with dark pigment; posterior margin of carapace, last 3 thoracic somites, and first 5 abdominal somites with dark posterior line; dorsal half of anterolateral plates of abdo-

179

men black; second and fifth abdominal somites with dark, rectangular, dorsal patches; sixth abdominal somite with dark posterolateral patches; telson with anterior and posterior black patches, posterior one on submedian carinae also; distal segment of exopod and distal half of endopod of uropod black.

*Size.*—Males, TL 21.9-101.4 mm; females, TL 21.3-118.9 mm. Other measurements of male, TL 66.1: carapace length, 15.4; cornea width, 3.8; rostral plate length, 2.3, width, 2.4; telson length, 12.0, width, 13.2.

*Discussion.*—R. W. Ingle of the British Museum (Natural History) kindly supplied sketches and notes on the types of *Squilla Dufresnii* White which show conclusively that it is conspecific with *S. prasinolineata* Dana. Ingle also included sketches of *Squilla prasinolineata:* Miers which show that Miers was dealing with *S. neglecta* Gibbes. In 1964 I examined the specimens reported by both White (1847) and Miers (1880) and I agree with Ingle's identifications.

Examination of material in the Vanderbilt Marine Museum has allowed a correction of Boone's record of *Squilla mantis* from Cuba (Manning, 1967b). Glassell (1934) corrected this identification to *S. empusa,* apparently without examining the specimen which proved to be *S. prasinolineata.*

*S. prasinolineata* and *S. neglecta* actually share only one important morphological feature, the presence of five teeth on the raptorial claw. *S. prasinolineata* lacks an anterior bifurcation on the median carina of the carapace and has a short, blunt lateral process on the fifth thoracic somite. *S. neglecta* has a well-developed anterior bifurcation on the median carina of the carapace and a broad spatulate lobe on the fifth thoracic somite.

*S. prasinolineata* also resembles *Squilla empusa* Say but that species has six teeth on the raptorial claw, a prominent bifurcation on the median carina of the carapace, and a sharp, anteriorly-curved lateral process on the fifth thoracic somite.

*S. obtusa* Holthuis and *S. surinamica* Holthuis both lack the anterior bifurcation of the median carina of the carapace and both have posterolaterally rounded processes on the sixth and seventh thoracic somites, as in *S. prasinolineata,* but both can be distinguished from the latter species by the anteriorly curved lateral process on the fifth thoracic somite and the claw which is armed with six teeth. There are also differences in color pattern, for *S. prasinolineata* lacks the dark crescents on the telson that are characteristic of the other two species.

*Remarks.*—There are no marked secondary sexual differences in this species.

*Ontogeny.*—Unknown.

*Type.*—Not extant.

*Type-locality.*—Rio de Janeiro, Brazil.

*Distribution.*—South Florida to Santa Catarina State, Brazil, including Yucatan, Jamaica, Cuba, and St. Thomas, in shallow water.

180

*Squilla neglecta* Gibbes, 1850
Figures 50b, 51

*Squilla neglecta* Gibbes, in Tuomey, 1848, p. xvi [listed; *nomen nudum*].—
Gibbes, 1850, p. 200 [p. 35 on separate].—Miers, 1880, p. 23 [listed].—
Howard, 1883, p. 294 [listed].—Bigelow, 1894, p. 510 [key].—Kemp,
1913, p. 201 [listed].— Lunz, 1933, p. 1, pl. 1.—Chace, 1954, p. 449.—
Lemos de Castro, 1955, p. 22, text-fig. 17, pl. 6, fig. 38, pl. 15, fig. 49.—
Springer & Bullis, 1956, p. 23.—Manning, 1959, p. 20; 1961, p. 13
[key]; 1966, p. 363, text-fig. 2.—Bullis & Thompson, 1965, p. 13 [listed].
—Manning, 1968, p. 129 [listed].
*Squilla prasinolineata:* Miers, 1880, p. 19, pl. 2, fig. 10.—Ives, 1891, p. 185
[discussion].—Kemp, 1913, p. 201 [listed].—Manning, 1959, p. 20
[not *S. prasinolineata* Dana].
*Chloridella neglecta:* Lunz, 1935, p. 154, text-fig. 4; 1937, p. 10, text-fig. 3.

*Previous records.*—NORTH CAROLINA: Beaufort (Lunz, 1935.)—
SOUTH CAROLINA: Charleston (Gibbes, 1850; Lunz, 1933, 1935); no
other data (Gibbes, in Tuomey, 1848; Howard, 1883).—FLORIDA: N
of Key West (Springer & Bullis, 1956; Manning, 1959); Lemon Bay
(Manning, 1959); Sanibel Id. (Lunz, 1937; Chace, 1954; Manning, 1959).
—BRAZIL: São Sebastião, Santos, Isla da Moela, Praia Grande (Lemos
de Castro, 1955); Anchieta (Manning, 1966); no other data (Miers,
1880). Bullis & Thompson (1965) record the species at one OREGON and
one COMBAT station.

*Material.*—NORTH CAROLINA: 1 ♂, 98.3; off Cape Hatteras; ALBA-
TROSS Sta. 2283; USNM 7225.—2 ♂, 61.8-99.0; 1 ♀, 57.6; same; ALBA-
TROSS Sta. 2282; USNM 8785.—SOUTH CAROLINA: 3 ♂, 58.1-94.2;
1 dam. ♀; Isle of Palms; washed up on beach; 7 January 1936; G. R. Lunz,
Jr.; USNM 76005.—1 ♂, 68.5; off Charleston Harbor; 14 November 1934;
USNM 76030.—1 ♂, 105.0; off Kiawak Id.; 9 September 1943; G. R.
Lunz, Jr.; USNM 81691.—2 ♂, 97.2-108.6; Rockville; 31 June 1947; T.
K. Ellis; USNM 84233.—1 ♂, 98.2; 1 ♀, 95.5; same; IM.—GEORGIA:
1 ♂, 42.5; S of Sapelo Id.; 27 July 1961; USNM 111151.—FLORIDA:
2 ♀, 105.6-111.0; Fernandina, Nassau Co.; ALBATROSS; USNM 110837.—
1 ♀, 91.9; W coast of Florida; Summer 1959; MCSN.—1 ♂, 105.0; off
Cape Romano, Collier Co.; OREGON Sta. 993; USNM 97536.—1 ♂, 96.2;
off Boca Grande, Charlotte Co.; 24 May 1959; W. Saenz; UMML 32.1227.
—1 ♂, 92.2; same; RMNH 369.—1 ♀, 96.4; Lemon Bay bridge, Placida,
Charlotte Co.; CHML.—1 dam. ♂, 74.8; Pensacola, Santa Rosa Co.;
from fish stomach; USNM 14110.—TEXAS: 1 ♂, 80.7; 0.5 mi. off beach
at point 1.5-2.0 mi. SW of entrance to Aransas Pass; ca. 5 m; 1940; H.
S. Ladd; USNM 113321.—1 ♂, 98.2; 1 ♀, 85.5; S of Port Aransas; 11 m;
15 May 1961; H. Compton; USNM 110386.—BRAZIL: 1 ♂, 71.1; no
other data; BMNH 1958.8.13.1.—1 ♂, 98.7; Rio de Janeiro; MCZ 643.—
1 ♀, 118.9; same; MCZ.—1 dam. ♂, CL 16.3; Carra Point, São Sebastião,
São Paulo; IOSP.—1 ♂, 85.5; near Moela Id., Santos; 14 July 1950;
USNM 111152.—1 ♂, 66.5; Santos; Fishing Biology Service Staff; May

1959; IOSP.—1 ♂, 71.8; SW of Anchieta village; 0.6 m; CALYPSO Sta. 93; 30 November 1961; USNM 111051.—1 ♂, 72.7; same; MNHNP.

*Diagnosis.*—Rostral plate with median carina; median carina of carapace with anterior bifurcation; dactylus of raptorial claw with 5 teeth; mandibular palp absent; 5 epipods present; lateral process of fifth thoracic somite a spatulate spine, rounded laterally; telson denticles 2-4, 5-7, 1.

*Description.*—Eye of moderate size, cornea set obliquely on stalk; corneal and peduncular axes of eye subequal; ocular scales truncate, dorsal surface sinuate, lateral angles acute but rounded; anterior margin of ophthalmic somite scarcely produced forward, rounded. Corneal indices of specimens of various sizes follow:

| Sex | TL | CL | CI |
|-----|-----|-----|-----|
| ♂ | 42.5 | 9.6 | 417 |
| ♂ | 71.8 | 16.0 | 471 |
| ♂ | 72.7 | 16.4 | 482 |
| ♀ | 83.6 | 17.0 | 421 |
| ♀ | 85.5 | 18.2 | 478 |
| ♂ | 98.2 | 21.8 | 484 |
| ♀ | 105.6 | 22.4 | 477 |
| ♀ | 111.0 | 23.6 | 481 |

Antennular peduncle shorter than carapace; antennular processes with sinuous margins, apex blunt, not spined.

Rostral plate truncate, broader than long, apex transverse, not rounded; distinct median carina present.

Carapace roughened, with strong carinae; median carina, anterior to cervical groove, bifurcate at both ends; distance from dorsal pit to anterior bifurcation subequal to distance from bifurcation to anterior margin; intermediate carinae falling short of anterior margin; anterolateral spines strong, extending well past base of rostral plate; posterolateral angles broadly rounded, not angled anteriorly; posterior margin projecting along midline.

Raptorial claw stout, dactylus armed with 5 teeth, outer margin sinuate; dorsal ridge of carpus subdivided into several rounded tubercles or irregular, terminating in obtuse lobe; basal segment of claw with strong, projecting spine.

Mandibular palp absent; 5 epipods present.

Last 3 thoracic somites with submedian and intermediate carinae, intermediates present as tubercles on fifth somite; lateral process of fifth somite a broad, spatulate spine, apex rounded; lateral processes of sixth and seventh somites each with small anterior tubercle (occasionally absent), larger posterior lobes broadly rounded; ventral keel on eighth somite ovoid, apex flattened.

Submedian abdominal carinae divergent on each somite; abdominal carinae spined as follows: submedian, 6; intermediate, (3) 4-6; lateral,

182

(1-2) 3-6; marginal, 1-5; sixth somite with sharp spine in front of articulation of uropod.

Telson flattened, about as long as broad, with sharp marginal teeth, prelateral lobes reduced but present; median carina very sharp, with long apical spine; denticles 2-4, 5-7, 1, submedians truncate, intermediates triangular, inner submedian the largest; ventral surface with inconspicuous postanal keel.

Outer margin of penultimate segment of exopod of uropods with 7-9 movable spines, last not extending to midlength of distal segment.

*Color.*—Carinae and posterior margins of carapace and body segment outlined in dark pigment; each abdominal somite with diffuse patch of dark pigment; carinae and curved rows of pits on telson dark; inner margin of exopod of uropod dark.

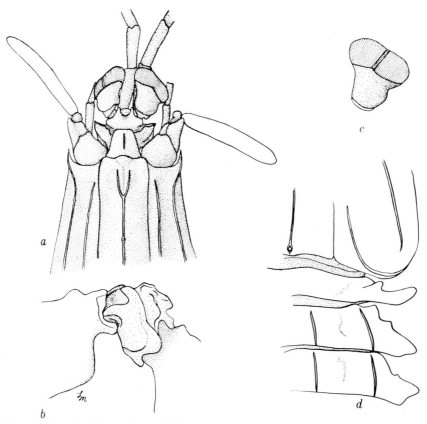

FIGURE 51. *Squilla neglecta* Gibbes, male, TL 71.8, CALYPSO Sta. 93. *a,* anterior portion of body; *b,* carpus of claw; *c,* eye; *d,* lateral processes of fifth to seventh thoracic somites (setae omitted; from Manning, 1966).

183

*Size.*—Males, TL 42.5-108.6 mm; females, TL 57.6-118.9 mm. Other measurements of male, TL 98.2: carapace length, 21.8; cornea width, 4.5; rostral plate length, 2.7, width, 3.5; telson length, 16.8, width, 16.7.

*Discussion.*—*S. neglecta* resembles *S. empusa* Say in many features, but differs in having a smaller eye (compare Corneal Indices of similar sizes), in lacking a mandibular palp, in having the anterior bifurcation of the median carina of the carapace nearer the anterior margin than the dorsal pit, in having five rather than six teeth on the raptorial claw, and in having a spatulate lateral process on the fifth thoracic somite. Comparison of specimens will also show numerous other differences of less importance.

*S. neglecta* resembles *S. prasinolineata* Dana, and has been confused with it by Miers (1880) and others, but the two species are quite distinct. *S. prasinolineata* lacks an anterior bifurcation on the median carina of the carapace, has a mandibular palp, has an evenly rounded instead of a sinuous dactylus on the raptorial claw, and has characteristic dark patches on the telson and abdomen that are completely lacking in *S. neglecta.*

The identity of *S. prasinolineata:* Miers could not have been settled without the aid of R. W. Ingle, British Museum (Natural History), who supplied sketches of Miers' material and who correctly identified it as *S. neglecta.* Miers' specimen was subsequently examined by me (1964).

Manning (1966) noted some differences between specimens from Brazil and those from the Gulf of Mexico, including (1) the anterior tubercle on the lateral process of the seventh thoracic somite is absent in northern specimens, present in southern; (2) all of the lateral abdominal carinae are usually spined in Gulf specimens, whereas the carinae of the first two somites are unarmed in those from Brazil; (3) the submedian abdominal carinae are less divergent in Brazilian specimens than in those from the Gulf. These differences are probably populational and are not enough to even consider separation of specimens from the two areas, which in all other respects are very similar.

*Remarks.*—*S. neglecta* has a typical Florida disjunct distributional pattern. It is found on the northeastern and entire western coasts but is absent from the southeastern coast. *Coronis excavatrix* Brooks is the only other stomatopod with a similar distribution pattern; it too has a parallel in Brazil, but in the case of *Coronis* the Brazilian form is a distinct species.

No secondary sexual differences were noted in the present material.

*Ontogeny.*—Unknown.

*Type.*—Not extant.

*Type-locality.*—Charleston Harbor, South Carolina.

*Distribution.*—North and South Carolina, Georgia, northeastern and western Florida and Texas in the Gulf of Mexico, and southern Brazil; no specimens have been recorded from intermediate areas. Littoral to 64 m, usually in shallow water. The species has not previously been recorded as far west as Texas in the Gulf of Mexico.

*Squilla surinamica* Holthuis, 1959
Figures 48b, 52

*Squilla surinamica* Holthuis, 1959, p. 184, text-figs. 76e-g, pl. 8, fig. 5, pl. 9, fig. 5.—Manning, 1961, p. 14 [key]; 1968, p. 129 [listed].

*Previous records.*—SURINAM: Holthuis, 1959.

*Material.*—SURINAM: 1 ♂, 38.7; COQUETTE Sta. 6; paratype; USNM 103217.—1 ♂, 34.7; COQUETTE Sta. 1; paratype; USNM 103218.—3 ♂, 34.9-38.8; 1 ♀, 34.9; COQUETTE Sta. 159; paratypes; USNM 103219.—15 ♂, 33.6-43.8; 7 ♀, 34.6-40.9; COQUETTE Sta. 144; paratypes; USNM 103220.—2 ♂, 38.1-40.5; 2 ♀, 37.3-41.2; same; UMML 32.1707.—9 ♂, 26.2-40.2; 2 ♀, 31.3-33.3; COQUETTE Sta. 2; paratypes; USNM 103221.

*Diagnosis.*—Rostral plate with median carina; median carina of carapace lacking anterior bifurcation; dactylus of raptorial claw with 6 teeth; mandibular palp and 5 epipods present; lateral process of fifth thoracic somite a broad, anteriorly-curved spine, lateral processes of next 2 somites bilobed, rounded posterolaterally; telson denticles 2-4 (5), 7-10, 1.

*Description.*—Eye of moderate size; cornea set transversely on and only slightly broader than stalk; stalk broad through most of its length, slightly narrowed proximally; ocular scales trapezoidal, acute but rounded laterally; anterior margin of ophthalmic somite projecting anteriorly, emarginate along midline. Corneal indices of 9 specimens of various sizes follow:

| Sex | TL | CL | CI |
|---|---|---|---|
| ♂ | 26.6 | 5.8 | 446 |
| ♀ | 31.3 | 6.9 | 460 |
| ♂ | 35.4 | 7.5 | 441 |
| ♀ | 33.3 | 7.5 | 469 |
| ♂ | 34.9 | 8.4 | 442 |
| ♀ | 38.7 | 8.6 | 430 |
| ♂ | 39.6 | 8.7 | 458 |
| ♂ | 38.7 | 8.9 | 494 |
| ♂ | 43.1 | 9.2 | 460 |

Antennular peduncle shorter than carapace; antennular processes produced into acute but rounded projections directed anterolaterally.

Rostral plate short, broader than long, lateral margins converging on rounded or flattened apex; median carina present on anterior third.

Carapace smooth, carinae poorly marked; median carina, anterior to cervical groove, not bifurcate at either end; intermediate carinae not extending to anterior margin; anterolateral spines, although strong, not extending to level of rostral base; posterolateral margins angled anteriorly.

185

Raptorial claw short, stout, dactylus armed with 6 teeth, outer margin of dactylus sinuate; dorsal ridge of carpus undivided.

Mandibular palp and 5 epipods present.

Last 3 thoracic somites with submedian and intermediate carinae, submedians poorly-marked; lateral process of fifth somite a thick spine, apex curved anteriorly; lateral process of sixth somite bilobed, anterior lobe small, rounded, posterior lobe large, margins rounded, apex convex; lateral process of seventh somite similar to sixth, anterior lobe more obtuse, less distinct; ventral keel on eighth somite low, rounded ventrally, angled posteriorly.

Abdominal somites with poorly-marked submedian carinae, subparallel on first 4 somites, divergent on fifth; abdominal carinae spined as follows: submedian, 5-6; intermediate, 4-6; lateral, (3) 4-6; marginal, 1-5; area between lateral and marginal carinae swollen in large males; sixth somite with ventral spine in front of articulation of each uropod.

Telson slightly inflated, slightly broader than long; prelateral lobes present; dorsal carinae of marginal teeth swollen in males larger than TL

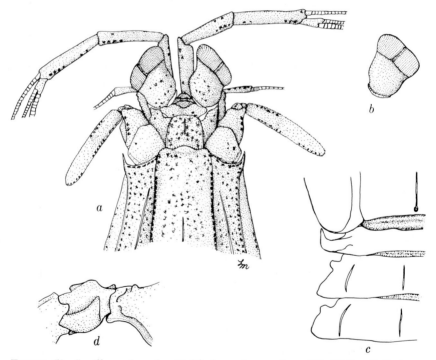

FIGURE 52. *Squilla surinamica* Holthuis, male paratype, TL 38.8, off Surinam. *a,* anterior portion of body; *b,* eye; *c,* lateral processes of fifth to seventh thoracic somites; *d,* carpus of claw (setae omitted).

186

37.0; denticles 2-4(5), 7-10, 1, outer submedian denticle larger than remainder; postanal keel absent.

Uropods with 7-8 movable spines on lateral margin of penultimate segment of exopod, distal spine not extending to midlength of distal segment.

*Color.*—Entire animal very dark, with many scattered dark chromatophores; gastric grooves and posterior margin of carapace outlined with dark pigment; last 3 thoracic and first 5 abdominal somites with dark posterior line and diffuse dorsal patch of dark chromatophores; telson with proximal pair of crescents, mesial surface convex; pits of telson dark; distal portions of uropods with scattered dark chromatophores.

*Size.*—Males, TL 26.2-43.8 mm; females, TL 31.3-40.9 mm. Other measurements of male, TL 40.5: carapace length, 8.8; cornea width, 2.0; rostral plate length, 1.3, width, 1.6; telson length, 7.8, width, 8.1.

*Discussion.*—This small species resembles *S. obtusa* Holthuis but differs from it and all other western Atlantic species in the shape of the eye, in which the cornea is scarcely broader than the stalk. *S. surinamica* further differs from *S. obtusa* in having a short carina on the rostral plate, and it agrees with *S. obtusa* and differs from *S. lijdingi* in the rounded posterolateral edges of the lateral processes of the sixth and seventh thoracic somites.

*Remarks.*—There is noticeable sexual dimorphism in this species. Males smaller than TL 37.0 mm closely resemble females, but in larger males the carinae of the telson become inflated. Also, in large males the lateral and marginal carinae of the abdomen become inflated; in some specimens most of the area between these carinae is inflated.

*Ontogeny.*—Unknown.

*Type.*—The holotype and a series of paratypes are in the Rijksmuseum van Natuurlijke Historie, Leiden. A series of paratypes is also present in the U.S. National Museum and the Institute of Marine Sciences, University of Miami.

*Type-locality.*—Off Surinam.

*Distribution.*—Known only from off Surinam, on mud bottom in depths of 26-27 m.

## Squilla obtusa Holthuis, 1959
### Figures 53a, 54

Squilla intermedia: Bigelow, 1901, p. 159.—Chace, 1954, p. 449 [part; Puerto Rico specimens only].—Manning, 1959, p. 19 [part; Curaçao specimens only] [not *S. intermedia* Bigelow, 1893].
Squilla brasiliensis: Manning, 1959, p. 18 [part; Trinidad specimens only] [not *S. brasiliensis* Calman, 1917].

187

*Squilla obtusa* Holthuis, 1959, p. 186, text-figs. 76h-j, pl. 9, figs. 3-4.—
Manning, 1961, p. 28, pl. 7.— Bullis & Thompson, 1965, p. 13 [listed].—
Manning, 1966, p. 365, text-fig. 3; 1968, p. 129 [listed].

*Previous records.*—PUERTO RICO: Bigelow, 1901; Chace, 1954.—
TRINIDAD: Manning, 1959, 1961; Bullis & Thompson, 1965.—
CURAÇAO: Manning, 1959.—SURINAM: Holthuis, 1959.—BRAZIL:
off Salvador (Manning, 1966).

*Material.*—PUERTO RICO: 1 abdomen; Mayaguez; FISH HAWK;
20 January 1899; USNM 64777.—2 ♂, 47.5-54.8; Mayaguez Harbor;
FISH HAWK Sta. 6059 (Porto Rico Sta. 131); USNM 64805.—1 ♂, 48.7;
same; FISH HAWK Sta. 6058 (Porto Rico Sta. 130); USNM 64804.—1
juv. ♀; same; FISH HAWK Sta. 6061 (Porto Rico Sta. 133); USNM 64806.
1 ♂, 46.2; off Mayaguez; dredge; 2 May 1963; G. Warmke, R. LaLlave;
USNM 111149.—COLOMBIA: 1 ♂, 50.4; 1 ♀, 49.0; Gulf of Darien;
08°15′N, 76°54′W to 08°16.5′N, 76°55′W; 46-51 m; ATLANTIS Sta. 254;
9 February 1960; USNM 105005.—3 ♂, 46.7-55.0; 4 ♀, 38.5-47.5;
OREGON Sta. 4846; USNM 119226.—2 ♀, 52.2-55.7; OREGON Sta. 4847;
USNM 119227.—1 ♂, 39.5; OREGON Sta. 4851; USNM 119228.—1 ♂,
43.3; OREGON Sta. 4852; USNM 119229.—2 ♂, 53.7-54.5; 3 ♀, 44.6-53.6;
OREGON Sta. 4857; USNM 119230.—1 ♂, 46.4; OREGON Sta. 4864;
USNM 119231.—1 ♂, 56.8; 1 ♀, 60.2; OREGON Sta. 4868; USNM 119232.
1 ♂, 55.5; OREGON Sta. 4898; USNM 119233.—VENEZUELA: 1 brk.
♀, CL 12.3; OREGON Sta. 4405 (probably erroneous, depth 391 m);
USNM 119224.—1 ♂, 53.7; 1 ♀, 48.4; OREGON Sta. 4492; USNM 119225.
—CURAÇAO: 1 ♂, 59.0; Curaçao; ALBATROSS; 1884; USNM 21480.—
TRINIDAD: 2 ♂, 44.0; near Trinidad; ALBATROSS Stats. 2121, 2122;
USNM 111150.—2 ♂, 54.5-62.1; 2 ♀, 54.7-56.5; Trinidad; OREGON Sta.
2208; USNM.—1 ♂, 49.0; 1 ♀, 65.9; same; OREGON Sta. 2209; UMML
32.1884.—1 ♀, 46.8; same; MNRJ.—SURINAM: 1 ♀, 55.2; off Surinam;
COQUETTE Sta. 144; paratype; USNM 103222.—BRAZIL: 1 brk. ♀; off
Salvador; 13°01.5′S, 38°24′W; 49-51 m; CALYPSO Sta. 54; 24 November
1961; MNHNP.—7 ♂, 31.9-72.8; 4 ♀, 30.0-59.5; same; 12°49.7′S,
38°31.4′W; 20-30 m; CALYPSO Sta. 62; 26 November 1961; MNHNP.—
1 ♀, 82.5; same; 13°28′S, 34°48.5′W; 39 m; CALYPSO Sta. 67; 26 November 1961; USNM 111054.

*Diagnosis.*—Rostral plate elongate, without median carina; median carina
of carapace without anterior bifurcation; dactylus of raptorial claw with
6 teeth; mandibular palp and 5 epipods present; lateral process of fifth
thoracic somite an acute spine, curved forward, lateral processes of next
somites posterolaterally rounded; telson denticles 3-5, 8-9, 1.

*Description.*—Eye large, cornea set obliquely on stalk, corneal axis greater
than peduncular; ocular scales truncate, rounded laterally; anterior margin
of ophthalmic somite produced forward, flattened apically. Corneal indices
of 11 specimens of various sizes follow:

188

| Sex | TL | CL | CI |
|------|------|------|-----|
| ♀ | 40.6 | 8.8 | 352 |
| ♂ | 44.0 | 9.1 | 350 |
| ♂ | 44.0 | 10.6 | 353 |
| ♂ | 47.5 | 11.0 | 367 |
| ♀ | 46.8 | 11.3 | 389 |
| ♂ | 54.8 | 12.6 | 382 |
| ♀ | 59.5 | 13.2 | 357 |
| ♂ | 59.0 | 14.3 | 421 |
| ♀ | 65.9 | 14.9 | 413 |
| ♂ | 72.8 | 16.4 | 364 |
| ♀ | 82.5 | 18.3 | 398 |

FIGURE 53. Color patterns of: *a*, *S. obtusa* Holthuis, female, TL 54.7, OREGON Sta. 2208; *b*, *S. lijdingi* Holthuis, male, TL 91.7, OREGON Sta. 2231; *c*, *S. chydaea* Manning, female paratype, TL 123.0, OREGON Sta. 2827.

189

Antennular peduncle about as long as carapace; antennular processes with acute apices directed obliquely forward.

Median carina of carapace, in front of cervical groove, not bifurcate at either end, interrupted portions of bifurcation present in some specimens; intermediate carinae extend almost to anterior margin; anterolateral spines strong but not extending to base of rostral plate; posterolateral margins angled anteriorly.

Raptorial claw slender, dactylus armed with 6 teeth; outer margin of

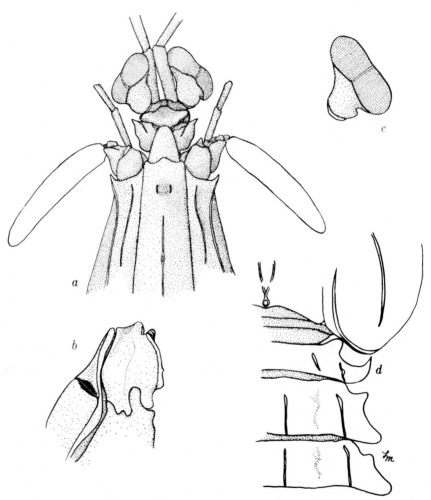

FIGURE 54. *Squilla obtusa* Holthuis, female, TL 82.5, CALYPSO Sta. 67. *a*, anterior portion of body; *b*, carpus of raptorial claw; *c*, eye; *d*, lateral processes of fifth to seventh thoracic somites (setae omitted; from Manning, 1966).

190

dactylus evenly curved; dorsal ridge of carpus uneven.

Mandibular palp and 5 epipods present.

Last 3 thoracic somites with submedian and intermediate carinae, intermediates also present on fifth somite; lateral process of fifth somite a broad sharp spine, apex curved forward; lateral processes of sixth and seventh somites bilobed, anterior lobe of sixth acute, that of seventh obtuse, posterior lobe of both processes larger than anterior and broadly rounded; ventral keel on eighth somite elongate, thin, inclined posteriorly.

Submedian abdominal carinae slightly divergent on each somite; second to fifth abdominal somites with inconspicuous median carinule, submedians and intermediates anteriorly notched; abdominal carinae spined as follows: submedian, 5-6; intermediate (2) 3-6; lateral, (1) 2-6; marginal, (1) 2-5; sixth somite with ventral spine in front of articulation of uropod.

Telson slightly inflated, length and width subequal, prelateral lobes present; denticles 3-5, 8-9, 1; ventral surface with inconspicuous postanal tubercle.

Uropod with 8-9 movable spines on outer margin of penultimate segment of exopod, last not extending to midlength of distal segment.

*Color.*—Carinae and grooves of carapace and posterior margins of carapace and body segments outlined in dark pigment; second and fifth abdominal somites with dark rectangular patch; anterolateral angle of sixth abdominal somite black; anterior half of telson with pair of dark crescents, inner side convex; apex of median carina and bases of teeth dark; distal segment of exopod and distal half of endopod of uropods dark; merus of raptorial claw with distal vertical line of dark pigment.

*Size.*—Males, TL 31.9-72.8 mm; females, TL 30.0-82.5 mm. Other measurements of female, TL 82.5: carapace length, 18.3; cornea width, 4.6; rostral plate length, 2.6, width, 2.6; telson length, 15.0, width, 15.1.

*Discussion.*—This small species resembles *S. brasiliensis* Calman, *S. lijdingi* Holthuis, and *S. chydaea* Manning, the latter two of which also have the pair of dark crescents on the telson, but differs primarily in having rounded posterolateral angles on the lateral processes of the sixth and seventh thoracic somites. In the western Atlantic, *S. rugosa* Bigelow, *S. discors* Manning, *S. deceptrix,* n. sp., and *S. grenadensis,* n. sp., also have a dark vertical bar on the merus of the raptorial claw.

The anteriorly curved spine on the fifth thoracic somite and the presence of 6 teeth on the raptorial claw will distinguish this species from *S. prasinolineata* Dana.

*S. surinamica* Holthuis also has obtuse posterolateral margins on the lateral processes of the sixth and seventh thoracic somites, but that species has a rostral carina and much broader eyestalks, with the cornea barely larger than the stalk.

Although there are no closely related species in the eastern Pacific, *S. parva* Bigelow from that area resembles *S. obtusa* in several features. *S. parva* also has an elongate rostral plate, but it has a well-marked median carina. As in *S. obtusa, parva* has rounded posterolateral angles on the lateral processes of the sixth and seventh thoracic somites and lacks the anterior bifurcation on the median carina of the carapace. No specimens of *parva* show a trace of the paired crescents on the telson that are present in *obtusa;* only *S. panamensis* Bigelow and *S. bigelowi* Schmitt of the eastern Pacific species of *Squilla* are decorated with the crescents.

*Remarks.*—Both Bigelow (1901) and Manning (1959) misidentified small specimens of this species, mistaking them for juveniles of larger forms.

Males about TL 50 mm and larger all have the marginal carinae of the telson noticeably swollen; none of the larger females shows any swelling.

The shape of the rostral plate is variable, particularly in small specimens, where the apex may be sharp, rounded off, or even truncate.

*Ontogeny.*—Unknown.

*Type.*—Rijksmuseum van Natuurlijke Historie, Leiden. A female paratype is in the U. S. National Museum.

*Type-locality.*—Surinam.

*Distribution.*—Puerto Rico; coast of South America from the Gulf of Darien, off Colombia, to off Salvador, Brazil, including Venezuela, Curaçao, Trinidad, and Surinam; 13-182 m, usually less than 90 m. The specimen from OREGON Sta. 4405 was probably mislabelled, for the depth at which that station was made (391 m) is far greater than all others.

### Squilla lijdingi Holthuis, 1959
### Figures 53b, 55

Squilla lijdingi Holthuis, 1959, p. 181, text-figs. 76b-d, pl. 9, figs. 1-2.— Manning, 1961, p. 26, pl. 6.— Bullis & Thompson, 1965, p. 13 [listed].— Manning, 1968, p. 129 [listed].
Squilla brasiliensis: Manning, 1959, p. 18 [part; not *S. brasiliensis* Calman].

*Previous records.*—TRINIDAD, VENEZUELA, and BRITISH GUIANA: (all Manning, 1959, 1961).—SURINAM: Holthuis, 1959; Manning, 1961.—FRENCH GUIANA: Manning, 1961. Bullis & Thompson (1965) list 11 OREGON Stations at which this species was taken.

*Material.*—COLOMBIA: 1 ♀, 117.5; OREGON Sta. 4844; USNM 119220. —2 ♀, 91.0-117.5; OREGON Sta. 4856; USNM 119221.—1 brk. ♀, CL 22.3; OREGON Sta. 4857; USNM 119222.—VENEZUELA: 2 ♂, 80.5-81.5; OREGON Sta. 4467; USNM 119219.—1 ♀, 81.6; ANTILLAS; March 1953; UMML 32.1152.—TOBAGO: 4 ♂, 82.4-113.2; 6 ♀, 77.7-107.4; OREGON Sta. 5025; USNM 119223.—TRINIDAD: 12 ♂, 34.2-85.5;

11 ♀, 50.0-133.0; ALBATROSS Stats. 2121-2122; USNM 6905.—1 ♂, 37.7; Trinidad; W. O. Crosby; USNM 110882.— 1 ♂, 62.3; OREGON Sta. 2207; UMML 32.1878.—BRITISH GUIANA: 1 ♂, 91.4; OREGON Sta. 2221; USNM 111157.—4 ♀, 53.0-97.3; OREGON Sta. 2226; USNM 111134.— 1 ♂, 65.6; 9 ♀, 61.3-95.8; OREGON Sta. 2228; USNM 111130.—2 ♀, 90.0-91.7; OREGON Sta. 2231; UMML 32.1879.—2 ♂, 63.1-65.7; OREGON Sta. 2250; USNM 111131.—SURINAM: 2 ♀, 78.3-88.5; off Surinam; summer 1960; H. Lijding; UMML 32.1873.—1 ♂, 61.4; OREGON Sta. 2272; USNM 103501.—1 ♀, 85.3; OREGON Sta. 2276; USNM 111132.— 2 ♂, 52.4-63.0; 5 ♀, 54.0-76.5; OREGON Sta. 2327; UMML 32.1877 (1 ♀ to BMNH; 1 ♀ to MNRJ).—1 ♂, 80.5; OREGON Sta. 4171; USNM 119216.—2 ♂, 64.2-72.5; 2 ♀, 51.5-79.1; OREGON Sta. 4179; USNM 119217.—2 ♀, 46.9-85.5; COQUETTE; 1957; paratypes; USNM 103223. —4 ♂, 56.0-67.8; 5 ♀, 52.2-64.2; COQUETTE Sta. 218; paratypes; USNM 103224.—2 ♂, 47.0-48.5; COQUETTE Sta. 144; paratypes; USNM 103225. —1 ♂, 66.7; COQUETTE Sta. 284; paratype; USNM 103226.—1 ♂, 60.1;

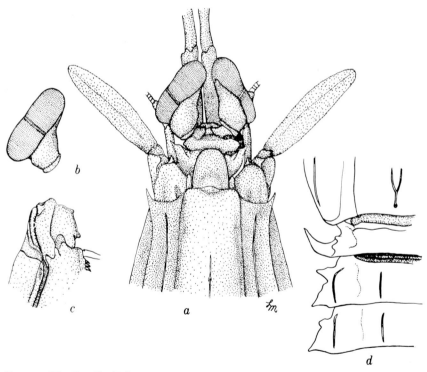

FIGURE 55. *Squilla lijdingi* Holthuis, male, TL 85.5, off Trinidad. *a,* anterior portion of body; *b,* eye; *c,* carpus of claw; *d,* lateral processes of fifth to seventh thoracic somites (setae omitted).

193

COQUETTE Sta. 33; paratype; USNM 103227.—1 ♂, 48.7; COQUETTE Sta. 15; paratype; USNM 103228.—2 ♀, 47.8-70.5; COQUETTE Sta. 20; paratypes; USNM 103229.—1 ♀, 71.2; COQUETTE Sta. 283; paratype; USNM 103230.—1 ♂, 51.0; 1 ♀, 61.8; COQUETTE Sta. 21; paratypes; USNM 103231.—1 ♂, 38.2; COQUETTE Sta. 159; paratype; USNM 103232.— 7 ♂, 22.0-42.7; 7 ♀, 34.2-48.7; COQUETTE Sta. 2; paratypes; USNM 103233.—1 ♀, 63.7; COQUETTE Sta. 209; paratype; USNM 103234.— 2 ♀, 64.3-91.4; COQUETTE Sta. 36; paratypes; USNM 103235.—FRENCH GUIANA: 1 ♂ abdomen; Cayenne; 05°03′N, 51°50′W; DANA Sta. 1176; 17 November 1921; UZM.—1 ♀, 54.2; OREGON Sta. 2306; UMML 32.1880.—6 ♂, 42.3-57.0; 3 ♀, 44.3-53.3; OREGON Stats. 2307 A&B; USNM 111133.—2 ♂, 50.7-52.5; 2 ♀, 43.4-47.6; same; IM.—BRAZIL: 1 ♀, 71.0; Amapa State; OREGON Sta. 4209; USNM 119492.—2 ♀, 69.0-88.5; same; OREGON Sta. 4211; USNM 119218.

*Diagnosis.*—Rostral plate subquadrate, without median carina; median carina of carapace without anterior bifurcation; dactylus of raptorial claw with 6 teeth; mandibular palp present; 5 epipods present; lateral process of fifth thoracic somite a slender, anteriorly-curved spine, lateral processes of next 2 somites sharp posteriorly; telson denticles 2-3, 8-10, 1, marginal teeth not markedly elongate.

*Description.*—Eye large, cornea set very obliquely on stalk, corneal axis much greater than peduncular; ocular scales truncate but rounded laterally; anterior margin of ophthalmic somite produced forward, apex rounded or slightly flattened. Corneal indices of 12 specimens of various sizes follow:

| Sex | TL | CL | CI |
|---|---|---|---|
| ♂ | 52.4 | 11.2 | 260 |
| ♀ | 54.0 | 11.6 | 269 |
| ♀ | 58.9 | 12.3 | 279 |
| ♂ | 63.0 | 13.2 | 248 |
| ♀ | 61.5 | 13.3 | 277 |
| ♀ | 67.6 | 14.1 | 266 |
| ♀ | 76.5 | 16.8 | 300 |
| ♀ | 97.3 | 19.4 | 334 |
| ♂ | 90.0 | 19.8 | 309 |
| ♂ | 91.7 | 20.0 | 294 |
| ♀ | 107.0 | 23.4 | 339 |
| ♀ | 133.0 | 26.1 | 363 |

Antennular peduncle longer than carapace; antennular processes produced into acute spines directed anterolaterally.

Rostral plate short, broader than long, apex rounded or truncate, median carina absent.

Median carina of carapace, anterior to cervical groove, without definite bifurcation at either end; faint indications of anterior bifurcation may be

194

present; anterior portion of intermediate carinae curved outward, carinae not extending to anterior margin; anterolateral spines strong but not extending to base of rostral plate; posterolateral margin with prominent anterior angles.

Raptorial claw large, slender; dactylus with 6 teeth, outer margin of dactylus evenly convex; dorsal ridge of carpus irregular.

Mandibular palp and 5 epipods present.

Last 3 thoracic somites with submedian and intermediate carinae; lateral process of fifth thoracic somite a strong, anteriorly-curved spine, apex sharp; lateral process of sixth somite bilobed, anterior lobe small, very sharp, posterior lobe much larger, posterolateral angle sharp; lateral process of seventh somite similar to that of sixth but anterior lobe more obtuse; ventral keel on eighth thoracic somite rounded, inclined posteriorly.

Submedian abdominal carinae slightly divergent on each somite; second abdominal somite with submedian carinae anteriorly notched; second to fifth abdominal somites with interrupted median carinule, notched intermediate carinae, and anterior tubercle between intermediate and lateral carinae; abdominal carinae spined as follows: submedian, (4)5-6; intermediate, (1-2) 3-6; lateral, 1-6; marginal, 1-5; sixth somite with prominent spine in front of articulation of uropod.

Telson as broad as long (relation variable) with sharp marginal teeth, prelateral lobes present; bases of marginal teeth only faintly swollen in large males; denticles 2-3, 8-10, 1, outer submedian denticle the largest; ventral surface with short postanal keel.

Outer margin of penultimate segment of uropodal exopod with 7-8 movable spines, last not extending to midlength of distal segment.

*Color.*—Posterior margin of carapace, last 3 thoracic somites, and first 5 abdominal somites lined with dark pigment; second abdominal somite with dark, rectangular, median dorsal patch; telson with dark crescent on either side of anterior portion of median carina, inner margin of crescent convex; inner side of last 2 segments of uropodal exopod and distal half of endopod dark.

*Size.*—Males, TL 22.0-113.2 mm; females, TL 34.2-133.0 mm. Other measurements of female, TL 95.8: carapace length, 20.5; cornea width, 5.9; rostral plate length, 3.0, width, 3.8; telson length, 18.7, width, 18.3.

*Discussion.*—Holthuis (1959) was the first to point out the differences between this species and *S. brasiliensis* Calman. Calman's species, which does not occur north of Cabo Frio, Brazil, has a median carina on the rostral plate, a noticeable anterior bifurcation on the median carina of the carapace, and has black rectangles rather than crescents on the anterior half of the telson; in addition, adult males of *S. brasiliensis* have the telson margins swollen, a secondary sexual feature that is lacking in *S. lijdingi.* The absence of an anterior bifurcation on the median carina of the carapace and the evenly curved dactylus of the raptorial claw will distinguish

195

*S. lijdingi* from *S. empusa* Say, which has a prominent bifurcation as well as a sinuous dactylus on the claw. The sharp lateral processes of the sixth and seventh thoracic somites, among other features, will distinguish *S. lijdingi* from *S. obtusa* Holthuis and *S. surinamica* Holthuis.

*S. lijdingi* is the southern counterpart of *S. chydaea* Manning from the Gulf of Mexico, to which it is very closely related. *S. chydaea,* however, has a very elongate rostral plate, with a faint median carina, at least traces of an anterior bifurcation on the median carina of the carapace, and a noticeably smaller eye. The Corneal Index alone can be used to separate specimens of these two species. Specimens of *S. lijdingi* under 70 mm TL have a CI of less than 300, whereas specimens of *S. chydaea* at a similar size have a CI of well over 300; a similar difference is found in larger specimens.

*S. tiburonensis* Schmitt is the eastern Pacific analogue of *S. lijdingi.* It is similar morphologically, differing chiefly by lacking anterior tubercles on the lateral processes of the sixth and seventh thoracic somites and by lacking the dark crescents on the telson, which are replaced by dark quadrangles.

Two other eastern Pacific species, *S. panamensis* Bigelow and *S. bigelowi* Schmitt, resemble *S. lijdingi* in general morphology but differ in having rounded posterolateral margins on the sixth and seventh thoracic somites and *S. bigelowi* has the prelateral lobes of the telson produced into sharp spines.

*Ontogeny.*—Unknown.

*Type.*—Rijksmuseum van Natuurlijke Historie, Leiden. A series of paratypes has also been deposited in the U. S. National Museum.

*Type-locality.*—Surinam.

*Distribution.*—Atlantic coast of northern South America, including Colombia, Trinidad, Tobago, Venezuela, the Guianas, and Amapa State, Brazil; 9-182 m (usually 100 m or less), on mud and shell.

## *Squilla chydaea* Manning, 1962
## Figures 53c, 56

*Squilla brasiliensis:* Springer & Bullis, 1956, p. 22.—Manning, 1959, p. 18 [part; not *S. brasiliensis* Calman].
*Squilla chydaea* Manning, 1962c, p. 215.—Dawson, 1963, p. 7; 1965, p. 14. —Bullis & Thompson, 1965, p. 13 [listed].—Manning, 1968, p. 129 [listed].

*Previous records.*—GULF OF MEXICO: W. Florida to Mexico (Springer & Bullis, 1956; Manning, 1959, 1962c; Bullis & Thompson, 1965); Alabama, Louisiana (Dawson, 1965).

*Material.*—FLORIDA: 1 ♀, 79.7; off Cape Canaveral; SILVER BAY Sta. 2732; USNM 111154.—1 ♂, 108.0; off Palm Beach; 182 m; grey mud;

196

August 1951; TRITON, Thompson-McGinty; USNM 119168.—1 ♂, 40.7; main channel off Key West; 4 October 1940; ANSP 4344.—2 ♀, 101.5-113.6; off Tortugas; 73 m; 7 August 1931; W. L. Schmitt; USNM 113323. —4 ♂, 63.5-89.4; 1 ♀, 66.0; off Tortugas; 24°22-23′N, 82°50-53′W; 89 m; GERDA Sta. 568; 12 April 1965; UMML.—13 ♂, 70.3-94.4; 21 ♀, 43.5-107.2; same; 24°23-24′N, 82°56-57′W; 86-96 m; GERDA Sta. 572; 13 April 1965; USNM 119165.—11 ♂, 60.6-90.0; 12 ♀, 35.5-95.5; same; 24°23-24′N, 82°54-57′W; 81-87 m; GERDA Sta. 573; 13 April 1965; USNM 119166.—1 ♀, 58.7; Tortugas; 31 m; 31 January 1957; F. Durant; CHML.—1 ♂, ca. 29.3; same; 29 January 1957; CHML.—1 ♂, 108.0; NW of Cape San Blas; OREGON Sta. 944; paratype; USNM 96401.—1 ♂, 97.5; S of Santa Rosa Id.; OREGON Sta. 332; holotype; USNM 92387.— MISSISSIPPI: 1 ♀, 123.4; PELICAN Sta. 11; paratype; USNM 109441.— MISSISSIPPI DELTA AREA: 1 ♀, 71.6; OREGON Sta. 88; paratype; USNM 91439.—1 ♀, 104.5; OREGON Sta. 90; paratype; USNM 92384.— 1 ♀, 104.6; OREGON Sta. 101; paratype; USNM 91440.—1 ♀, 69.9; OREGON Sta. 103; paratype; USNM 91438.—2 ♀, 79.1-82.5; OREGON Sta. 107; paratypes; USNM 91441.—2 ♂, 85.6-104.1; OREGON Sta. 340; paratypes; USNM 92388.—2 ♂, 78.4-108.5; OREGON Sta. 342; paratypes; USNM 92389.—1 ♂, 74.5; OREGON Sta. 843; paratype; USNM 96278.— 1 ♂, 60.6; 3 ♀, 69.4-81.0; OREGON Sta. 845; paratypes; USNM 96287. —1 dam. ♂ ; OREGON Sta. 2203; paratype; UMML 32.2169.—1 ♂, 124.5; 1 ♀, 123.0; OREGON Sta. 2827; paratypes; UMML 32.2168.—1 ♂, 74.6; OREGON Sta. 4583; USNM 119167.—LOUISIANA: 3 ♂, 76.4-92.6; 1 ♀, 105.4; PELICAN Sta. 69-6; paratypes; USNM 109439 (1 ♂, 78.3, to BMNH).—1 ♀, 74.7; PELICAN Sta. 84-3; paratype; USNM 109437.—1 ♀, 117.1; PELICAN Sta. 85-6; paratype; USNM 109440.—1 damaged ♂, ca. 87.0; PELICAN Sta. 96-1; paratype; USNM 109436.—1 ♂, 61.7; 1 ♀, 59.3; S of Grand Isle; 37 m; 20 November 1959; C. E. Dawson; paratypes; USNM 104748.—1 ♂, 53.2; 3 ♀, 44.6-62.5; same; not types; GCRL.— 1 ♀, 72.0; same; 1 November 1969; GCRL I61:364.—1 ♂, 112.5; SILVER BAY Sta. 181; paratype; UMML 32.2167.—1 brk. ♀; SILVER BAY Sta. 182; paratype; UMML 32.2166.—1 ♂, 70.7; OREGON Sta. 2862; USNM 111155.—1 ♂, 77.6; off Isles Dernières; OREGON Sta. 3769; USNM 111153. —TEXAS: 1 ♂, 68.3; 2 ♀, 69.0-71.7; PELICAN Sta. 39; paratypes; USNM 109438.—2 ♂, 70.2-74.9; 1 ♀, 82.8; PELICAN Sta. 42; paratypes; USNM 110868.—1 ♂, 76.7; 1 ♀, 70.2; SE of Port Aransas; 25 m; 14 March 1961; H. Compton; paratypes; USNM 107022.—1 ♂, 79.0; 1 ♀, 66.5; 50 mi E of Port Aransas; 92m; A. W. Anderson; USNM 92126.—1 ♂, 85.1; OREGON Sta. 156; paratype; USNM 92385.—1 ♀, 108.1; OREGON Sta. 159; paratype; USNM 92386.—1 damaged ♂, 65.0; 2 ♀, 52.4-108.2; OREGON Sta. 162; paratypes; USNM 91947.—1 ♂, 70.0; OREGON Sta. 1082; USNM 111156.—1 ♂, 71.3; OREGON Sta. 1084; Tulane Univ.— 2 ♂, 55.5-69.7; 9 ♀, 54.1-72.4; OREGON Sta. 3968; USNM 113322.— 2 ♂, 63.5-69.2; 2 ♀, 69.1-72.2; same; IM.—MEXICO: 1 ♀, 78.7; OREGON

Sta. 662; paratype; USNM 94465.—2 ♀, 86.5-88.7; Barra del Tampico; 12 September 1963; U. Barron; USNM 113322.—2 ♀, 88.6-89.7; same; LB 1768.—1 damaged ♂ ; Punta Delgada, Veracruz; 122 m; from stomach of red snapper; 13-14 November 1960; D. Fuentes; ITV.—1 ♂, 71.0; 1 damaged ♀, 67.0; 1 damaged juv.; between Abra and Puntilla, Veracruz; 116 m; from the stomach of red snapper; 26-28 November 1960; D. Fuentes; ITV.—1 ♂, 88.7; 1 ♀, 92.0; Chachalacas, Veracruz; 14 March 1959; C. Garcia; ITV.

*Diagnosis.*—Rostral plate elongate, with faint median carina; median carina of carapace with poorly-developed anterior bifurcation; dactylus of raptorial claw with 6 teeth; mandibular palp present; 5 epipods present; lateral process of fifth thoracic somite a slender, anteriorly-curved spine, lateral processes of next 2 somites sharp posteriorly; telson denticles 2-5, 7-11, 1, marginal teeth elongate.

*Description.*—Eye large, cornea set very obliquely on stalk; corneal axis greater than peduncular; ocular scales obliquely truncate, usually rounded laterally, occasionally acute and sharp; anterior margin of ophthalmic somite projecting, faintly emarginate medially. Corneal indices of 10 specimens of various sizes follow:

| Sex | TL | CL | CI |
|---|---|---|---|
| ♀ | 52.4 | 11.5 | 328 |
| ♀ | 69.0 | 14.5 | 354 |
| ♂ | 68.3 | 14.9 | 373 |
| ♀ | 71.7 | 16.5 | 413 |
| ♀ | 78.7 | 17.6 | 383 |
| ♀ | 82.8 | 18.2 | 350 |
| ♀ | 105.4 | 22.7 | 413 |
| ♂ | 112.5 | 25.3 | 372 |
| ♀ | 123.0 | 25.7 | 402 |
| ♂ | 124.5 | 27.2 | 406 |

Antennular peduncle longer than carapace; antennular processes produced into acute spines directed anterolaterally.

Rostral plate elongate, longer than broad, lateral margins converging on rounded apex; faint median carina, rarely entire, present.

Median carina of carapace, anterior to cervical groove, not bifurcate posteriorly, anterior bifurcation present, poorly-marked, branches of carina not extending to anterior margin; intermediate carinae not extending to anterior margin; anterolateral spines strong but not extending to base of rostral plate; posterolateral margin sharply angled anteriorly.

Raptorial claw large; dactylus with 6 teeth, outer margin of dactylus a simple curve; dorsal ridge of carpus uneven.

Mandibular palp and 5 epipods present.

Last 3 thoracic somites with well-developed submedian and intermediate

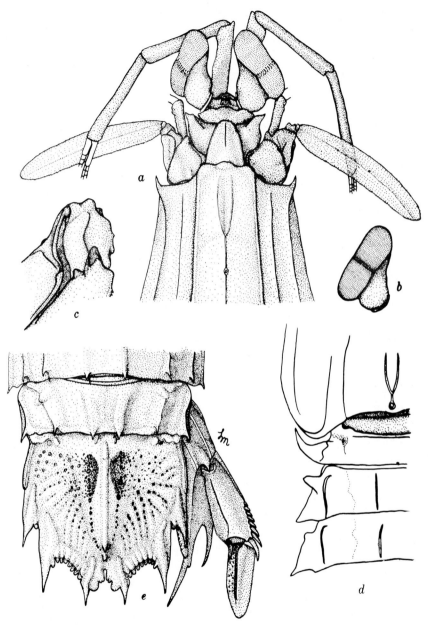

Figure 56. *Squilla chydaea* Manning, male holotype, TL 97.5, Oregon Sta. 332. *a*, anterior portion of body; *b*, eye; *c*, carpus of claw; *d*, lateral processes of fifth to seventh thoracic somites; *e*, last two abdominal somites, telson, and uropod (setae omitted).

carinae; lateral process of fifth somite a sharp, anteriorly-curved spine; lateral process of sixth somite bilobed, anterior lobe small, blunt, usually rounded, posterior lobe large, apex sharp posterolaterally; lateral process of seventh somite bilobed, anterior lobe less prominent than on sixth somite, posterior lobe larger than anterior, acute posterolaterally; ventral keel on eighth somite low, rounded.

Submedian abdominal carinae subparallel on anterior somites, slightly divergent on posterior somites; second to fifth abdominal somites each with median tubercle, anterior notch on intermediate carinae, and anterior tubercle between intermediate and lateral carinae; lateral carinae appear longitudinally sulcate, groove not well-marked; abdominal carinae spined as follows: submedian, (4) 5-6; intermediate, (1) 2-6; lateral, 1-6; marginal, 1-5; sixth somite with sharp spine in front of articulation of each uropod.

Telson longer than broad, length and width occasionally subequal, with sharp, elongate, marginal teeth; prelateral lobes present; denticles 2-5, 7-11, 1, outer submedian denticle larger than the remainder; carinae of marginal teeth swollen in large males; ventral surface with short postanal keel.

Uropods with 7-9 movable spines on outer margin of penultimate segment of exopod, last spine not extending to midlength of distal segment.

*Color.*—Gastric grooves and posterior margin of carapace, last 3 thoracic and first 5 abdominal somites lined with dark pigment; second abdominal somite with single rectangular black patch or 2 submedian patches on dorsal surface; telson with pair of anterior black crescents, convex mesially; uropod with inner margin of distal segment of exopod and distal portion of endopod dark.

*Size.*—Males, TL 29.3-124.5 mm; females, TL 35.5-123.4 mm. Other measurements of male, TL 85.1: carapace length, 19.0; cornea width, 5.5; rostral plate length, 3.2, width, 2.7; telson length, 16.5, width, 15.9.

*Discussion.*—*S. chydaea* closely resembles *S. lijdingi* Holthuis from the northern coast of South America, but differs in that the eyes are smaller (CI usually greater than 300) and slenderer, the rostral plate is slenderer and has a faint median carina, the median carina of the carapace has a broken but distinct anterior bifurcation, and the teeth of the telson are markedly elongate. The anterior lobe of the lateral process of the sixth thoracic somite is more rounded in *S. chydaea* than in *S. lijdingi*. Mature males of *S. chydaea* also seem to have more prominent swollen marginal carinae on the telson, although neither *S. chydaea* nor *S. lijdingi* have the telson margin between the marginal spines swollen as in *S. brasiliensis*. The slender eyes, slender rostral plate, and color pattern, with the prominent black crescents on the telson, among other features, will distinguish *S. lijdingi* from *S. empusa,* with which it is associated in the Gulf of Mexico.

200

*S. chydaea* is similar to two eastern Pacific species, *S. panamensis* Bigelow and *S. bigelowi* Schmitt, both of which have black crescents on the telson. *S. panamensis* has the posterior lobes of the lateral processes of the sixth and seventh thoracic somites rounded, and *S. bigelowi* has spined prelateral lobes.

This species, which occurs primarily in the Gulf of Mexico, was confused with *S. brasiliensis* and the related species from northern South America, *S. lijdingi,* until Holthuis (1959) described the latter. The ranges of *S. chydaea* and *S. lijdingi* are not now known to overlap, and, although they appear to be distinct species, additional collections from Central America may well show that they are terminal populations of a single species.

*Ontogeny.*—Unknown.

*Type.*—The holotype is in the U. S. National Museum. A series of paratypes has been deposited in the collection of the Institute of Marine Sciences, University of Miami, and one lot of paratypes is in the British Museum.

*Type-locality.*—Northern Gulf of Mexico, off Santa Rosa Island, Florida, 30°02.5′N, 86°53′W, in 109 m.

*Distribution.*—Off Cape Canaveral, Florida; Gulf of Mexico, from Tortugas, Florida to Veracruz, Mexico, including numerous records from off Mississippi, Louisiana, and Texas; 24-366 m, usually in less than 200 m depth. Its depth range is deeper than that of *S. empusa* and shallower than that of *S. edentata,* both of which may occur together with it in the Gulf of Mexico.

## *Squilla empusa* Say, 1818
Figures 57a, 58-59

*Squilla empusa* Say, 1818, p. 250.—Gould, in Hitchcock, 1833, p. 563 [listed].—H. Milne-Edwards, 1837, p. 525 [discussion].—DeKay, 1844, p. 32, pl. 13, fig. 54.—White, 1847, p. 84.—Gibbes, in Tuomey, 1848, p. xvi [listed].—Gibbes, 1850, p. 199 [p. 35 on separate]; 1850a, pp. 25, 28 [part; listed].—Leidy, 1855, p. 18.—Verrill, 1873, pp. 369, 377, 434, 439, 452, 468, 515.—Smith, 1873, p. 536, pl. 8, fig. 36 [ontogeny].— Verrill, *et al.,* 1873, p. 551.—Brooks, 1878, p. 143, pls. 9-13 [ontogeny]. Coues & Yarrow, 1878, p. 298.—Miers, 1880, p. 23, pl. 2, fig. 12 [part; African records = *S. calmani* Holthuis].—Faxon, 1882, p. 17, pl. 7, figs. 17-19, pl. 8, figs. 1-4 [ontogeny].—Preudhomme de Borre, 1882, p. cxi [listed].—Howard, 1883, p. 274 [listed].—R. Rathbun, 1883, pp. 121, 130; in Goode, 1884, p. 823, pl. 274 [general account].—Kingsley 1884, p. 66, text-figs. 82, 83, 85.—Brooks, 1885, p. 10 [biology].—Norman, 1886, p. 10 [listed].—Brooks, 1886, p. 25, pl. 1, figs. 4-5 [larvae], pl. 2 fig. 7; 1886a, p. 166 [biology].—M. J. Rathbun, 1892, p. 88.—M. J. Rathbun, in Evermann, 1892, p. 90.—R. Rathbun, 1893, p. 823, pl. 274 [general account; reprinted from 1884].—Bigelow, 1893, p. 102.—Sharp, 1893, p. 107.—Stebbing, 1893, p. 283 [discussion].—Bigelow, 1894, p. 525.—

201

Faxon, 1896, p. 165.—Wilson, 1900, p. 352 [listed].—Arnold, 1901, p. 288, text-fig. p. 288 [popular account].—Paulmeier, 1905, p. 149.—Sheldon, 1905, p. 335 [discussion; larva].—Mayer, 1905, pp. 95, 175, text-fig. 65 [popular account].—Norman, 1905, p. 11 [listed].—Kemp, 1913, p. 200 [listed].—Pratt, 1916, p. 384, fig. 615.—Kemp & Chopra, 1921, p. 298 [listed].—Parisi, 1922, p. 91.—Allee, 1923, p. 180.—Bigelow, 1941, p. 399, text-fig. 1.—Gurney, 1946, p. 160 [larval forms: listed].—Hedgpeth, 1950, p. 76.—Gunter, 1950, p. 12.—Miner, 1950, p. 532, pl. 170, color pl. X,2.—Anonymous, 1952, p. 2.—Blake, 1953, p. 29 [fossil].—Hedgpeth, 1953, p. 159; 1954, p. 211.—Chace, 1954, p. 449.—Perez-Farfante, 1954, p. 98.—Hildebrand, 1954, p. 261; 1955, p. 189.—Parker, 1956, pp. 311, 319, 323, 345, 354, text-fig. 16b, pl. 1, figs. 17a-b [ecology].—Menzel, 1956, p. 46 [listed].—Springer & Bullis, 1956, p. 22.—Manning, 1959, p. 19.—Holthuis, 1959, p. 177, text-fig. 76a, pl. 8, figs. 3-4.—Parker, 1960, p. 322 [ecology].—Wass, 1961, p. XII-19.—Milne & Milne, 1961, p. 422 [discussion].—Manning, 1961, p. 20, pl. 4, figs. 4-5.—Tabb & Manning, 1962, p. 594.—Lee & McFarland, 1963, p. 126 [physiology].—Dawson, 1963, p. 7 [listed].—Hessler, 1964, pp. 8, 51 *et seq.*, figs. 44, 45, 47 [morphology].—Smith, 1964, p. 127, pl. 17, fig. 32.—Dawson, 1965, p. 14.—Bullis & Thompson, 1965, p. 13 [listed].—Manning, 1968, p. 129 [listed], figs. 5a, 7a-b.
not *Squilla empusa* de Haan, 1844, pl. 51, fig. 6 [ = *Pseudosquilla haani* Holthuis].
(?)*Squilla mantis:* Howard, 1883, p. 294 [listed].—Guppy, 1894, p. 115 [listed] [not *S. mantis* (Linnaeus)].
not *Squilla empusa:* Osorio, 1889, p. 138.—Büttikofer, 1890, pp. 466, 487.—Osorio, 1898, p. 194.—Jurich, 1904, p. 8, pl. 1, fig. 3.—Johnston, 1906, p. 862.—Balss, 1916, p. 50.—Holthuis, 1941, p. 31 [ = *S. calmani* Holthuis, 1959].
*Chloridella empusa:* M. J. Rathbun, 1899, p. 628; 1905, p. 29.—Cary & Spaulding, 1909, p. 13.—Fowler, 1912, pp. 303, 539, pls. 91, 95, fig. 2; 1913, p. 64.—Summer, Osburn & Cole, 1913, p. 137 [biology].—Fish, 1925, p. 153, text-fig. 54 [larvae].—Cowles, 1930, p. 354.—M. J. Rathbun, 1935, p. 119 [fossil].—Pratt, 1935, p. 445, fig. 615.—Lunz, 1935, p. 157, text-fig. 6; 1937, p. 8.—Richards, 1938, p. 216, text-fig. 37 [popular account].—Berry, 1939, p. 466 [fossil].—Townes, 1939, p. 174.—Reed, 1941, p. 46 [discussion].—Pearse, Humm, & Wharton, 1942, p. 185.—Anonymous, 1942, p. 5.—Brown, 1948, p. 372.—Behre, 1950, p. 19.
not *Squilla empusa:* Moreira, 1903a, p. 60 [p. 5 on separate]; 1905, p. 125 [p. 5 on separate].—Andrade Ramos, 1951, p. 141 [ = *Squilla brasiliensis* Calman].
*Squilla empura:* Torralbas, 1917, p. 621 [p. 81 on separate] [erroneous spelling of *empusa*].
not *Squilla empusa:* Glassell, 1934, p. 454 [misidentification, = *S. prasinolineata* Dana].
*Chloridella* Amos, 1959, fig. on p. 8.

*Previous records.*—MASSACHUSETTS: Vineyard Sound (Verrill & Smith, 1873); Woods Hole (Allee, 1923; Smith, 1964); no specific locality (Gould, *in* Hitchcock, 1833; Sharp, 1893; Rathbun, 1905; Bigelow, 1941).—RHODE ISLAND: Say, 1818; White, 1847; Leidy, 1855; Miers, 1880; Rathbun, 1883.; Sharp, 1893; Rathbun, 1905.—CONNECTICUT: New Haven Harbor (Lunz, 1937); Stonington (Rathbun, 1883); no specific locality (Rathbun, 1905).—NEW YORK: Cold Spring Harbor

(Paulmeier, 1905); Long Island (Townes, 1939); Staten Island (Fowler, 1912); New York City (DeKay, 1844); Eastchester (Parisi, 1922).—NEW JERSEY: Delaware Bay (Lunz, 1937); Corson's Inlet, Stone Harbor, Anglesea, Point Pleasant, Cape May (all Fowler, 1913); no specific locality (Richards, 1938). — DELAWARE: Amos, 1959. — MARYLAND: Long Beach (Holthuis, 1959).—CHESAPEAKE BAY: Parisi, 1922; Cowles, 1930; Anon., 1952; Wass, 1961.—VIRGINIA: Fort Wool, Old Point Comfort (Brooks, 1878); Chincoteague Inlet, Wallops Beach (Fowler, 1912); Wallops Id. (Fowler, 1913); Chincoteague Id. (Holthuis, 1959). — CAROLINAS: Lunz, 1935. — NORTH CAROLINA: Beaufort (Brooks, 1885, 1886, 1886a); Fort Macon and Bird Shoal, Beaufort (Pearse, Humm, & Wharton, 1942); no specific locality (Coues & Yarrow, 1878; Sharp, 1893).—SOUTH CAROLINA: Charleston Harbor (Gibbes, 1850); no specific locality (Gibbes, in Tuomey, 1848; Howard, 1883; Sharp, 1893).—GEORGIA: mouth of Wilmington River and St. Simon Sound (Lunz, 1937).—FLORIDA: E Florida (Say, 1818); numerous localities between Fernandina and Cape San Blas (Manning, 1959); Matanzas Inlet (Lunz, 1937); Everglades National Park (Tabb & Manning, 1962); Sanibel Id. (Lunz, 1937; Chace, 1954); Pensacola (Bigelow, 1893; Lunz, 1937; Chace, 1954).—GULF OF MEXICO: Sharp, 1893; Hedgpeth, 1953, 1954; Springer & Bullis, 1956; Manning, 1959. — ALABAMA: Dawson, 1965. — MISSISSIPPI: Dawson, 1965.—MISSISSIPPI DELTA: Parker, 1956, 1960.—LOUISIANA: Grand Isle (Anon., 1942; Behre, 1950; Chace, 1954; Manning, 1959); Cameron and Breton Id. (Manning, 1959); no specific locality (Cary & Spaulding, 1909; Hildebrand, 1954; Dawson, 1965).—TEXAS: Galveston (Rathbun, 1892; Rathbun in Evermann, 1892; Chace, 1954; Manning, 1959; Lee & McFarland, 1963); Rockport (Manning, 1959); Aransas Bay (Gunter, 1950); Aransas Pass (Manning, 1959); Corpus Christi Bay (Manning, 1959); no specific locality (Reed, 1941; Hedgpeth, 1950, 1953, 1954; Hildebrand, 1954).—MEXICO: Campeche area (Faxon, 1896; Chace, 1954; Hildebrand, 1955; Springer & Bullis, 1956; Manning, 1959).—CUBA: Gulf of Batabano (Perez-Farfante, 1954); no specific locality (Torralbas, 1917).—JAMAICA: Port Royal Cays and Kingston Harbor (Rathbun, 1899); no specific locality (White, 1847; Miers, 1880).—SOUTH AMERICA: Sharp, 1893.—TRINIDAD: Manning, 1959, 1961.—VENEZUELA: (?) Guppy, 1894; Holthuis, 1959.—SURINAM: Holthuis, 1959; Manning, 1959, 1961.—FRENCH GUIANA: Manning, 1959. Bullis & Thompson (1965) list 38 U.S. Fish and Wildlife stations at which this species was taken.

*Material.*—?BERMUDA: 2 ♂; YPM 4491.—NEW ENGLAND: 2 ♂, 5 ♀; no other data; YPM 4486.—12 ♂, 22 ♀; same; YPM 4487.—MAINE: 1 ♂; Casco Bay; YPM 4480.—MASSACHUSETTS: 1 ♂; Boston Bay, Sandwich; USNM 92785.—1 ♀; Boston Bay, 10 mi. off Sandwich; MCZ 12068.—1 dry ♀; Cape Cod; USNM 102314.—1 ♂; Pocasset;

USNM.—1 ♂; Marion, Buzzard's Bay; MCZ 7882.—1 ♂, 1 ♀; Buzzard's Bay; USNM 21475.—4 ♀; Buzzard's Bay; Fɪsʜ Hᴀᴡᴋ Sta. 955; USNM 21476.—1 ♂; same; Fɪsʜ Hᴀᴡᴋ Sta. 961; YPM 4418.—1 ♀; same; USNM 36617.—1 ♂; New Bedford; ANSP 31.—1 ♂; same; USNM 40124.—2 ♂; same; USNM 2517.—1 ♂, 2 ♀; Acushnet River, New Bedford; USNM 22593.—1 ♀; same; USNM 22796.—1 ♂; Vineyard Sound; YPM 4431.—1 ♀; same; USNM 14115.—1 ♂; Woods Hole; YPM 4421.—1 ♀; same; USNM 3752.—1 ♂, 1 ♀; same; USNM 76020.— RHODE ISLAND: 1 brk. ♂; MCZ 7861.—1 dry ♂; lectotype; BMNH 158c.—1 dry ♂; ANSP 3182.— 1 brk. ♀; off Halfway Rock, Newport; Fɪsʜ Hᴀᴡᴋ Sta. 883; YPM 4419.— 1 juv. ♂, 1 brk. ♀; Newport; BMNH 1911.12.22.1-2.—1 ♀; S end of Hope Id., Narragansett Bay; Fɪsʜ Hᴀᴡᴋ Sta. 820; YPM 4467.—1 spec.; Narragansett Bay; Fɪsʜ Hᴀᴡᴋ Sta. 882; USNM 5139.—1 ♂; Greenwich Bay; YPM 4476.—1 ♂, 1 ♀; Providence River; USNM 13007.—1 ♂; same; USNM 21477.—1 ♀; same; USNM 3689.—1 ♀; Point Judith Pond; USNM 76012.—CONNECTICUT: 1 ♀; AMNH 10126.—1 ♂; Stonington; USNM 4118.—1 ♂; Westbrook; USNM. —1 ♀; Guilford; YPM 4478.—1 ♂; (?)Stony Creek; YPM 4489.—1 ♂; Indian Neck, Branford; YPM 4461.—1 ♂; New Haven; USNM 2253.— 1 ♂; same; MNHNP.—1 ♂; same; BMNH 1898.5.7.853.—2 ♀; same; YPM 1248.—1 ♂, 1 ♀; same; YPM 4463.—1 ♀; same; YPM 4464.— 1 ♂; same; YPM 4465.—1 ♂; S end of New Haven; YPM 4466.—1 ♀; off Morris Cove, New Haven; YPM 4477.—1 ♂; New Haven Harbor; YPM 4488.—1 ♀; same; YPM 4469.—15 ♂, 9 ♀; Bridgeport; YPM 4930.— LONG ISLAND SOUND: 1 ♀; Duck Id.; YPM 4462.—1 ♂; no other data; YPM 4481.—3 ♂, 1 ♀; same; YPM 387.—1 ♀; Fɪsʜ Hᴀᴡᴋ Sta. 1555; USNM 21474.—1 ♀; Fɪsʜ Hᴀᴡᴋ Sta. 1560; USNM 19638.—1 ♀; Fɪsʜ Hᴀᴡᴋ Sta. 1727; USNM 26182.—1 ♀; Fɪsʜ Hᴀᴡᴋ Sta. 1755: USNM 26183. — 1 ♂; Fɪsʜ Hᴀᴡᴋ Sta. 1768; USNM 75997. — NEW YORK: 2 ♀; Long Island; USNM 2518.—1 ♀; East Hampton, L.I.; AMNH 7494.—1 ♂; Great Peconic Bay, L.I.; AMNH 3516.—1 ♂; Peconic Bay, L.I.; AMNH 2088.—1 ♀; Cold Spring Harbor, L.I.; USNM 47286.—1 ♀; Inner Harbor, Cold Spring, L.I.; AMNH 6256.—1 ♂, 2 ♀; Lloyd's Neck, L.I.; AMNH 6689.—1 ♀; Huntington Bay, L.I.; USNM 11302.—2 spec.; Great South Bay, L.I.; USNM 9153.—1 ♂; Point Lookout, Long Beach, L.I.; AMNH 7267.—1 ♀; Gravesend Bay, L.I.; USNM 21603.—2 ♂; New York Harbor; AMNH 321.—1 ♂; (?) Eastchester Bay; AMNH 320.—1 ♀; Stapleton, Staten Id.; USNM 3243.—1 ♀; Hudson River, Piermont; AMNH 2074.—1 ♂; Westchester Co.; AMNH 2083.— NEW JERSEY: 1 ♀; no other data; AMNH 9904.—1 brk. ♀; Ventnor; ANSP 4820.—1 dry ♂; Cape May; ANSP 3596.—1 ♂, 1 ♀; Cape May Point; ANSP 4871.—1 ♂; Cape May Harbor; ANSP 4870.—1 ♂; Delaware Bay; YPM 4438.—1 ♂, 1 ♀; (?) Brandywine; USNM 76432.—1 ♀; off New England Creek; USNM 63189.—MARYLAND: 1 brk. ♂; Solomon's Id.; USNM 76011.—CHESAPEAKE BAY: 1 ♂; no specific locality; USNM 3383.—2 ♂, 1 ♀; same; BMNH 94.10.17.1-3—2 ♂, 7 ♀; near Barren Id.; USNM 5140.—2 ♀; Fɪsʜ Hᴀᴡᴋ Sta. 1058; USNM 5144.—3 ♂, 2 ♀; Fɪsʜ Hᴀᴡᴋ Sta. 1075; USNM 4488.—3 ♂, 1 ♀; Fɪsʜ Hᴀᴡᴋ Sta. 8396; USNM 110862.—1 ♀; Fɪsʜ Hᴀᴡᴋ Sta. 8511; USNM 4490.—3 ♂; Fɪsʜ Hᴀᴡᴋ Sta. 8341; USNM 75995.— 1 ♂, 1 ♀; same; USNM 110848.—1 ♀; Fɪsʜ Hᴀᴡᴋ Sta. 8342; USNM 75998.—1 ♀; Fɪsʜ Hᴀᴡᴋ Sta. 8345; USNM.—1 ♂, 1 ♀; Fɪsʜ Hᴀᴡᴋ Sta. 8353; USNM 110858.—1 ♂, 2 ♀; Fɪsʜ Hᴀᴡᴋ Sta. 8359; USNM 110860.—1 ♂, 2 ♀; Fɪsʜ Hᴀᴡᴋ Sta. 8361; USNM 110859.—1 ♀; Fɪsʜ Hᴀᴡᴋ Sta. 8366; USNM 75999.—1 ♂; Fɪsʜ Hᴀᴡᴋ Sta. 8372; USNM 75994.—2 brk. spec.; Fɪsʜ Hᴀᴡᴋ Sta. 8388; USNM 110861.—5 ♂, Fɪsʜ Hᴀᴡᴋ Sta. 8394; USNM 76000.—1 ♀, Fɪsʜ Hᴀᴡᴋ Sta. 8395; USNM 75996.—3 ♀; Fɪsʜ Hᴀᴡᴋ Sta. 8396; USNM 110862.—1 ♀; Fɪsʜ Hᴀᴡᴋ Sta. 8511; USNM 76015.—3 ♀; Fɪsʜ Hᴀᴡᴋ Sta. 8523; USNM.—1 ♀; Fɪsʜ Hᴀᴡᴋ Sta. 8599; USNM 76014.—1 ♂; same; USNM 110849.— 1 ♂; Fɪsʜ Hᴀᴡᴋ Sta. 8603; USNM 110830.—VIRGINIA: 3 ♂; Wallops Beach; ANSP 4872.—3 ♀; Crisfield to Cape Charles; USNM 57132.—4 ♀; Rappahannock River; USNM 57146.—1 ♂; Hampton Roads; USNM 110834.—1 ♀; Old Point Comfort; MCZ 7845.—1 brk. ♀; Grand View Beach, near Fort Monroe; USNM 110829.-—1 dry ♂; Virginia Beach; USNM 88061.—NORTH CAROLINA: 1 ♀; off Cape Hatteras; Aʟʙᴀᴛʀᴏss Sta. 2283; USNM 8786. —2 ♂; same Aʟʀᴀᴛʀᴏss Sta. 2285; USNM 8789.—1 ♂, 1 ♀; same; Cᴏᴍ- ʙᴀᴛ Sta. 375; USNM 101725.—1 ♀; same; Cᴏᴍʙᴀᴛ Sta. 378; USNM.— 1 ♀; Ocracoke; ANSP 4277.—1 ♀; S of Ocracoke Inlet; Pᴇʟɪᴄᴀɴ Sta. 188- 8; USNM 110828.—1 ♀; off Cape Lookout; Pᴇʟɪᴄᴀɴ Sta. 192-1; USNM 110842.—1 ♀; Cape Lookout; AMNH 2075.—1 ♂, 2 ♀; Fort Macon; YPM 4460.—1 ♂, 1 ♀; same; USNM 81680.—1 ♀; Beaufort; ANSP 34.— 1 ♀; same; MCZ 7850.—1 ♂; same; USNM 85529.—2 ♂; same; USNM 110835.—2 ♂, 16 ♀; off Morehead City; Sɪʟᴠᴇʀ Bᴀʏ Sta. 2165; USNM.— 1 ♀; Morehead City; USNM 26110.—2 ♀; Core Creek, Newport River, Morehead City; USNM 110315.—SOUTH CAROLINA: 1 ♀; off Cape Romain; Pᴇʟɪᴄᴀɴ Sta. 182-16; USNM 110864.—1 ♀; off Caper's Id.; USNM 76006.—1 ♂; 2 mi. off Caper's Id.; AMNH 6974.—1 ♀; same; MCZ 9638.—3 brk. ♂; off Cape Fear River; Fɪsʜ Hᴀᴡᴋ Stats. 8278-8280; USNM 110836.—1 ♂; E end Sullivan's Id., Charleston; USNM 3165.— 3 ♂, 1 ♀, 1 fragment; Charleston Harbor; USNM 3152.—1 ♂, 2 ♀; off Charleston Harbor; YPM 4479.— 1 ♀; Charleston; USNM 3235.—1 ♀; same; USNM 4091.—1 ♀; same; USNM 4638.—1 ♂, 2 ♀; same; MCZ 569.—1 ♀; same; MCZ 582.—1 ♂, 2 ♀; Blackfish Banks, 12 mi. off Charleston; USNM 5141.—1 ♀; Hilton Head; ANSP 33.—GEORGIA: 7 ♂, 10 ♀; mouth of Wilmington River; YPM 4485.— 1 ♀; (?) near Green Id., Chatham Co.; USNM 40054.—3 ♂; Brunswick; USNM.—1 ♀; Jekyll Id.; UMML 32.1463.—3 ♀; same; USNM.—1 ♀; same; USNM.—4 ♀; W of Jekyll Id.; USNM.—1 ♂; Jekyll Sound; USNM.—2 ♂; St. Simon's

Sound; YPM 4484.—1 ♀; same; USNM.—2 ♂; (?) Mollclark Creek; USNM 110831.—FLORIDA: 1 ♂; no specific locality; MCZ 563.—2 ♂; 3 ♀; same; MCZ 567.—1 ♀; same; MCZ 7881.—1 ♀; same; UZM.—1 ♂, 1 ♀; Fernandina; USNM 66611.—1 ♂, 1 ♀; same; USNM 76016.—1 ♂; same; USNM 76002.—1 ♀; St. Augustine; PELICAN Sta. 202-6; USNM 110825.—4 ♂, 1 ♀; St. Augustine; UZM.—1 dry ♂; same; USNM 61044. —2 dry ♂; same; USNM 62058.—1 ♂; same; USNM.—1 ♂, 4 ♀; off S shore, St. Augustine; YPM 4472.—2 ♂; SE of St. Augustine; PELICAN Sta. 203-1; USNM 110844.—9 ♂, 5 ♀; off Matanzas Inlet; YPM 4482.— 1 ♂; SE of Cape Canaveral; PELICAN Sta. 206A-3; USNM 110843.—2 ♀; same; PELICAN Sta. 206A-5; USNM 110826.—2 ♂; same; SILVER BAY Sta. 5096; USNM.—1 ♀; off Melbourne; PELICAN Sta. 169-1; USNM 110827. —1 ♂; Indian Creek, Miami Beach; UMML 32.812.—3 ♂, 2 ♀; Biscayne Bay; USNM.—1 ♂; same; USNM.—1 ♀; same; USNM 64917.—1 ♂; Hawk Channel, Marathon; USNM.—1 brk. spec.; off Key West; FISH HAWK Sta. 7272; USNM 110863.—1 ♂; Key West; FISH HAWK Sta. 7273; USNM 75993.—1 ♂, 1 ♀; Tortugas shrimp grounds; USNM.—1 ♂, 2 ♀; same; USNM.—1 ♀; same; CHML.—9 ♂, 25 ♀; same; CHML.—6 ♂, 3 ♀; same; UMML 32.1191.—2 ♂, 3 ♀; same; UMML 32.1195.—2 ♂, 7 ♀; same; UMML 32.1190 (1 ♂, 1 ♀; BMNH).—4 ♀; same; USNM.—2 ♂; same; USNM.—4 ♂; same; USNM (1 ♂, 1 ♀; same lot; MMF).—13 ♂, 22 ♀; same; USNM.—9 ♂, 9 ♀; same; USNM.—1 ♀; same; USNM.—1 ♂; same; USNM 99889.—2 ♂; same; USNM 99887.—2 ♀; same; OREGON Sta. 562; USNM 93680.—25 ♂, 32 ♀; same; USNM (1 ♂, 1 ♀; same lot; MNRJ.—2 ♂; off Tortugas; GERDA Sta. 566; UMML.—1 ♀; Florida Bay; USNM.—1 ♀; Conchie Channel, Florida Bay; UMML 32.1362.—6 ♂, 4 ♀; Sandy Key Basin, Florida Bay; USNM.—1 ♂; off Flamingo; UMML 32.1699.—1 ♂; same; UMML 32.1172.—1 ♂; Buttonwood Canal, Flamingo; USNM.—1 juv. ♂; Oyster Bay; USNM.—1 ♀; W Whitewater Bay; USNM.—1 ♀; Slagle's Ditch, Cape Sable; USNM.—1 ♀; off East Cape Sable; USNM.—1 ♂; off Naples; OREGON Sta. 982; USNM 97535.—1 ♂, 1 ♀; same; SILVER BAY Sta. 67; USNM 101617.—1 ♀; off Marco; USNM. —1 ♀; Captiva Key; MCZ 581.—1 ♂; off Sanibel Id.; USNM.—3 ♂; same; ANSP 4406.—1 ♀; same; MCZ 10411.—1 ♀; same; FISH HAWK Sta. 7795; USNM 76003.—4 ♂, 2 ♀; 4 mi. off Englewood; USNM 76018.—1 ♀; off Port Tampa; USNM 26109.—1 juv. ♂; same; USNM 26181.—2 ♂, 2 ♀; Tampa Bay; FISH HAWK Sta. 7109; USNM 25647.—2 ♂; Cedar Key; USNM 6472.—2 ♀; off Apalachee Bay; USNM.—1 ♂, 1 ♀; Apalachicola Bay; FSBCML.—1 ♀; N Higgins Shoal, Apalachicola Bay; FSBCML.— 2 ♂; Lone Pine, Apalachicola Bay; FSBCML.—1 ♂, 1 ♀; off Cape St. George; OREGON Sta. 274; USNM 92392.—2 ♀; off Cape San Blas; USNM 91422.—2 ♂, 1 ♀; East Pass Channel, Choctawahtchee Bay; USNM 99888.—1 ♂; Pensacola; USNM 4029.—1 ♂; same; ANSP 4694.—2 ♂, 3 ♀; same; MCZ 10636.—101 ♂, 93 ♀; N of Pensacola Bay Bridge; USNM. —2 ♀; off Barrancas Point, Pensacola Bay; YPM 4483.—2 ♂; Pensacola

Bay; YPM 4439.—1♂, 2♀; same; YPM 4468.—3♂, 10♀; same; YPM 4470.—3♂, 4♀; same; YPM 4471.—1♂, 8♀; same; YPM 4473.—3♂, 5♀; same; YPM 4474.—7♂, 10♀; same; YPM 4475.—GULF OF MEXICO: 1 dry ♂; ANSP 8.—ALABAMA: 1♀; Mobile Bay; USNM 81678.—1♀; near Cedar Point; USNM 81682.—1♀; OREGON Sta. 188; USNM 92391.—1♂; OREGON Sta. 185; USNM 91931.—1♀; OREGON Sta. 382; GCRL.—MISSISSIPPI: 1♀; no other data; USNM 87376.— 1♂; E end Mississippi Sound; USNM 85537.—21♂, 17♀; Ocean Springs to Horn Id.; USNM.—1♂, 4♀; ¾ mi. N of Horn Id.; USNM 92439.— 1♀; Ship Id.; MCZ 7849.—1brk. ♀; same; YPM 4437.—1♂; W of Deer Id., off Biloxi; USNM.—1♀; off Biloxi-Ocean Springs Causeway; USNM. —OFF MISSISSIPPI DELTA: 1♂; PELICAN Sta. 1; USNM 110823.— 1♂, 2♀; PELICAN Sta. 69-6; USNM 110486.—1♂; PELICAN Sta. 96-1; USNM 110845.—1♂; OREGON Sta. 110; USNM 91942.—1♀; OREGON Sta. 123; USNM 92390.—1♂; OREGON Sta. 342; USNM 92393.—1♀; OREGON Sta. 845; USNM 96288.—1♂, 1♀; OREGON Sta. 3713; USNM. —LOUISIANA: 1♂, 1♀; no specific locality; USNM 110833.—2♀; Breton Id.; USNM 64241.—1 brk. ♂, 1♀; Barataria Bay; USNM 110832. —1♂; off Isles Dernières; PELICAN Sta. 81-3; USNM 110841.—2♂, same; OREGON Sta. 3768; USNM.—1♂; same OREGON Sta. 3769; USNM. —5♂; S of Vermilion Bay; OREGON Sta. 3776; USNM.—1♀; Grand Isle; PELICAN Sta. 20; USNM 110824.—2♀; S of Grand Isle; GCRL I61:363. —1♀; same; GCRL I61:365.—1 juv. ♂; same; USNM 103754.—3♂, 3♀; Chauvin; AMNH 9812.—1 brk. ♀; Cameron; USNM 30575.—1 juv. ♂; same; USNM 33114.—TEXAS: 2♂, 1♀; no specific locality; BMNH 1952.5.27.1-3.—1 juv. ♂; Sabine Pass; USNM 101085.—1♂; Galveston Bay; USNM 17139.—1 brk. ♂; Galveston; USNM 76013.—3♂, 5♀; same; USNM 81681.—2♂, 2♀; same; USNM 81693.—2♀; same; MCZ 7852.—1♂; same; USNM.—21♂, 12♀; same; USNM.—5♂, 5♀; same; IM.—1♂, 1♀; Freeport; USNM.—1♀; same; USNM.—3♂, 3♀; Rockport; USNM 81683.—1♂, 1♀; Aransas Pass; USNM 81677.—2♂; Corpus Christi Bay; USNM 72184.—1♂; same; USNM 80510.—1♂; Corpus Christi; USNM 21473.—2♂; same; USNM 76023.—1♀; (?) Pass Cabello; USNM 76028.—MEXICO: 1♀; Gulf coast of Mexico; USNM 110822.—1♂, 2♀; (?) off Hut's Bayou; PELICAN, col.; USNM.—1♂, 1♀; Tamaulipas, Tampico; USNM.—1♀; Tampico; LB 1145-167.—2♂ (2 lots); same; USNM.—1♂; between Antigua and Chachalacas, Veracruz; ITV.—2♀; Estario de Tacoma, Tuxpan, Veracruz; LB 2497-605.— 1♀; Tuxpan, Veracruz; LB 2490-608.—2♂; Rio Pantepec, Tuxpan, Veracruz; USNM.—1♂; same; USNM.—1♀; Veracruz; USNM.—1♂; NNE of Alvarado, Veracruz; ITV.—3♂, 1♀; off Ciudad del Carmen; ITV.— 1♂; mouth of Rio San Pedro, Campeche; ITV.—1♂, 1♀; Punta de la Disciplina, Campeche; ITV.—1♂; same; ITV.—1♂, 5♀; off Campeche; OREGON Sta. 720; USNM 94466.—2♀; NW of Arcas, Campeche; LB 774n-216.—14♂, 10♀; Campeche chrimp grounds; USNM.—16♂, 12♀;

FIGURE 57. Color patterns of: *a, S. empusa* Say, male, TL 129.0, OREGON Sta. 4306; *b, S. brasiliensis* Calman, female, TL 105.0, CALYPSO Sta. 150.

same; USNM.—NICARAGUA: 1 brk. ♂; OREGON Sta. 1902; USNM.—CUBA: 2 ♂; off Havana; USNM 58667.—JAMAICA: 1 ♂; Kingston Harbor; USNM 21240.—2 dry spec.; no specific data; BMNH 158a-158b.—SOUTH AMERICA: 1 ♀; no specific data; OREGON, col.; USNM.—2 brk. spec.; no specific data; ANSP 39.—COLOMBIA: 1 ♂, 2 ♀; OREGON Sta. 4886; USNM.—VENEZUELA: 11 ♂, 10 ♀; OREGON Sta. 4493; USNM.—1 ♂; Bar Santa Rosa, Caracas; UCVMB 5009.—1 ♀; Gulf of Venezuela; UCVMB 5013.—1 ♂, 2 ♀; same; UCVMB 5014.—TRINIDAD: 4 ♂, 1 ♀; OREGON Sta. 2208; USNM.—1 ♀; OREGON Sta. 2209; UMML.—1 ♂; Trinidad; BMNH 1940.vii.8.7.—1 ♂, 1 ♀; Trinidad, Tobago and British Guiana; USNM 110852.—BRITISH GUIANA: 3 ♂, 4 ♀; CAPE ST. MARY, col.; USNM.—2 ♂, 1 ♀; same; USNM.—1 ♂; no specific data;

BMNH 1949.2.28.49.—5 ♂, 5 ♀; same; USNM.—5 ♂, 4 ♀; Oregon Sta. 4306; USNM.—SURINAM: 3 ♂, 5 ♀; no specific data; USNM.—1 ♂; Coquette Sta. 140; USNM 103216.—1 ♀; Coquette Stats. 281-282; USNM 103215.—2 ♀; Coquette Sta. 297; USNM 103214.—3 ♂, 1 ♀; Oregon Sta. 2279; USNM.—1 ♂, 1 ♀; Oregon Sta. 4170; USNM.—3 ♂, 5 ♀; Oregon Sta. 4171; USNM.—1 ♂, 2 ♀; Oregon Sta. 2327; UMML.

*Diagnosis.*—Rostral plate broad, subquadrate or trapezoidal, with median carina; median carina of carapace with anterior bifurcation open for most of distance between pit and anterior margin; dactylus of raptorial claw with 6 teeth, outer margin sinuate; lateral process of fifth thoracic somite a sharp, anteriorly-curved spine; lateral processes of next 2 somites acute posteriorly; telson denticles 3-5, 6-9, 1.

*Description.*—Eye large, cornea set very obliquely on stalk; ocular scales rounded or obtusely angled laterally; anterior margin of ophthalmic somite produced forward, usually emarginate, occasionally flattened or provided with an apical spinule. Corneal indices of 18 specimens follow:

| Sex | TL | CL | CI |
|-----|------|------|-----|
| ♂ | 31.9 | 7.1 | 296 |
| ♀ | 39.7 | 8.4 | 350 |
| ♂ | 42.0 | 9.6 | 369 |
| ♀ | 55.3 | 11.8 | 358 |
| ♀ | 59.0 | 13.0 | 361 |
| ♂ | 60.0 | 14.7 | 387 |
| ♀ | 81.0 | 17.0 | 395 |
| ♀ | 93.5 | 18.2 | 387 |
| ♂ | 92.1 | 19.9 | 442 |
| ♂ | 100.0 | 21.1 | 440 |
| ♀ | 102.5 | 21.1 | 414 |
| ♂ | 105.0 | 22.3 | 455 |
| ♀ | 109.7 | 23.3 | 424 |
| ♂ | 126.6 | 25.9 | 425 |
| ♀ | 129.7 | 27.7 | 454 |
| ♂ | 132.9 | 28.8 | 465 |
| ♀ | 133.9 | 29.9 | 490 |
| ♀ | 151.7 | 32.9 | 491 |

Antennular processes tapering to blunt spines directed anterolaterally; antennular peduncle as long as or slightly shorter than carapace.

Rostral plate subquadrate or trapezoidal, broader than long, lateral margins tapering to blunt, rounded, or transverse apex; median carina present.

Carapace minutely punctate, with prominent carinae; median carina, anterior to cervical groove, bifurcate anteriorly only; distance from dorsal pit to bifurcation much less than distance from bifurcation to anterior margin; anterolateral spines strong, extending to or slightly beyond base of rostral plate; posterolateral margin angled anteriorly.

Dactylus of claw with 6 teeth, outer margin of dactylus sinuate; dorsal ridge of carpus with 2-3 irregular tubercles.

Mandibular palp and 5 epipods present.

Last 3 thoracic somites with submedian and intermediate carinae, submedians divergent on each somite; intermediate carinae represented by tubercles on fifth somite; lateral process of fifth somite a slender, sharp

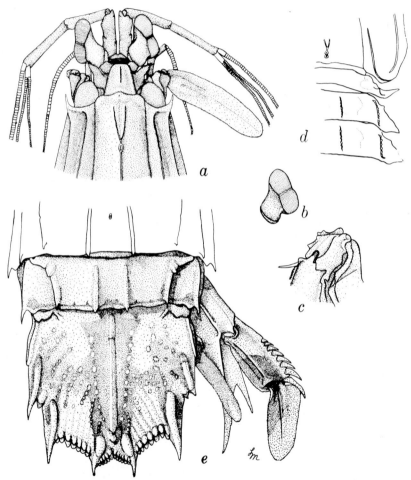

FIGURE 58. *Squilla empusa* Say, female, TL 96.5, Florida Bay. *a,* anterior portion of body; *b,* eye; *c,* carpus of claw; *d,* lateral processes of fifth to seventh thoracic somites; *e,* last two abdominal somites, telson, and uropod (setae omitted).

210

spine, curved slightly forward; lateral processes of next 2 somites bilobed, each with small anterior lobe, sharper and larger on sixth somite than seventh, and large posterior lobes, posterolateral apices sharp, acute; ventral keel on eighth somite subtriangular, apex rounded.

Submedian abdominal carinae posteriorly divergent on each somite; second to fifth somites with anterior tubercle between intermediate and lateral carinae; abdominal carinae spined as follows: submedian, (4)5-6; intermediate, (2)3-6; lateral, 1-6; marginal, 1-5; sixth somite with sharp spine on each side in front of articulation of uropod.

Telson about as broad as long, with sharp marginal spines; prelateral lobes present; denticles rounded, 3-5, 6-9, 1; ventral surface of telson with postanal keel.

Uropod with 7-8 (11) graded movable spines on outer margin of proximal segment of exopod, last not extending to midlength of distal segment; lobe on outer margin of inner spine of basal prolongation inconspicuous.

*Color.*—Last 3 thoracic and first 5 abdominal somites each with dark posterior line; second abdominal somite usually with median rectangular black patch; bases of submedian teeth of telson dark; distal half of penultimate segment and proximal half of distal segment of uropodal exopod dark, distal half of endopod dark.

In life, background color white; carinae of body orange-yellow; distal portions of merus and propodus of claw yellow; eyes, light green. Bigelow (1941) also recorded color in life of specimens from Woods Hole.

*Size.*—Males, TL 28.7-165.0 mm; females, TL 33.8-185.0 mm. Other measurements of female, TL 96.6: carapace length, 20.8; cornea width, 5.0; rostral plate length, 3.0, width, 3.8; telson length, 19.0, width, 19.3.

*Discussion.*—*Squilla empusa,* the most common American species of *Squilla,* can readily be distinguished from all other species in the genus by the following features: (1) the anterior bifurcation of the median carina of the carapace is open from most of the distance between the dorsal pit and the anterior margin; (2) the outer margin of the dactylus of the claw is sinuate; (3) the lateral process of the fifth thoracic somite is a single, slender, curved spine: and (4) the lateral processes of the next two somites are bilobed, with the anterior lobe much smaller than the posterior and the posterior with a very acute posterolateral apex.

*S. neglecta* Gibbes and *S. prasinolineata* Dana both resemble *S. empusa* in general appearance and color pattern and either of the first two species may occur together with *S. empusa.* Both of these species, however, have but five teeth on the dactylus of the claw. *S. neglecta* further differs in having a shorter anterior bifurcation on the median carina of the carapace and in having a spatulate lateral process on the fifth thoracic somite. *S. prasinolineata,* which has no anterior bifurcation on the median carina of the median carina of the carapace, has a short, straight lateral process on the fifth thoracic somite.

211

*S. aculeata* Bigelow from the eastern Pacific and the similar West African *S. calmani* Holthuis both resemble *S. empusa* in most features, but both can be distinguished from *S. empusa* by the presence of a prominent spine on the basis of the raptorial claw.

All of the other large species of *Squilla* in the western Atlantic have an evenly convex dactylus on the raptorial claw.

Comparison of northern and southern populations of *S. empusa* reveals some minor differences. In specimens from northern South American, in the southern part of the range, (1) the rostral plate is slenderer, more elongate, with the lateral margins subparallel; (2) the anterior margin of the ophthalmic somite is not always emarginate but often is provided with an apical tubercle; (3) the posterolateral apices of the lateral processes of the sixth and seventh thoracic somites are sharper, more nearly spiniform; (4) the submedian carinae of the fourth abdominal somite are rarely provided with spines; (5) the telson appears flatter and the marginal spines are more elongate; and (6) the postanal keel is longer. Some of these features also can be observed in specimens from the northern part of the range of the species. For example, in one lot from Tortugas, Florida some specimens had the anterior margin of the ophthalmic somite flattened, in most it was medially emarginate, and, in a few specimens, the anterior margin was provided with a small spinule. In most northern specimens the submedian carinae of the fourth abdominal somite are armed, but in a few specimens the spines are lacking. The Corneal Indices of specimens from both northern and southern portions of the range overlap completely, at all sizes, and give no indication of divergence of the populations. In my opinion, these differences lie within the expected range of variation of the species and are not of sufficient importance to warrant recognition of separate subspecies at this time.

Fossil *S. empusa* have been reported from Pleistocene formations in Maryland by Rathbun (1935) and Blake (1953). The latter author discussed the presence of burrow-like tunnels, two inches in diameter and two feet or more in length, in formations in which *S. empusa* occurred and tentatively attributed these burrows to that species.

*Remarks.*—Bigelow (1941) was the first to record sexual dimorphism in this species. The marginal carinae of the abdomen in adult males, TL in excess of 140 mm, are noticeably more inflated than in females; the margins of the telson may also be slightly inflated in large males. The dimorphism is not nearly as well-marked as in some related species, including *S. edentata* (Lunz) and *S. intermedia* Bigelow.

Bigelow's account included some general observations on behavior and color pattern.

*S. aculeata* Bigelow is the eastern Pacific analogue of *S. empusa;* the major difference between these species has been noted above.

*S. empusa* is very abundant in moderate depths off the southern coast

212

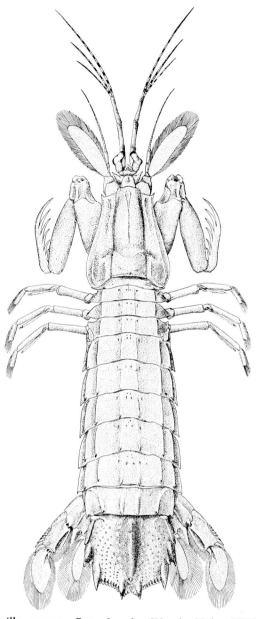

Figure 59. *Squilla empusa* Say, female, Woods Hole, USNM 3752. The figure errs in showing spines on the submedian carinae of the third abdominal somite (from R. Rathbun, 1884).

213

of Florida, particularly in the shrimp grounds northwest of Key West. There it is one of the characteristic components of the "trash" fishery. It is so abundant off Florida that the dactyli of the raptorial claws are sold as ornamental jewelry.

*S. empusa* is the only western Atlantic stomatopod to have been recorded from waters of low salinity, although from the habitat of *Nannosquilla grayi* (Chace) one might infer that it also occurs in low salinities. It has been taken in brackish water (salinity 16-30 ppt) in Chesapeake Bay (Cowles, 1930), off Texas (Gunter, 1950), and in south Florida (Tabb & Manning, 1962). Experimental data on the osmoregulatory abilities of *S. empusa* have been presented by Lee & McFarland (1963).

*S. empusa* occurs in relatively shallow waters, in depths of 40 m or less, although there are some collections extending its depth range to 154 m. Specimens were taken at greater depths at OREGON Sta. 382 (346-382 m, off Alabama) and OREGON Sta. 1902 (246 m, off Nicaragua). The occurrence of this species at these depths needs to be verified.

The record from Bermuda also should be verified. This lot, in the Peabody Museum at Yale, includes two specimens as well as two labels. One label states: Bermuda; G. Brown Goode; 1876-7; A. E. Verrill; 1903. The other label states: Outer Island; 1903; A. E. Verrill. It seems possible that at least one of the two specimens was taken by Goode at Bermuda, but the species has not been found there since. The source of Verrill's specimen is unknown; I cannot find the exact locality of the "Outer Island."

*Ontogeny.*—Larvae identified as those of *S. empusa* have been recorded by Brooks (1878, 1886), Faxon (1882) and Fish (1925); Faxon summarized Brooks' observations. Brooks' account of 1878 was based on larvae taken at Fort Wool, Virginia; the identification may well be correct, for *S. empusa* is the only *Squilla* known to occur in that area as an adult. Larvae of *S. empusa* have not been reared in the laboratory.

*Type.*—British Museum (Natural History). The dry specimen (BMNH 158c) from Rhode Island donated to the British Museum by Say and first recorded in that collection by White (1847) is here selected as the lectotype. Although Rathbun (1935) stated that the type was in the Academy of Natural Sciences at Philadelphia, a search there revealed none of Say's specimens of *Squilla*.

*Type-locality.*—Coasts of Rhode Island and Florida, here restricted to Rhode Island.

*Distribution.*—Western Atlantic, from Bermuda (?) and Maine southward through the Gulf of Mexico to Surinam, in depths between 0 and 154 m, generally in 40 m or less.

West African records of *S. empusa* actually refer to *S. calmani* Holthuis.

Specimens recorded from French Guiana by Manning (1959) should read Surinam.

214

## Squilla brasiliensis Calman, 1917
Figures 57b, 60, 61a-b

*Squilla panamensis* var. C Bigelow, 1894, p. 529.
*Squilla empusa:* Moreira, 1903a, p. 60 [p. 5 on separate]; 1905, p. 125 [p. 5 on separate].—Andrade Ramos, 1951, p. 141 [not *S. empusa* Say].
*Squilla brasiliensis* Calman, 1917, p. 139, text-figs. 1-3.—Hansen, 1921, p. 7.—Andrade Ramos, 1951, p. 142 [listed].—Lemos de Castro, 1955, p. 18, text-fig. 16A, pl. 4, fig. 30, pl. 5, fig. 37, pl. 14, figs. 47-48.—Manning, 1966, p. 361, text-fig. 1; 1968, p. 129 [listed].
*Squilla braziliensis:* Kemp & Chopra, 1921, p. 298 [listed].
not *Squilla brasiliensis:* Andrade Ramos, 1951, p. 145, pl. 1, fig. 1 [possibly undescribed].
not *Squilla brasiliensis:* Springer & Bullis, 1956, p. 22.—Manning, 1959, p. 18 [ = *S. chydaea* Manning and *S. lijdingi* Holthuis].

*Previous records.*—BRAZIL: Cabo Frio (Calman, 1917; Lemos de Castro, 1955); Ilha Rasa, Ponta de Guaratiba, and Ilha Grande, Sepetiba Bay (Moreira, 1903, 1905; Lemos de Castro, 1955); off Rio de Janeiro, Ilha de São Sebastião, Santos, São Francisco do Sul, Porto Alegre, and Rio Grande (Manning, 1966).—URUGUAY: Hansen, 1921; Manning, 1966.

*Material.*—BRAZIL: 1 ♀, 113.3; nr. Cabo Frio; 73 m; TERRA NOVA Sta. 42; holotype; BMNH 1917.3.1.1-7.—3 ♂, 39.2-76.9; 2 ♀, 37.6-46.9; data as in holotype; paratypes; same Reg. no.—1 ♂, 124.5; Cabo Frio; 21 October 1955; UMML 32.1745.—1 ♂, 92.6; off Cabo Frio; ALBATROSS Sta. 2762; USNM 18469.—1 damaged ♂, ca. 58.0; Rio de Janeiro; from stomach of *Genypterus brasiliensis;* BMNH 1906.6.9.86-87.—1 ♂, 63.7; SE of Rio de Janeiro; 23°06.5'S, 42°50'W; 63 m; CALYPSO Sta. 105; 2 December 1961; MNHNP.—1 ♂, 68.3; off Rio de Janeiro; 23°07.5'S, 43°11.5'W; 54 m; CALYPSO Sta. 108; 7 December 1961; USNM 111053. —1 ♂, 150.0; S of Rio de Janeiro; 23°27.5'S, 43°23'W; 100 m; CALYPSO Sta. 109; 8 December 1961; MNHNP.—1 brk. ♂, ca. 125.0; Ilha Rasa, Rio de Janeiro; P. S. Moreira; July 1959; IOSP.—3 ♂ (1 small, soft), 122.7-138.0; 4 ♀, ca. 53.0-101.2; off Ilha de São Sebastião; 23°43.5'S, 44°57'W; 46 m; CALYPSO Sta. 130; 10 December 1961; MNHNP.—1 ♂, 137.0; same; 24°06.5'S, 45°29'W; 48 m; CALYPSO Sta. 136; 11 December 1961; MNHNP.—1 ♂, 140.0, 1 ♀, 100.0; Santos; May 1959; USNM 119163.—1 damaged ♂, ca. 56.5; SE of Santos; 24°43'S, 45°10'W; 97-100 m; CALYPSO Sta. 138; 11 December 1961; MNHNP.—1 ♂, 114.0; 1 ♀, 115.0; E of São Francisco do Sul; 26°34'S, 47°22'W; 100 m; CALYPSO Sta. 145; 15 December 1961; MNHNP.—5 ♂, 46.7-127.3; 8 ♀, 54.0-110.0; E of Porto Alegre; 30°40'S, 49°35'W; 135-141 m; CALYPSO Sta. 150; 17 December 1961; USNM 111059.—5 ♂, ca. 45.0-69.0; 6 ♀, ca. 56.0-78.0; NE of Rio Grande; 31°24'S, 50°36'W; 66 m; CALYPSO Sta. 152; 17 December 1961; MNHNP.—URUGUAY: 2 ♂, 75.0-77.5; 34° 19'S, 52°57'W; 57 m; CALYPSO Sta. 157; 21 December 1961; USNM 111057.—15 ♂, 50.0-133.5; 12 ♀, 49.0-117.0; E of Maldonado; 35°05'S,

52°33'W; 115 m; CALYPSO Sta. 160; 21 December 1961; MNHNP.

*Diagnosis.*—Rostral plate elongate, with short median carina; median carina of carapace with anterior bifurcation; dactylus of raptorial claw with 6 teeth; mandibular palp present; 5 epipods present; lateral process of fifth thoracic somite a sharp anteriorly-curved spine; lateral process of next 2 somites acute posteriorly; telson denticles 4, 8-10, 1.

*Description.*—Eye large, cornea set very obliquely on stalk, corneal axis greater than peduncular; ocular scales acute laterally, almost spiniform, sinuous dorsally; anterior margin of ophthalmic somite produced forward, emarginate along midline. Corneal indices for 11 specimens follow:

| Sex | TL | CL | CI |
|---|---|---|---|
| ♂ | 63.7 | 13.5 | 355 |
| ♂ | 68.3 | 14.2 | 364 |
| ♀ | 78.2 | 16.3 | 354 |
| ♀ | 87.8 | 18.8 | 392 |
| ♂ | 92.6 | 19.6 | 369 |
| ♀ | 101.2 | 21.0 | 362 |
| ♂ | 114.0 | 23.8 | 425 |
| ♂ | 140.0 | 30.1 | 470 |
| ♀ | 149.0 | 30.2 | 458 |
| ♂ | 138.0 | 30.5 | 492 |
| ♂ | 150.0 | 32.4 | 463 |

Antennular peduncle subequal to or slightly longer than carapace; antennular processes produced into slender, forwardly-directed spines.

Rostral plate elongate, longer than broad; lateral margins faintly convex, tapering slightly to rounded apex; short median carina present.

Median carina of carapace, anterior to cervical groove, with anterior bifurcation only, distance from dorsal pit to bifurcation less than or equal to distance from bifurcation to anterior margin; intermediate carinae extending almost to anterior margin; anterolateral margins straight, not concave; anterolateral spines strong but not extending to base of rostral plate; posterolateral margins angled anteriorly.

Raptorial claw large, dactylus armed with 6 teeth, outer margin of dactylus evenly curved; dorsal ridge of carpus irregular, usually with 3 sharp tubercles.

Mandibular palp and 5 epipods present.

Last 3 thoracic somites with well-marked submedian and intermediate carinae, intermediates more prominent; lateral process of fifth thoracic somite a thick, sharp, forwardly curved spine; lateral process of sixth

216

somite bilobed, anterior lobe small, acute, posterior very large, acute, sharp apex directed posterolaterally; lateral process of next somite similar, anterior lobe inconspicuous; ventral keel on eighth thoracic somite cordiform.

Submedian abdominal carinae faintly divergent on each somite; second to fifth abdominal somites with anterior tubercle between intermediate and lateral carinae; lateral and marginal carinae swollen in adult male; abdominal carinae spined as follows: submedian, 4-6; intermediate, (1) 2-6; lateral, 1-6; marginal, 1-5; sixth somite with sharp spine in front of articulation of uropod.

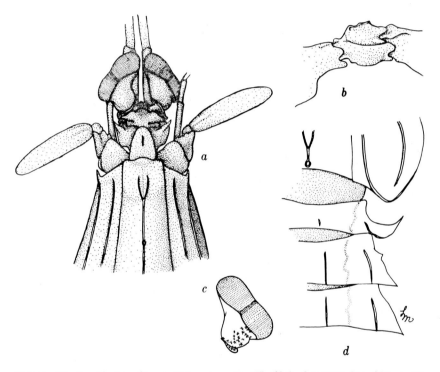

FIGURE 60. *Squilla brasiliensis* Calman, male, TL 68.3, CALYPSO Sta. 108. *a*, anterior portion of body; *b*, carpus of raptorial claw; *c*, eye; *d*, lateral processes of fifth to seventh thoracic somites (setae omitted; from Manning, 1966).

Telson slightly broader than long, with sharp marginal teeth, prelateral lobes present; entire posterior margin swollen in adult males; denticles 2-4, 8-10, 1, outer submedian denticle largest; ventral surface with prominent postanal keel.

217

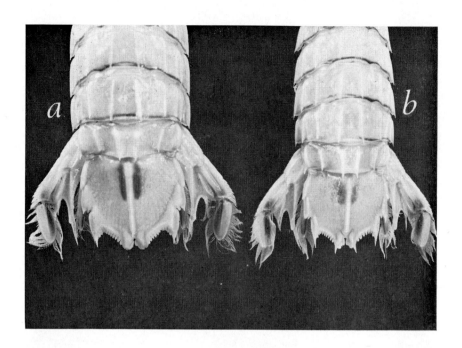

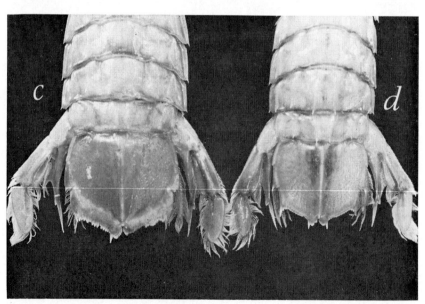

FIGURE 61. Telsons of: *S. brasiliensis* Calman, *a,* male, TL 127.3, CALYPSO Sta. 150; *b,* female, TL 110.0, same. *S. edentata edentata* (Lunz), *c,* male, TL 153.0, OREGON Sta. 2827; *d,* female, TL ca. 175.0, same.

Uropod with 6-7 movable spines on outer margin of penultimate segment of exopod, last not extending to midlength of distal segment.

*Color.*—Carinae and grooves of carapace outlined in dark pigment, last 3 thoracic and first 5 abdominal somites each with posterior black line; second abdominal somite with dark rectangular patch of black pigment; telson with anterior pair of rectangular black spots; uropods with endopod dark distally, exopod with dark spot at articulation of distal segment, inner half of distal segment black.

*Size.*—Males, TL 39.2-150.0 mm; females, TL 37.6-149.0 mm. Other measurements of male, TL 92.6: carapace length, 19.6; cornea width, 5.3; rostral plate length, 3.5, width, 3.0; telson length, 18.0, width, 18.5.

*Discussion.*—*S. brasiliensis* closely resembles the other western Atlantic species of *Squilla* with six teeth on the claw and acute posterolateral angles on the lateral processes of the sixth and seventh thoracic somites, *S. lijdingi* Holthius, *S. chydaea* Manning, *S. edentata* (Lunz), *S. caribaea* Manning, *S. intermedia* Bigelow, and *S. empusa* Say. It differs from *S. lijdingi* and *S. chydaea* in having a rostral carina, a distinct anterior bifurcation on the median carina of the carapace, and in having the entire posterior margin of the telson of adult males swollen dorsally. It differs from *S. edentata, S. caribaea,* and *S. intermedia* in that the rostral plate is elongate, the submedian carinae of the fourth abdominal somite are spined, and the lateral process of the fifth thoracic somite is not as sharply curved. *S. empusa* has a short rostral plate, the anterior bifurcation on the median carina of the carapace open for a much greater distance, and a sinuous outer margin on the dactylus of the raptorial claw. The straight, sloping anterolateral margins of the carapace in *S. brasiliensis* are not found in any other American species of *Squilla.*

The color pattern of *S. brasiliensis* also differs from that of the five species mentioned above. On the anterior portion of the telson there is a pair of rectangular black spots. In *S. chydaea, S. lijdingi,* and *S. edentata,* as well as in the similar *S. panamensis* Bigelow from the eastern Pacific, the anterior dark spots on the telson are crescent-shaped. The color pattern of *S. intermedia* is not known and *S. empusa* completely lacks any anterior dark markings on the telson. The color in *S. brasiliensis* is more similar to that found in the eastern Atlantic *S. mantis* (Linnaeus), which has square black spots on the telson and differs in morphological features as well; as in *S. empusa, S. mantis* has a sinuate dactylus on the raptorial claw.

*S. panamensis* Bigelow may be considered the eastern Pacific analogue of *S. brasiliensis.* These species are similar in general morphology, but differ in that the lateral processes of the sixth and seventh thoracic somites are posteriorly rounded in *S. panamensis* and that the anterior black crescents on the telson found in *panamensis* are replaced by black rectangles in *brasiliensis*. Both species have a much stouter propodus on the raptorial claw than related species such as *S. lijdingi* and *S. chydaea,* and both have sexual dimorphism well-marked in adults.

Until Holthuis (1959) pointed out the distinctness of his *S. lijdingi* from *S. brasiliensis,* several workers, including myself, had confused *S. brasiliensis, S. chydaea,* and *S. lijdingi.* R. W. Ingle of the British Museum (Natural History) also helped clear up the confusion by supplying me with sketches and notes on the types of *S. brasiliensis.*

The specimen described and illustrated by Andrade Ramos (1951, p. 145, pl. 1, fig. 1) is apparently an undescribed species; the telson resembles that of *Cloridopsis dubia* in several respects. M. Vannucci of the Instituto Oceanografico at São Paulo informed me that the specimen could not be located.

*Remarks.*—Sexual dimorphism in this species is reflected in the swollen posterior margin of the telson in large males; the marginal swelling is interrupted at each marginal tooth. The carinae of the sixth abdominal somite and the lateral and marginal carinae of the abdomen are also inflated in males but not females. The dimorphic features are first noticeable in males around TL 120 mm.

*Ontogeny.*—Unknown.

*Type.*—British Museum (Natural History).

*Type-locality.*—Cabo Frio, Brazil.

*Distribution.*—This is a southern species, known from numerous localities between Cabo Frio, Brazil and Uruguay, in depths between 46 and 141 m.

*Squilla edentata edentata* (Lunz, 1937)
Figures 61c-d, 62

*Squilla intermedia* Bigelow, 1893, p. 102 [part]; 1894, p. 530 [part].—Chace, 1954, p. 449.—Springer & Bullis, 1956, p. 22.—Manning, 1959, p. 19 [part] [reference to Gulf of Mexico specimens only].
*Chloridella edentata* Lunz, 1937, p. 14, text-figs. 7-10.—Manning, 1968, p. 129 [listed].
*Squilla edentata:* Chace, 1954, p. 449.—Manning, 1959, p. 19; 1961, p. 22 [part].—Dawson, 1963, p. 7 [listed]; 1965, p. 14.—Bullis & Thompson, 1965, p. 13 [part; listed].

*Previous records.*—NORTHERN GULF OF MEXICO: Springer & Bullis, 1956; Chace, 1954; Manning, 1959.—FLORIDA: SW of Pensacola (Lunz, 1937). — ALABAMA: Dawson, 1963, 1965. — MISSISSIPPI DELTA: Bigelow, 1893, 1894. The specimens from northern South America reported by Manning (1961) are herein referred to *S. edentata australis,* new subspecies, and *S. caribaea,* new species, both of which are described below. Bullis & Thompson (1965) recorded the occurrence of

220

this species at 27 OREGON and six SILVER BAY Stations; these records are referable to *S. edentata, S. e. australis,* n. ssp., and *S. caribaea,* n. sp.

*Material.*—SOUTH CAROLINA: 1 ♀, 85.6; PELICAN Sta. 182-8; USNM 110839.—FLORIDA: 2 ♀, 67.0-121.0; off Cape Canaveral; SILVER BAY Sta. 2039; USNM 119193.—1 ♀, 119.0; same; SILVER BAY Sta. 2725; USNM 119192.—1 ♀, 118.5; SILVER BAY Sta. 2731; USNM 119194.— 4 ♂, 99.6-116.8; 1 ♀, 137.5; same; SILVER BAY Sta. 2732; USNM 119254. —1 ♂, 31.2; same; PELICAN Sta. 204-4; USNM 110880.—1 ♂, 68.8; 1 ♀, 111.1; same; PELICAN Sta. 205-5; USNM 110840.—1 ♂, 82.0; 1 ♀, 95.3; off Ft. Pierce; SILVER BAY Sta. 1968; USNM 119188.— 1 ♂, 30.0; off Cape St. George; ALBATROSS Sta. 2402; USNM 9768.—2 ♂, 98.0-105.5; 2 ♀, 111.8-148.0; same; OREGON Sta. 273; USNM 92395.— 1 ♂, 136.1; off Panama City; OREGON Sta. 945; USNM 96402.—1 ♂, 105.5; same; OREGON Sta. 1382; USNM 99503.—1 ♂, 44.0; 65 mi WSW of Pensacola; 29°16′N, 87°54′W; 218-237 m; ATLANTIS Sta. 2377; 24 March 1935; holotype; YPM 4405.—2 ♂, 91.5-94.2; 3 ♀, 130.6-142.3; same; OREGON Sta. 27; USNM 91095.—2 ♀, 124.8-129.0; same; OREGON Sta. 326; USNM 92396.—1 ♀, 69.5; same; OREGON Sta. 4943; USNM 119186.—1 ♀, 113.0; S of Santa Rosa Id.; OREGON Sta. 332; USNM 92397.—MISSISSIPPI DELTA AREA: 1 ♂, 108.3; ALBATROSS Sta. 2378; paralectotype of *S. intermedia* Bigelow; USNM 9658.—1 ♂, 162.4; ORE-GON Sta. 60; USNM 91132.—1 ♀, 156.0; OREGON Sta. 103; USNM 91442. —3 ♂, 129.5-156.0; 2 ♀, 131.3-136.5; OREGON Sta. 2203; USNM 119255. —1 ♂, 129.0; 1 ♀,148.0; same; RMNH 382.—1 ♂, 85.1; OREGON Sta. 2799; USNM 119187.—3 ♂, 145.6-161.2; 5 ♀, 132.6-ca. 175.0; OREGON Sta. 2827; USNM 120346.—2 ♂, 128.3-144.8; 3 ♀, 119.2-145.0; OREGON Sta. 3201; USNM 119190.—1 ♂, 74.6; OREGON Sta. 3649; USNM 119191. —1 brk. ♀, ca. 49.0; OREGON Sta. 4002; USNM 119185.—2 ♀, 142.8-145.2; OREGON Sta. 4583; USNM 119189.—1 dry ♀; PELICAN Sta. 9; USNM 110883.—1 ♂, 56.8; PELICAN Sta. 96-1; USNM 110871.— LOUISIANA: 1 ♂, 104.2; 2 ♀, 115.7-142.0; PELICAN Sta. 4; USNM 110855.—2 ♂, 113.7-114.5; 2 ♀, 83.7-111.0; same; IM.—TEXAS: 1 ♂, 130.0; OREGON Sta. 158; USNM 92394.

*Diagnosis.*—Ocular scales rounded laterally; median carina of carapace composed of 2 subparallel lines of pits, distance from dorsal pit to anterior bifurcation equal to or less than distance from bifurcation to anterior margin; dactylus of claw with 6 teeth; lateral process of fifth thoracic somite a sickle-shaped spine; abdomen usually broader than carapace; telson inflated, broader than long, median carina convex in lateral view; denticles 2-5, 11-15, 1, outer submedian denticle the largest.

*Description.*—Eye large, cornea set very obliquely on stalk; ocular scales rounded laterally, occasionally serrate, not spiniform; anterior margin of opthalmic somite usually medially emarginate, projecting forward. Corneal indices for specimens of various sizes follow:

| CL in mm | n | CI mean | CI range |
|---|---|---|---|
| 7-8 | 1 | 338 | . . . . . . |
| 9-10 | 1 | 343 | . . . . . . |
| 11-12 | 1 | 390 | . . . . . . |
| 13-14 | 1 | 372 | . . . . . . |
| 15-16 | 2 | 372 | 370-373 |
| 17-18 | 3 | 398 | 385-414 |
| 19-20 | 3 | 385 | 376-402 |
| 21-22 | 2 | 407 | 393-420 |
| 23-24 | 3 | 421 | 416-428 |
| 25-26 | 10 | 428 | 402-460 |
| 27-28 | 2 | 439 | 437-441 |
| 29-30 | 7 | 433 | 416-448 |
| 31-32 | 3 | 449 | 430-465 |
| 33-34 | 6 | 463 | 445-487 |
| 35-36 | 3 | 474 | 452-489 |
| 37-38 | 2 | 481 | 470-492 |

Antennular processes produced into sharp spines directed obliquely forward; antennular peduncles slightly shorter than carapace.

Rostral plate broader than long, appearing elongate, subtriangular; upturned lateral margins slope to truncate or slightly rounded apex; sharp median carina present on anterior half.

Carapace with prominent, rough carinae; median carina composed of 2 subparallel rows of pits, appearing bicarinate under high magnification; median carina, anterior to cervical groove, bifurcate at either end, posterior bifurcation faint; distance from dorsal pit to anterior bifurcation less than distance from bifurcation to anterior margin; intermediate carinae usually not extending to anterior margin; anterolateral spines strong; projecting more dorsally than anteriorly, but not extending to base of rostral plate; posterolateral margin with anterolateral angle.

Raptorial claw large; dactylus with 6 teeth, outer margin of dactylus evenly curved; dorsal ridge of carpus with 2 sharp tubercles.

Mandibular palp present; 5 epipods present.

Last 3 thoracic somites with pitted, rugose submedian and intermediate carinae; fifth somite with short irregular carinae, those on last 3 somites well-developed; lateral process of fifth somite an acute, sickle-shaped spine, base broad, apex directed anteriorly; lateral processes of next 2 somites bilobed, anterior lobes small, posterior lobes much larger, posterior apices acute; anterior lobe of process of sixth somite truncate, more prominent than that of seventh somite, which is usually rounded; ventral keel of eighth somite a rounded lobe, inclined posteriorly.

Greatest width of abdomen greater than carapace length; abdominal carinae, especially intermediates and laterals, pitted, rugose; submedian carinae subparallel on anterior somites, slightly divergent on posterior somites; intermediate and marginal carinae of last 3 somites as well as all carinae of sixth somite swollen in male; abdominal carinae spined as follows: submedian, 5-6; intermediate, (1) 2-6; lateral, 1-6; marginal, 1-5; under TL 100 mm, abdominal carinae spined as follows: submedian, 5-6;

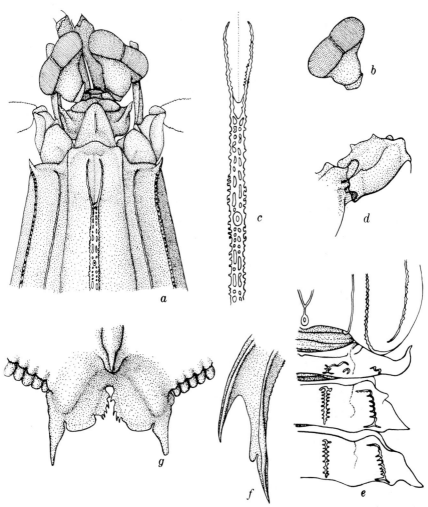

FIGURE 62. *Squilla edentata edentata* (Lunz), female, TL 95.3, SILVER BAY Sta. 1968. *a,* anterior portion of body; *b,* eye; *c,* anterior half of median carina of carapace, enlarged; *d,* carpus of claw; *e,* lateral processes of fifth to seventh thoracic somites; *f,* basal prolongation of uropod; *g,* submedian denticles of telson, enlarged (setae omitted).

223

intermediate, (2) 3-6; lateral, 1-6; marginal, 1-5; sixth somite with spine anterior to articulation of each uropod.

Telson very tumid, broader than long, median carina convex in lateral view at all sizes; marginal teeth short; prelateral lobes present, projecting laterally; median carina and margin of telson thickly swollen in adult male, TL 110 mm and over, swellings interrupted at intermediate denticle and prelateral lobe; denticles (1) 2-5 (6), 11-15, 1, outer submedian denticle large, rounded, remainder of denticles small, sharp; ventral surface of telson with prominent but short postanal keel.

Outer margin of penultimate segment of uropodal exopod with 6-8 graded, movable spines, last not extending to midlength of distal segment; large, rounded lobe present on outer margin of inner spine of basal prolongation.

*Color.*—Usually completely faded in preserved specimens, occasionally some specimens with posterior margin of carapace and body segments with dark posterior line. The following notes were made on freshly preserved material: Overall background color light orange, parts of body flushed with pink; carinae, gastric grooves, and posterior margin of carapace lined with dark posterior line. The following notes were made on freshly preserved grooves; last 3 thoracic and first 5 abdominal somites lined with dark pigment, submedian and intermediate carinae darker than body but lighter than margin; each somite with broad band of orange-pink; merus and carpus of raptorial claw outlined in orange-pink, propodus with dark line along pectinate margin; telson with proximal pair of orange crescents, open laterally, marginal denticles and carinae of female pink, male straw; endopod and inner spine of basal prolongation of uropod lined with brown, inner and distal portions of proximal segment and proximal half of distal segment of exopod brown.

*Size.*—Males, TL 30.0-162.4 mm; females, TL 49.0-175.0 mm. Other measurements of female, TL 85.6; carapace length, 17.8; cornea width, 4.5; rostral plate length, 2.1 width, 2.6; telson length, 15.0; width, 16.4.

*Discussion.*—The southern form of *S. edentata,* reported from off the coast of northern South America by Manning (1961), is described below as a new subspecies. The southern subspecies differs from the nominal subspecies in having a larger eye (compare Corneal Indices), and a more elongated rostral plate with a truncated apex. The males of the southern subspecies mature at a larger size than do those of *S. e. edentata.* See also remarks below under the discussion of the southern subspecies.

*S. edentata* can be distinguished from *S. intermedia* and *S. caribaea,* n. sp., by three features: (1) in *S. edentata* the median carina of the carapace is composed of 2 subparallel rows of pits; in the other two species the median carina is a simple, raised ridge, roughened but not appearing bicarinate; (2) in *S. edentata* the telson is broad, noticeably tumid, with the median

224

carina appearing convex in lateral view at all sizes; in the other two species the median carina is straight or sinuous in lateral view, but not convex; and (3) the lobe on the outer margin of the inner spine of the basal prolongation of the uropod is larger in *S. edentata* than in either of the other two species. *S. edentata* shares rounded ocular scales with *S. caribaea;* in *S. intermedia* the ocular scales are spined laterally. *S. edentata* further differs from *S. intermedia* in having the outer submedian denticle of the telson large, rounded, in having smaller eyes (compare Corneal Indices), and in the relative lengths of the portions of the median carina of the carapace anterior and posterior to the anterior bifurcation. In *S. edentata* the distance from the dorsal pit to the bifurcation is less than the distance from the bifurcation to the anterior margin; in *S. intermedia* the distance from the pit to the bifurcation is equal to or greater than the distance from the bifurcaiton to the anterior margin. Other differences between *S. edentata, S. caribaea,* and *S. intermedia* are discussed under the accounts of the latter species.

Bigelow (1894) illustrated the Gulf of Mexico syntype of *S. intermedia,* which was reidentified as *S. edentata* by Manning (1961), but did not show that the telson was damaged. The submedian denticles on the right side are broken; there are seven on the left side, of which the outer is much the largest. This is not clearly shown in his illustration.

*Remarks.—S. edentata* is the only one of this group of deepwater species including *S. intermedia* and *S. caribaea* in which any color pattern has been noted. The color pattern of *S. edentata* superficially resembles that of *S. lijdingi* and related species.

*S. edentata* is a shallower species than *S. intermedia;* although it has been taken in depths to 319 m, it generally occurs in depths of less than 200 m.

Of all the specimens examined, only six had the lateral carinae of the first abdominal somite provided with spines. In the southern subspecies these carinae are armed in all specimens.

*Ontogeny.*—Unknown.

*Type.*—The holotype, a young male, is in the Yale Peabody Museum.

*Type-locality.*—Northern Gulf of Mexico, 65 mi. off Pensacola, Florida, in 218-237 m; ATLANTIS Sta. 2377.

*Distribution.*—Off South Carolina, east Florida as far south as Fort Pierce, and northern Gulf of Mexico from Cape St. George to Texas, in depths between 55-319 m, generally 200 m or less.

**Squilla edentata australis,** new subspecies
Figure 63

*Squilla edentata:* Manning, 1961, p. 22, pl. 5 [part = *S. caribaea,* n. sp.].

*Previous records.*—BRITISH GUIANA: OREGON Sta. 1983.—SURIN-

225

AM: OREGON Stats. 2286, 2288.—FRENCH GUIANA: OREGON Stats. 2021, 2295 (all Manning, 1961). Material from other stations reported therein is here referred to *S. caribaea* which see; *S. e. australis* and *S. caribaea* were taken together at stats. 1983 and 2286.

*Holotype.*—1 ♂, 120.8; Venezuela; 10°54′N, 68°01′W; 182 m; OREGON Sta. 4446; 10 October 1963; USNM 119256.

*Paratypes.*—4 ♂, 88.9-137.7; 4 ♀, 138.5-147.5; Surinam; 07°26′N, 54°40′W; 173 m; OREGON Sta. 2288; 8 September 1958; USNM 119257.

*Other material.*—HONDURAS: 1 ♂, 129.2; 1 ♀, 98.4; OREGON Sta. 3626; USNM 120347.—PANAMA: 1 ♂, 125.4; 1 ♀, 151.0; OREGON Sta. 3588; USNM 119258.—1 ♀, 170.0; OREGON Sta. 3595; USNM 119259.—BRITISH GUIANA: 1 ♀, 177.7; OREGON Sta. 1983; UMML.—SURINAM: 1 ♂, 135.9; OREGON Sta. 2286; USNM 119260.—FRENCH GUIANA: 3 ♂, 87.4-136.2; OREGON Sta. 2295; UMML 32.1888.—1 ♂, 109.7; same; BMNH.—1 ♂, 134.6; OREGON Sta. 2021; USNM 119261.

*Diagnosis.*—As in *S. edentata edentata,* but eye larger, rostral plate more elongate, with truncate apex, and males maturing at TL 135 mm rather than 110 mm. Corneal indices for specimens of various sizes follow:

| CL in mm | n | CI mean | CI range |
|---|---|---|---|
| 15-16 | 1 | 370 | . . . . . . |
| 17-18 | - | . . . | . . . . . . . |
| 19-20 | 2 | 359 | 356-361 |
| 21-22 | - | . . . | . . . . . . . |
| 23-24 | 1 | 377 | . . . . . . |
| 25-26 | 3 | 375 | 360-392 |
| 27-28 | 1 | 416 | . . . . . . |
| 29-30 | 6 | 412 | 396-430 |
| 31-32 | 4 | 419 | 414-427 |
| 33-34 | 2 | 442 | 437-447 |
| 35-36 | 1 | 448 | . . . . . . |

Abdominal carinae spined as follows: submedian, 5-6; intermediate, (2) 3-6; lateral, 1-6; marginal, 1-5; telson denticles (3) 4-6, (12) 13-15 (18), 1.

*Color.*—Faded in most specimens; some specimens with trace of dark spot at base of telson and at the articulation of the segments of the uropodal exopod; uropodal endopod dark.

*Size.*—Males, TL 87.4-137.7 mm; females, TL 98.4-170.0 mm. Other measurements of male, TL 129.2 mm: carapace length, 30.2; cornea width, 7.3; rostral plate length, 4.1, width, 4.3; telson length, 26.2, width, 28.9.

*Discussion.*—*S. edentata australis* is the southern form of the species. It differs from the nominate subspecies, known only from the Carolinas, NE Florida, and the northern Gulf of Mexico, in having noticeably smaller eyes (compare Corneal Indices), a more elongate rostral plate, with a truncated apex, and in having the males exhibit sexual maturity at a different size. Males of the new subspecies mature at a TL of 135 mm or

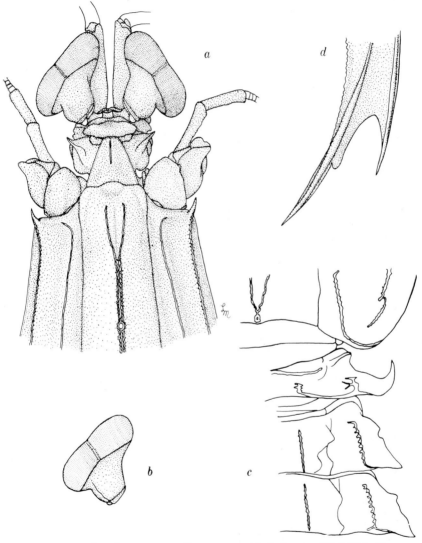

FIGURE 63. *Squilla edentata australis*, n. ssp., male holotype, TL 120.8, OREGON Sta. 4446. *a*, anterior portion of body; *b*, eye; *c*, lateral processes of fifth to seventh thoracic somites; *d*, basal prolongation of uropod (setae omitted).

over, whereas males of the nominate subspecies exhibit sexual dimorphism at TL 110 mm. The southern subspecies generally has fewer of the intermediate abdominal carinae armed posteriorly; the carinae of the second somite are rarely spined in *S. e. australis,* always armed in *S. e. edentata.* In all other respects, *S. e. australis* is similar to the nominate subspecies. The broad, tumid telson and broad abdomen will immediately distinguish specimens of *S. e. australis* from those of *S. caribaea* with which it may be associated; both species have been taken together at two OREGON stations, as noted under the discussion of *S. caribaea.*

*Remarks.*—The swollen lateral and marginal abdominal carinae, swollen carinae of the sixth abdominal somite, and swollen margin of the telson are as well marked in this subspecies as in *S. edentata* proper.

The figures given by Manning (1961) for *S. edentata* are actually of two species; the male telson is that of *S. edentata australis,* whereas the female telson is that of *S. caribaea.* The telson of female *S. e. australis* is much broader than that of *S. caribaea.*

*Etymology.*—The subspecific name is derived from the Latin, "australis," meaning southern.

*Type.*—U.S. National Museum.

*Type-locality.*—Off Venezuela, 10°54′N, 68°01′W, OREGON Sta. 4446, in 182 m.

*Distribution.*—Southern Caribbean Sea and adjacent coast of northern South America, including off Honduras, Panama, Venezuela, British Guiana, Surinam, and French Guiana, in depths between 173-273 m.

*Squilla intermedia* Bigelow, 1893
Figures 64, 65a-b

*Squilla intermedia* Bigelow, 1893, p. 102 [part; Bahama specimen only]; 1894, p. 530, text-fig. 19 [part].—Faxon, 1895, p. 237.—Kemp, 1913, p. 201 [listed].—Chace, 1954, p. 449 [part].—Manning, 1959, p. 19 [part]; 1961, p. 25 [discussion; lectotype selection].—Bullis & Thompson, 1965, p. 13 [listed].—Manning, 1968, p. 129 [listed].
not *Squilla intermedia:* Bigelow, 1901, p. 159 [ = *S. obtusa* Holthuis].—Springer & Bullis, 1956, p. 22 [ = *S. edentata* (Lunz)].—Ingle, 1960, p. 573, text-figs. 11-12 [Nigeria; undescribed].
not *Squilla affinis* var. *intermedia* Nobili, 1903, p. 39 [Indo-West Pacific; identity uncertain].

*Previous records.*—BAHAMAS: Bigelow, 1893, 1894; Chace, 1954; Manning, 1959.—FLORIDA: Manning, 1959.—The remainder of the records in the literature are referrable to other species as noted in the synonymy. Bullis & Thompson (1965) recorded specimens at 14 U.S. Fish and Wildlife stations.

*Material.*—BAHAMAS: 1 ♀, 105.2; N of Little Bahama Bank; ALBATROSS

228

Sta. 2655; lectotype; USNM 11543.—1 ♀, 121.7; Santaren Channel; SIL-
VER BAY Sta. 2468; USNM 119207.—1 ♀, 153.0; same; SILVER BAY Sta.
2458; USNM 119205.—FLORIDA: 1 ♀, 48.3; E of Fort Pierce; 27°25′N,
78°37.5-41.0′W; 291-309 m; GERDA Sta. 251; 5 February 1964; USNM
119204.—1 ♂, 111.0; 2 ♀, 100.0-101.5; same; 27°34-37′N, 78°56-58′W;
464-491 m; GERDA Sta. 256; 6 February 1964; USNM 120350.—4 ♂,
108.9-127.8; same; COMBAT Sta. 235; USNM 101619.—1 ♂, 83.4; same;
COMBAT Sta. 237; USNM 101727.—1 ♂, 115.2; 1 ♀, 104.8; same; COM-
BAT Sta. 238; UMML 32.1177.—HONDURAS: 1 ♀, 140.4; OREGON Sta.
3627; USNM 119214.—CUBA: 1 fragment; 23°01′N, 83°14′W; 346 m;
BLAKE Sta. 23; 1877-1878; MCZ.—1 ♀, 53.6; 22°46′N, 78°45′W; 355-
410 m; ATLANTIS Sta. 2982-C; 11 March 1938; MCZ.—1 ♀, 83.7 mm;
23°11′N, 79°08′W; 428-473 m; ATLANTIS Sta. 2983; 12 March 1938;
MCZ.—4 ♀, 103.5-121.0; OREGON Sta. 1341; USNM 119209.—1 ♂,
130.3; OREGON Sta. 1343; USNM 98676.—4 ♂, 60.5-135.2; OREGON Sta.
1344; USNM 101618.—DOMINICAN REPUBLIC: 1 brk. ♀; SILVER
BAY Sta. 5166; USNM 119206.—PUERTO RICO: 1 ♂, 123.5; 18°35′30″
N, 65°23′54″W; 546 m; Johnson-Smithsonian Deep-Sea Exped. Sta. 81;
26 February 1933; USNM 119208.—1 ♀, 116.0; OREGON Sta. 2652;
USNM 119212.—1 ♀, 111.5; OREGON Sta. 2658; USNM 119211.—1 ♂,
124.0; same; RMNH 383.—1 ♂, 123.0; OREGON Sta. 2664; UMML.—
1 ♂, 123.3; OREGON Sta. 2639; USNM 119213.—VIRGIN ISLANDS:

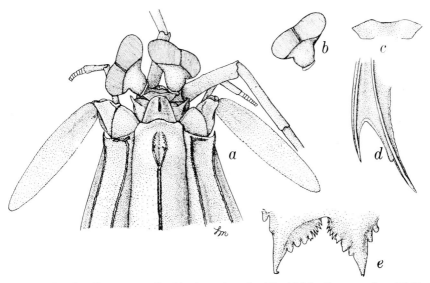

FIGURE 64. *Squilla intermedia* Bigelow, female, TL 116.0, OREGON Sta. 2652.
*a*, anterior portion of body; *b*, eye; *c*, ocular scales, enlarged; *d*, basal prolongation of uropod; *e*, submedian denticles of telson, enlarged (setae omitted).

229

1 ♂, 123.6; 1 ♀, ca. 122.0; OREGON Sta. 2606; USNM 119215.—BAR-
BADOS: 1 ♂, 97.8; OREGON Sta. 5018; USNM 119210.

*Diagnosis.*—Ocular scales with spiniform apices; median carina of carapace
simple, distance from dorsal pit to bifurcation usually greater than distance
from bifurcation to anterior margin; dactylus of claw with 6 teeth; lateral
process of fifth thoracic somite a sickle-shaped spine; abdomen broader
than carapace; telson broader than long, median carina sinuous in lateral
view; denticles usually 6-8, 13-15, 1, submedian denticles subequal.

*Description.*—Eye large, cornea set very obliquely on stalk; ocular scales
spiniform laterally; anterior margin of ophthalmic somite projecting, medially emarginate. Corneal indices of specimens of various sizes follow:

| CL in mm | n | CI mean | CI range |
|---|---|---|---|
| 11-12 | 1 | 316 | ... ... |
| 13-14 | 1 | 307 | ... ... |
| 15-16 | - | ... | ... ... |
| 17-18 | - | ... | ... ... |
| 19-20 | 1 | 337 | ... ... |
| 21-22 | 2 | 357 | 354-360 |
| 23-24 | 5 | 340 | 332-357 |
| 25-26 | 4 | 364 | 345-378 |
| 27-28 | 10 | 364 | 351-381 |
| 29-30 | 1 | 378 | ... ... |
| 31-32 | 2 | 383 | 372-394 |
| 33-34 | 1 | 400 | ... ... |

Antennular processes produced into sharp spines directed obliquely
forward; antennular peduncle longer than carapace and rostral plate
combined.

Rostral plate subtriangular, broader than long, with sharp median carina;
lateral margins faintly sinuous, sloping to truncate apex.

Carinae of carapace very sharp, occasionally with lateral pits; median
carina simple, not composed of 2 subparallel lines of pits; on median carina,
distance from dorsal pit to anterior bifurcation greater than or equal to
(usually greater than) distance from bifurcation to anterior margin; intermediate carinae extending to anterior margin; anterolateral spines strong
but not extending to base of rostral plate; posterolateral margin of carapace
strongly angled anteriorly.

Raptorial claw large, extending, when folded, from level of eyes to
median posterior margin of carapace; dactylus with 6 teeth, outer margin of
dactylus evenly convex; dorsal ridge of carpus with 2 sharp tubercles.

Mandibular palp and 5 epipods present.

Last 3 thoracic somites with strong submedian and intermediate carinae;
lateral process of fifth somite a broad, sickle-shaped spine, apex directed

230

anteriorly; lateral process of sixth somite bilobed, anterior lobe a small, sharp tubercle, posterior lobe much larger, apex acute, almost spiniform; lateral process of seventh somite at most faintly bilobed, anterior lobe usually absent, occasionally present as an obtuse prominence grading into larger posterior lobe, the latter almost spiniform posterolaterally; ventral keel on eighth somite a low, rounded lobe, inclined posteriorly.

Submedian abdominal carinae slightly divergent, divergence well-marked on fourth and fifth somites; submedians, although distinct, lower, less prominent than remainder of carinae; some pitting present on abdominal carinae, pits not as prominent as in *S. edentata;* abdominal carinae spined as follows; submedian, 5-6; intermediate, 1-6; lateral, 1-6; marginal, 1-6; sixth abdominal somite with prominent spine in front of articulation of each uropod; abdomen very broad, in adult male greatest width equal to length of carapace and rostral plate combined; marginal carinae inflated in large males.

Telson broader than long, appearing very broad, flattened, not inflated, median carina sinuous in lateral view, convex anteriorly, concave posteriorly; prelateral lobes present, well-marked dorsally, scarcely extending laterally beyond margin of lateral teeth; submedian and intermediate margin as well as carinae of marginal teeth noticeably swollen in adult males, swollen area interrupted at bases of marginal teeth; denticles small, rounded or triangular, 5-10, 10-18, 1, usually 6-8, 13-15, 1; outer submedian denticle similar to others in size; margin of intermediate area convex; ventral surface with prominent postanal keel.

Outer margin of proximal segment of uropodal exopod with 6-8 graded, movable spines, last short, not extending to midlength of distal segment; lobe on outer margin of inner spine of basal prolongation low, inconspicuous.

*Color.*—Faded in most specimens. In a male from off Barbados, OREGON Sta. 5018, the abdomen was marked with orange submedian carinae and with faint dark posterior lines on each somite; the radiating pits on the surface of the telson were dark. No specimens have shown traces of the dark pigment patterns on the abdomen, telson and uropods that are characteristic of many other species in the genus.

*Size.*—Males, TL 60.5-135.2 mm; females, TL 48.3-153.0 mm. Other measurements of female lectotype, TL 105.2: carapace length, 23.5; cornea width, 7.0; rostral plate length, 2.9, width 3.9; telson length, 22.3, width, 24.1.                                                           ,

*Discussion.*—*S. intermedia* most closely resembles *S. edentata* (Lunz); both species are of large size, with males showing marked sexual dimorphism, and both have a strongly-hooked lateral process on the fifth thoracic somite. It differs from *S. edentata* in the following features: (1) the eyes are larger (compare CI summaries); (2) the ocular scales are spiniform

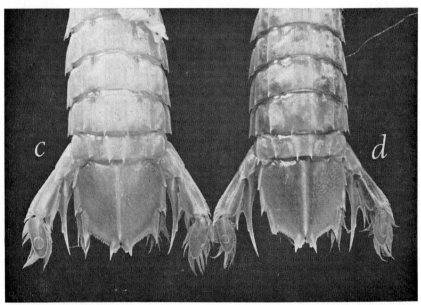

FIGURE 65. Telsons of: *S. intermedia* Bigelow, *a*, male, TL 111.0, GERDA Sta. 256; *b*, female, TL 101.5, same. *S. caribaea*, n. sp., *c*, male paratype, TL 138.0, OREGON Sta. 4860; *d*, female paratype, TL 130.0, same.

laterally rather than rounded; (3) the median carina of the carapace is a simple ridge, occasionally pitted, but not composed of two lines of pits; (4) on the median carina the distance from the dorsal pit to the anterior bifurcation is equal to or greater than (usually greater than) the distance from the bifurcation to the anterior margin; (5) the lateral process of the sixth thoracic somite has a smaller, sharper anterior lobe; (6) the lateral process of the seventh thoracic somite is rarely bilobed; (7) the submedian denticles of the telson are subequal in size; (8) the telson is flattened, not swollen; (9) the median carina of the telson is sinuous or straight, not convex in lateral view; and (10) the lobe on the outer margin of the inner spine of the basal prolongation of the uropod is much smaller.

*S. intermedia* also resembles *S. caribaea,* particularly in large size and general appearance and in having a simple median carina on the carapace, but *S. caribaea* differs in that: (1) the ocular scales are rounded; (2) the distance from the pit to the bifurcation on the median carina of the carapace is less than the distance from the bifurcation to the anterior margin; (3) the lateral process of the seventh thoracic somite is definitely bilobed; (4) the lateral process of the fifth thoracic somite is not as strongly hooked; (5) the prelateral lobe of the telson is more prominent, projecting beyond the outline of the telson margin; and (6) the intermediate margin of the telson is concave Also, *S. intermedia* is a broader species than *S. caribaea.* In adult males of the former (as in *S. edentata*) the width of the fifth abdominal somite is equal to the combined lengths of the carapace and rostral plate, whereas in *S. caribaea* it is shorter than the carapace. Males of *S. intermedia* mature at a much smaller size (TL 110.0) than do those of *S. caribaea* (TL over 130.0).

*S. intermedia:* Ingle from West Africa differs in having smaller eyes, a much larger ventral lobe on the eighth thoracic somite, no spines on the intermediate carinae of the first abdominal somite, fewer denticles on the telson and a more persistent color pattern. In my opinion Ingle's form represents an undescribed species.

*S. brasiliensis* resembles *S. intermedia* as well as *S. edentata* and *S. caribaea* in some respects, but the submedian carinae of the fourth abdominal somite are armed in that species, the intermediate area of the telson is much narrower, and the anterior margin of each lateral plate of the carapace is straight, not concave. Other differences are noted under the discussion of *S. brasiliensis.*

*S. intermedia* is the Atlantic analogue of the eastern Pacific *S. biformis* Bigelow; that species can be distinguished immediately by the presence of a spined postanal keel. Also, in adult males of *S. biformis* the marginal swellings on the telson are not interrupted at the bases of the intermediate marginal teeth.

*Remarks.*—In *S. intermedia* sexual dimorphism in the male is well-marked at a TL of 110 mm; maturity of males of this species is accompanied by a

233

marked swelling of the margin of the telson, with the tumid area interrupted at the bases of the intermediate marginal teeth, accompanied by an increase in the tumidity of the median carina of the telson and the carinae of the sixth abdominal somite as well as the lateral and marginal carinae of the last three abdominal somites. In dimorphism of the telson, *S. intermedia* resembles *S. edentata* rather than *S. caribaea;* in the latter species dimorphism is very poorly marked, even in the largest males examined. The telsons of a mature male and female of *S. intermedia* are shown in Figure 65a-b.

*S. intermedia* seems to be the northern counterpart of *S. caribaea;* both species were taken together off Honduras at OREGON Sta. 3627, in 364 m. In general, *S. caribaea* has been taken in somewhat shallower water than *S. intermedia. S. heptacantha* (Chace) is the only other western Atlantic species of *Squilla* that occurs in waters as deep as does *S. intermedia.*

According to the present Code, Nobili's *S. affinis var. intermedia* requires a new name. In view of the uncertain status of that species, no replacement name is proposed herein.

*Ontogeny.*—Unknown.

*Type.*—The female lectotype, selected by Manning (1961, p. 25), is in the U.S. National Museum. The paralectotype, a male from the Gulf of Mexico, is actually *S. edentata* (Lunz).

*Type-locality.*—N of Little Bahama Bank, 27°22′N, 78°07′30″W, 615 m, ALBATROSS Sta. 2655, 2 May 1886.

*Distribution.*—Western Atlantic, from the Bahamas, southeast Florida, Honduras, Cuba, Dominican Republic, Puerto Rico, and Barbados, in depths between 291 and 615 m, rarely under 300 m. It has not been recorded previously outside of the Bahamas and Florida.

The West African records should be referred to a new species.

## Squilla caribaea, new species
Figures 65c-d, 66

*Squilla edentata:* Manning, 1961, p. 22, pl. 5 [part; not *S. edentata* (Lunz)].

*Previous records.*—T R I N I D A D: OREGON Sta. 2351.—B R I T I S H GUIANA: OREGON Stats. 1983, 1985.—SURINAM: OREGON Stats. 2285, 2286 (all Manning, 1961). Material from other stations reported therein is here referred to *S. edentata australis,* new subspecies, which see; *S. caribaea* and *S. e. australis* occurred together at OREGON Stats. 1983, 2286.

*Holotype.*—1 ♂, 120.0; Colombia; 09°30′45″N, 76°25.5′W; 282 m; ALBATROSS Sta. 2143; 23 March 1884; USNM 6936.

*Paratypes.*—7 ♂, 78.2-138.8; 4 ♀, 94.5-130.0; Colombia; 11°09′N, 74°26′W; 282-291 m; OREGON Sta. 4860; 19 May 1964; USNM 120345.—1 ♂,

234

97.2; Venezuela; 11°53'N, 69°28'W; 391 m; Oregon Sta. 4405; 27 September 1963; USNM 113745.—1 ♀, 105.0; Trinidad; 11°30'N, 60°46'W; 364-437 m; Oregon Sta. 5028; 22 September 1964; USNM 113752.

*Other material.*—HONDURAS: 1 ♂, 127.8; Oregon Sta. 1870; USNM 113761.—3 ♀, 111.5-129.9; Oregon Sta. 1883; USNM 113753.—1 ♀, 108.8; Oregon Sta. 3627; USNM 119164.—NICARAGUA: 2 ♂, 101.8-129.5; 3 ♀, 125.4-135.5; Oregon Sta. 3574; USNM 113758.—PANAMA: 1 ♀, 137.0; Oregon Sta. 3585; USNM 113760.—7 ♂, 128.0-142.5; 1 ♀, 120.2; Oregon Sta. 3597; USNM 113759.—COLOMBIA: 2 ♂, 62.1-134.5; Oregon Sta. 4838; USNM 113749.—4 ♀, 63.5-150.7; Oregon Sta. 4858; USNM 113750.—2 ♀, 121.2-150.2; Oregon Sta. 4880; USNM 113747.—3 ♂, 129.5-142.6; Oregon Sta. 4911; USNM 113751.—VENEZUELA: 1 ♀, 174.0; Oregon Sta. 2782; USNM 113757.—1 ♀, 66.9; Oregon Sta. 4434; USNM 113746.—TRINIDAD: 1 ♂, 100.8; Oregon Sta. 2351; USNM 113756.—BRITISH GUIANA: 1 ♂, 104.1; 1 ♀, 137.7;

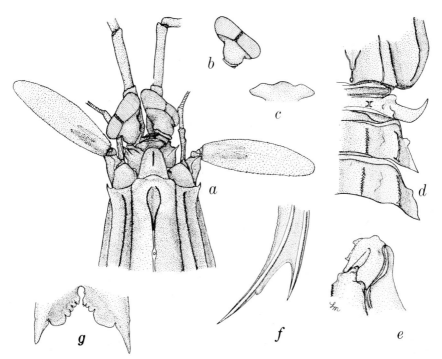

Figure 66. *Squilla caribaea*, n. sp., female, TL 108.8, Oregon Sta. 3627. *a,* anterior portion of body; *b,* eye; *c,* ocular scales, enlarged; *d,* lateral processes of fifth to seventh thoracic somites; *e,* carpus of claw; *f,* basal prolongation of uropod; *g,* submedian denticles of telson, enlarged (setae omitted).

OREGON Sta. 1983; UMML 32.1889.—1 ♂, 81.9; 2 ♀, 60.6-82.2; OREGON Sta. 1985; USNM 113754.—SURINAM: 2 ♂, 87.6-93.5; OREGON Sta. 2285; UMML 32.1887.—3 ♂, 114.5-128.1; OREGON Sta. 2286; USNM 113755.—1 ♂, 77.4; 5 ♀, ca. 75.0-ca. 155.0; OREGON Sta. 4302; USNM 113748.

*Diagnosis.*—Ocular scales rounded laterally; median carina of carapace simple, distance from pit to bifurcation less than distance from bifurcation to anterior margin; dactylus of claw with 6 teeth; lateral process of fifth thoracic somite a sharp, tapering, anteriorly-curved spine; abdominal width not greater than carapace length; telson as long as or longer than broad, median carina sinuous in lateral view; prelateral lobe projecting; denticles usually 5-8, 11-14, 1.

*Description.*—Eye large, cornea set obliquely on stalk; ocular scales rounded laterally; anterior margin of ophthalmic somite produced forward, medially emarginate. Corneal indices of specimens of various sizes follow:

| CL in mm | n | CI mean | CI range |
|---|---|---|---|
| 13-14 | 1 | 315 | . . . . . . |
| 15-16 | 3 | 352 | 349-359 |
| 17-18 | 4 | 341 | 327-354 |
| 19-20 | 2 | 357 | 355-358 |
| 21-22 | 3 | 362 | 343-390 |
| 23-24 | 2 | 356 | 347-364 |
| 25-26 | 5 | 370 | 362-378 |
| 27-28 | 3 | 366 | 366 |
| 29-30 | 8 | 379 | 371-389 |
| 31-32 | 8 | 378 | 366-401 |
| 33-34 | 4 | 392 | 382-405 |

Antennular processes produced into acute spines directed anterolaterally; antennular peduncle slightly shorter than carapace.

Rostral plate short, triangular, sinuous upturned lateral margins converging on rounded apex; short median carina present on anterior half.

Carapace with prominent, roughened carinae; median carina flanked by small pits, not composed of 2 rows of pits as in *S. edentata;* on median carina, distance from dorsal pit to anterior bifurcation less than distance from bifurcation to anterior margin; intermediate carinae extend to or almost to anterior margin; anterolateral spines strong but not extending to base of rostral plate; posterolateral margin with sharp anterior angle.

Raptorial claw large, dactylus with 6 teeth, outer margin of dactylus evenly curved; dorsal ridge of carpus with 2 sharp tubercles.

Mandibular palp and 5 epipods present.

Last thoracic somites with prominent, pitted submedian carinae; carinae of fifth somite short, obscure, those of last 3 somites well-developed; lateral process of fifth somite a sharp, tapering, anteriorly-curved spine, not sickle-shaped as in *S. edentata* and *S. intermedia;* lateral processes of next 2 somites bilobed, anterior lobes small, rounded, posterior lobes much larger, triangular, each with sharp posterolateral apex; posterior margin of process of seventh somite convex; ventral keel on eighth somite rounded, inclined posteriorly.

Abdominal carinae pitted, noticeably rough; submedian carinae divergent, particularly on posterior somites; carinae of abdomen not noticeably swollen in male; abdominal carinae spined as follows: submedian, 5-6; intermediate, 1-6; lateral, 1-6; marginal, 1-5; sixth somite with spine in front of articulation of each uropod; abdomen relatively slender, in adult males, greatest width not greater than median length of carapace.

Telson flattened, not inflated, as long as or longer than broad, appearing slender; prelateral lobes present, well-marked, projecting laterally beyond general outline of telson margin; median carina of telson straight or slightly sinuous in lateral view, not convex; telson margin slightly inflated in very large males (TL 140 mm or over), swellings interrupted at bases of intermediate and lateral teeth; intermediate denticles forming straight or concave line, not convex as in *S. edentata* and *S. intermedia;* denticles small, rounded, 2-11, 11-17, 1, usually 5-8, 11-14, 1; outer submedian denticle slightly larger than remainder or all submedian denticles subequal; ventral surface of telson with postanal keel, usually subdivided into tubercles posteriorly.

Uropods with 6-8 graded, movable spines on outer margin of proximal segment of exopod, last short, not extending to midlength of distal segment; lobe on inner spine of basal prolongation inconspicuous.

*Color.*—No well-marked color pattern is visible in any of the present specimens, although some show traces of dark pigment on the posterior margin of each abdominal somite.

*Size.*—Males, TL 62.1-142.6 mm; females, TL 60.6-174.0 mm. Other measurements of male holotype, TL 120.0: carapace length, 26.4; cornea width, 7.0; rostral plate length, 3.8, width, 3.8; telson length, 23.8, width, 23.8.

*Discussion.*—*S. caribaea* is a slenderer species than either *S. edentata* or *S. intermedia* which it closely resembles. *S. caribaea* has larger eyes than *S. edentata edentata* (compare corneal indices), but the eyes of *S. e. australis* are of similar size. It agrees with *S. edentata* and differs from *S. intermedia* in the relation of the distance from the dorsal pit to the bifurcation and the bifurcation to the anterior margin on the median carina of the carapace and in the rounded ocular scales, but differs from the former in the nature of the median carina of the carapace. In *S. edentata* (both subspecies) the median carina is composed of two lines of pits, whereas in *S. caribaea* the carina is a simple, single structure. It also differs from *S.*

*edentata* in that the lateral process of the fifth thoracic somite is not as strongly hooked and the intermediate carinae of the first abdominal somite usually terminate in spines. The telson of *S. caribaea* is not as inflated as in *S. edentata,* and the median carina is not convex in lateral view. The telson is also slenderer in the new species, which has a noticeably narrower intermediate area on the telson, with the intermediate margin concave; in this latter feature the new species differs from both *S. edentata* and *S. intermedia.*

In slenderness and in shape of the lateral process of the fifth thoracic somite, as well as in comparative narrowness of the intermediate area of the telson, *S. caribaea* resembles *S. brasiliensis.* That species, however, has a larger eye (compare CI), a straight anterior margin on each lateral plate of the carapace, and the submedian carinae of the fourth abdominal somite always terminate in spines.

*Remarks.*—Males of *S. caribaea* mature at a larger size than those of either *S. edentata* or *S. intermedia;* sexual differences can be seen in only the largest males (TL 140 mm) of *S. caribaea.* The abdominal carinae of *S. caribaea* are rarely swollen to the extent observed in either *S. edentata* or *S. intermedia.* The telsons of adult male and female *S. caribaea* are shown in Figure 65.

The submedian denticles of the telson in *S. caribaea* may be either subequal in size or with the outer submedian denticle the largest; in this respect *S. caribaea* is intermediate between *S. edentata* in which the outer denticle is always largest and *S. intermedia* in which the denticles are always subequal.

*S. caribaea,* the southern counterpart of *S. intermedia,* has been collected together with that species at OREGON Sta. 3627, off Honduras, in 364 m. The southern form generally occurs in shallower waters than does *S. intermedia.* *S. caribaea* has also been taken together with *S. edentata australis,* at OREGON Stats. 1983 (off British Guiana, in 228 m) and 2286 (off Surinam, in 191-218 m). The species are distinct wherever they occur together.

One specimen from OREGON Sta. 1983 was sent to the Museu Nacional, Rio de Janeiro, on exchange. The specimen is labeled *S. edentata* but may prove to be either *S. e. australis* or *S. caribaea.*

*Etymology.*—The name is derived from the general range of the species, the Caribbean Sea.

*Ontogeny.*—Unknown.

*Type.*—The male holotype and a series of paratypes are in the U. S. National Museum.

*Type-locality.*—Colombia, 09°30'45"N, 76°25.5'W, 282 m, ALBATROSS Sta. 2143, 23 March 1884.

238

*Distribution.*—Caribbean Sea, from off Honduras, Nicaragua, Panama, Colombia, Venezuela, Trinidad, and off British Guiana and Surinam, in depths between 190 and 437 m.

Family GONODACTYLIDAE Giesbrecht, 1910

Gonodactylinae Giesbrecht, 1910, p. 148.—Gurney, 1946, p. 134.—Manning, 1963a, p. 325.
Gonodactylidae: Manning, 1967, p. 238; 1968, pp. 109 [key], 137 [text].

*Definition.*—Propodi of third and fourth thoracic appendages longer than broad, lacking ventral ribbing; telson with distinct median carina; submedian marginal teeth always with movable apices, submedian denticles present or absent; no more than 2 intermediate denticles present.

*Type-genus.*—*Gonodactylus* Berthold, 1827.

*Number of Genera.*—Thirteen, of which six occur in the western Atlantic. The key given below will serve to distinguish the American genera.

*Ontogeny.*—Larva an erichthus, early stages with 2 maxillipeds and 5 pairs of pleopods; telson with no more than one intermediate denticle.

*Remarks.*—As pointed out by Manning (1968), the genera assigned to this family fall into two broad sections, one characterized by *Gonodactylus* and its allies and the other by *Pseudosquilla* and its allies. The genera related to *Gonodactylus* all have a subterminal ischiomeral articulation on the raptorial claw, with the merus projecting posteriorly beyond the articulation. In this section, only *Odontodactylus* has the claw armed with teeth. In general, genera of the *Gonodactylus* section have the claw inflated at the articulation of the propodus and dactylus, lack the pectinations of the propodus of the claw, and have a flattened dorsal plate on the basal segment of the antenna.

The genus *Hemisquilla* seems to be most closely related to the *Gonodactylus* section, but it differs in having a terminal ischiomeral articulation of the claw.

Two groups of genera now can be recognized within the *Pseudosquilla* section of the family. The first of these includes *Pseudosquilla, Pseudosquillopsis,* and *Parasquilla,* and the second includes *Coronidopsis, Manningia, Eurysquilla,* and *Eurysquilloides.* These groups of genera are more closely related to each other than to any of the other genera now included in the Gonodactylidae.

The first group of genera share several distinctive features. The raptorial claw is armed with three teeth, the abdomen is compact, and the telson is slender. *Pseudosquilla* differs from both of the other genera in having the dorsal channeled process on the antennal protopod, a slender raptorial claw with the propodus only partially pectinate, and the basal prolongation of the uropod produced into two spines. *Parasquilla* differs from the other

239

two genera in having a flattened, carinate abdomen and in having submedian denticles on the telson, but it shares with *Pseudosquillopsis* the bilobed eye, similar raptorial claw, and three-spined uropod. These combinations of characters can best be explained if a common origin for these genera is postulated.

The second group of genera shares several features with the first, including a sharp median carina on the telson and the presence of only two intermediate denticles on the telson, but these genera differ in several basic features. The body is depressed, loosely articulated, the raptorial claw is armed with more than three teeth, the telson is broad, and the uropod structure is basically different. This group will be discussed in greater detail under *Eurysquilla*.

Although some species of *Eurysquilla* resemble lysiosquillids in general facies, the slender maxillipeds and median carina of the telson of *Eurysquilla* will serve to distinguish species of that genus from all lysiosquillids.

The nomenclature of the carinae of the telson in *Gonodactylus*, *Pseudo-*

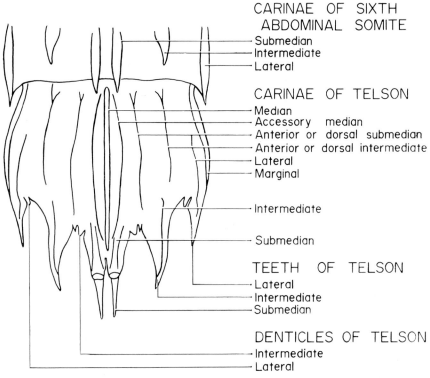

CARINAE OF SIXTH ABDOMINAL SOMITE
- Submedian
- Intermediate
- Lateral

CARINAE OF TELSON
- Median
- Accessory median
- Anterior or dorsal submedian
- Anterior or dorsal intermediate
- Lateral
- Marginal
- Intermediate
- Submedian

TEETH OF TELSON
- Lateral
- Intermediate
- Submedian

DENTICLES OF TELSON
- Intermediate
- Lateral

FIGURE 67. Terms used in the descriptive accounts of *Hemisquilla*, *Pseudosquilla*, and allied genera.

240

*squilla,* and related genera, stems from that adopted by Kemp (1913). This nomenclature is artificial in that the carina of one of the marginal teeth does not necessarily bear the same name as the tooth; thus, in Kemp's system, the lateral carina extends onto the intermediate tooth. In the carinal nomenclature used herein, the name of a dorsal carina is derived from that of the marginal tooth onto which it extends. A comparison of the present nomenclature with that of Kemp is given in Table 8.

TABLE 8

NOMENCLATURE OF CARINAE OF TELSON IN GONODACTYLIDAE

| Kemp, 1913 | Present | Remarks |
| --- | --- | --- |
| Submedian | Accessory median | Parallel to median carina |
| Second submedian | Second accessory median | Parallel to median carina (*Odontodactylus* only) |
| Intermediate | Submedian | On submedian tooth |
| First lateral | Intermediate | On intermediate tooth |
| Second lateral | Lateral | On lateral tooth |
| Marginal | Marginal | |

Figures explaining the terminology used in the accounts of *Pseudosquilla, Gonodactylus,* and *Odontodactylus,* are included under the account of each of those genera. The terminology used for *Eurysquilla, Hemisquilla,* and *Parasquilla* is most similar to that used for *Pseudosquilla* and is shown in Figure 67.

Manning (1968) has presented a key to all genera in the family. The following key will distinguish the American genera.

KEY TO THE AMERICAN GENERA OF THE FAMILY
GONODACTYLIDAE

1. Ischiomeral articulation terminal; merus grooved inferiorly throughout its length . . . . . . . . . . . . . . . . . . . . . . . . . . . . . . . . . . . . . . . . . . . . . . . . 2
1. Ischiomeral articulation subterminal, merus projecting posteriorly beyond articulation; inferior groove on merus incomplete . . . . . . . . . . 6

2. Dactylus of claw unarmed; sixth abdominal somite unarmed posteriorly . . . . . . . . . . . . . . . . . . . . . . . . . . . . . . . . . . . . . . *Hemisquilla,* p. 242.
2. Dactylus of claw with teeth; sixth abdominal somite with armed carinae or with posterior spines . . . . . . . . . . . . . . . . . . . . . . . . . . . . . . . . . 3

3. Inner spine of basal prolongation of uropod longer than outer; dactylus of claw with more than 4 teeth . . . . . . . . . . . . . . *Eurysquilla,* p. 248.
3. Outer spine of basal prolongation of uropod longer than or subequal to inner; dactylus of claw with 3 teeth . . . . . . . . . . . . . . . . . . . . . . . 4

241

4. Basal prolongation of uropod with 2 spines, inner margin unarmed
...................................... *Pseudosquilla*, p. 262.
4. Basal prolongation of uropod with 3 spines, proximal smallest .... 5

5. First 5 abdominal somites with prominent carinae; telson with submedian denticles ......................... *Parasquilla*, p. 278.
5. First 5 abdominal somites not carinate; submedian denticles absent
............................. [*Pseudosquillopsis* Serène, 1962].

6. Dactylus of claw with teeth; rostral plate without median spine ....
...................................... *Odontodactylus*, p. 284.
6. Dactylus of claw unarmed; rostral plate with slender median spine ...
...................................... *Gonodactylus*, p. 291.

## Genus *Hemisquilla* Hansen, 1895

*Hemisquilla* Hansen, 1895, p. 72.—Holthuis & Manning, 1964, p. 42 [for
inclusion on Official List].—Manning, 1966, p. 376; 1968, p. 138 [key].

*Definition.*—Dorsal surface of body smooth; cornea subglobular, set very
obliquely on stalk; rostral plate triangular, unarmed; antennal protopod
without papillae, dorsal surface with a flat plate on anterointernal border;
carapace narrowed anteriorly, without spines or longitudinal carinae; posterior margin of lateral plates with marginal carina which is not anteriorly
recurved; cervical groove not distinct across dorsum of carapace but well-
marked on lateral plates; thoracic somites without dorsal carinae; ventral
surface of eighth thoracic somite with ventrally directed spinule in males,
blunt prominence in females; raptorial claw short, dactylus unarmed, base
slightly inflated; propodus of claw with 2 proximal movable spines, first
the larger; upper margin of propodus minutely serrate proximally; dorsal
ridge of carpus terminating in blunt tooth; ischiomeral articulation terminal,
merus channeled ventrally throughout its length for reception of propodus;
mandibular palp and 5 epipods present; endopod of walking leg made up
of 1 segment; abdomen convex, slightly depressed, unarmed; articulated
anterolateral plates present; carinae absent on first 4 somites, fifth somite
with blunt carina at level of intermediate carina of sixth somite; sixth somite
with 8 longitudinal, flattened carinae; spine present ventrally on sixth
somite in front of articulation of uropod; telson with median carina and 1
pair of submedian carinae on dorsal surface, and 3 pairs of marginal teeth,
submedians with movable apices; submedian denticles absent, 1 or 2 intermediate and 1 lateral denticle present; ventral surface of telson unarmed,
not carinate; basal prolongation of uropod with inner spine, outer margin
convex, outer spine absent or reduced to a tubercle; basal segment of
uropod unarmed dorsally, proximal segment of exopod shorter than distal
and armed with movable spines on outer margin.

*Type-species.*—*Gonodactylus styliferus* H. Milne-Edwards, 1837, p. 530

242

(a subjective junior synonym of *Gonodactylus ensiger* Owen, 1832), by monotypy.

*Gender.*—Masculine.

*Number of species.*—Two.

*Nomenclature.*—The type-species was originally described in *Gonodactylus* and later transferred to *Pseudosquilla;* it was separated from the latter genus by Hansen in 1895. Although the second species, *H. braziliensis,* was described in 1903, it was not transferred to *Hemisquilla* until 1940. A petition to have the name placed on the Official List of Generic Names in Zoology is now pending (Holthuis & Manning, 1964).

*Remarks.*—Only two species are now placed in this genus. The Pacific species, *H. ensigera* (Owen, 1832), has recently been studied by W. Stephenson (1967) who has shown that three subspecies, one Californian, one Chilean, and one Australian, can be recognized.

The eye of *Hemisquilla* is very similar in shape to that of *Odontodactylus,* with the cornea almost globular and set very obliquely on the stalk. The cornea is divided by an oblique band of highly sensitive cells, and the corneal axis can be measured along or across this band. In all calculations of corneal indices given here and in subsequent accounts of this genus, the corneal diameter measured across the band will be used.

This is one of the few stomatopod genera in which any variation is found in the number of segments of the mandibular palp. In most stomatopods, the palp is either present and composed of three segments or absent, and its presence (or absence) is usually a reliable specific character. In both species of *Hemisquilla,* the palp is always present and it may be made up of one, two or three segments. The size and armature of the intermediate denticles of the telson are also variable; either one or two large lobes may be present and either or both may be armed with a small spinule.

The terminology of the carinae of the telson is shown in figure 67.

*Affinities.*—The exact affinities of this genus are uncertain. In general body form, *Hemisquilla* approaches both *Gonodactylus* and *Odontodactylus;* the latter also has the flattened plate on the dorsal surface of the antennal protopod. As in *Gonodactylus,* the dactylus of the raptorial claw is unarmed, but the basal portion of the dactylus is less inflated. The eyes and the basic structure of the uropod are similar in *Hemisquilla* and *Odontodactylus,* although in *Hemisquilla* the two spines of the basal prolongation of the uropod are poorly developed and the proximal segment of the uropod lacks a dorsal spine.

### *Hemisquilla braziliensis* (Moreira, 1903)
### Figures 68, 69

*Pseudosquilla braziliensis* Moreira, 1903, p. 60; 1903a, p. 5, 2 text-figs.; 1905, p. 125, pls. 1, 2 [p. 5 on separate dated 1906].—Balss, 1938, p. 129 [listed as *brasiliensis*].

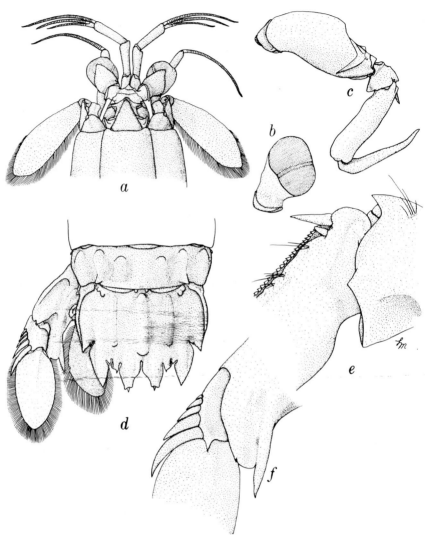

FIGURE 68. *Hemisquilla braziliensis* (Moreira), male, TL 140.0, CALYPSO Sta. 143. *a,* anterior portion of body; *b,* eye; *c,* raptorial claw; *d,* sixth abdominal somite, telson, and uropod; *e,* proximal portion of propodus of claw, enlarged; *f,* basal prolongation of uropod (*a-d* from Manning, 1966).

*Hemisquilla braziliensis:* Schmitt, 1940, p. 182, text-fig. 18b.—Andrade Ramos, 1951, pp. 142, 148, pl. 1, fig. 4.—Lemos de Castro, 1955, p. 31, pl. 9, fig. 41, pl. 15, fig. 50.—Serène, 1962, p. 7 [discussion].—Manning, 1966, p. 376, text-fig. 7.

*Previous records.*—BRAZIL: Espiritu Santo (Andrade Ramos, 1951); Ilha Rasa, Rio de Janeiro Bay (Moreira, 1903); off Rio de Janeiro (Manning, 1966); Ilha Grande, Sepetiba Bay (Moreira, 1903, 1905; Schmitt, 1940; Andrade Ramos, 1951; Lemos de Castro, 1955); Ilha de São Sebastião and Santos (Manning, 1966); south Brazil (Lemos de Castro, 1955).

*Material.*—1 ♂, 160.0; Cabo Frio, Rio de Janeiro; October 1955; UMML 32.1744.—1 ♂, 149.5; 1 ♀, 128.5; ESE of Ilha Rasa, Rio de Janeiro Bay; 100 m; C. Moreira; January 1903; syntypes; USNM 29327.—1 ♂, 153.5; same; not types; USNM 78009.—1 ♀, 87.0; same; July 1959; P. S. Moreira; USNM 111109.—1 ♂, 158.0; Ilha Rasa; USNM 48304.—1 ♂, 77.8; SE of Rio de Janeiro; 23°06.5′S, 42°50′W; 63 m; CALYPSO Sta. 105; 2 December 1961; MNHNP.—3 ♂, 136.0-146.0; Ilha de São Sebastião; 23°40′S, 45°01′W; 37 m; CALYPSO Sta. 129; 10 December 1961; MNHNP. —1 ♂, 163.0; same; 24°06.5′S, 45°29′W; 48 m; CALYPSO Sta. 136; 11 December 1961; MNHNP.—2 ♂, 148.0-155.0; off Santos; 24°18′S, 45°23′ W; 66 m; CALYPSO Sta. 137; 11 December 1961; MNHNP.—3 ♂, 140.0-152.0; same; 24°35.5′S, 46°31′W; 45 m; CALYPSO Sta. 143; 14 December 1961; USNM 111060.—1 ♂, 150.0; Alvaredo, Santa Catarina State; 7 June 1960; IOSP.

*Diagnosis.*—Eyes very large, almost cuboidal, cornea noticeably expanded laterally beyond stalk; no more than 2 intermediate denticles present on telson; intermediate and lateral denticles each usually armed with small spinule, lateral denticle always armed; uropod with 4-5 movable spines on outer margin of proximal segment of exopod.

*Description.*—Eye very large with cornea inflated, set obliquely on stalk; eyes extending past first segment of antennular peduncle; anterior margin of ophthalmic somite produced into obtuse angle; ocular scales compressed anteroposteriorly, lateral margins rounded.

Antennule short, stout, peduncle shorter than carapace, flagella shorter than or subequal to last 2 segments of peduncle.

Rostral plate triangular, longer than broad, without carinae, extending anteriorly to point of attachment of eyestalks.

Antennal scale large, ovate, about four-fifths carapace length; antennal peduncle with broad, flat plate on anterointernal angle of basal segment; antennal protopod without papillae.

Carapace smooth, short, about one-fifth total length.

Raptorial claw stout but small, propodus over one-half as long as carapace; merus stout, inflated; articulation of propodus and dactylus slightly inflated.

Lateral processes of sixth and seventh thoracic somites broadly rounded.

Posterolateral angles of abdominal somites all rounded, more acute anteriorly than posteriorly; greatest width of abdomen longer than carapace; carinae of sixth abdominal somite depressed, intermediates most prominent.

Telson noticeably flattened, median carina and submedian dorsal carinae prominent but not sharp; median carina with 1 pair of anterior tubercles; anterior margin, lateral to submedian carinae, with pair of tubercles, occasionally fused, on each side; ventral surface of telson without sharp carinae.

Uropods very broad, terminal segments ovate; basal segment of exopod with ventral spine at articulation of distal segment; outer spine of basal prolongation of uropod usually obsolete; last movable spine on outer margin of proximal segment of exopod short, not extending to midlength of distal segment.

*Color.*—Males with body metallic green, telson reddish; females with body dark, telson metallic green (Moreira, 1905). In preservative, some specimens show traces of dark pigment along inner surface of antennules and along sides of carapace and body; dactylus and dorsal depression of merus of raptorial claw blue; inner half of distal segment of exopod of uropod dark.

*Size.*—Males, TL 77.8-163.0 mm; only female examined, TL 128.5 mm. Measurements of selected specimens are given in Table 9.

*Discussion.*—*H. braziliensis* and *H. ensigera* are very similar in almost all features. Manning (1966) pointed out the following differences:

1. The cornea is comparatively larger in *H. braziliensis* than in *H. ensigera*, i.e., the diameter of the cornea measured across the band is noticeably larger than that measured along the band. The Corneal Length-Width Index (CoLWI) in *H. braziliensis* ranges from 929 to 985, mean 949, for 10 specimens with TL 78 to 163 mm, whereas in eight specimens of *H. ensigera* from Chile (TL 101-168) the CoLWI ranges from 703 to 833, mean 776. This difference in cornea size is very evident when specimens of similar size are compared.

2. The lateral denticle of the telson is always armed with an apical spinule in *H. braziliensis* and the intermediate denticles are usally armed too. In *H. ensigera*, none of the denticles is spined and the lateral denticle is usually obsolete.

3. The carine of the sixth abdominal somite and telson are usually sharper and more prominent in *H. ensigera* than in *H. braziliensis*, but this feature is somewhat variable.

The first two of these features are more reliable than the third and can be used to identify specimens even when the locality is unknown.

*Remarks.*—Although Moreira (1905, p. 125) noted that his species differed from *P. stylifera* (= *H. ensigera*) in lacking denticles on the inner

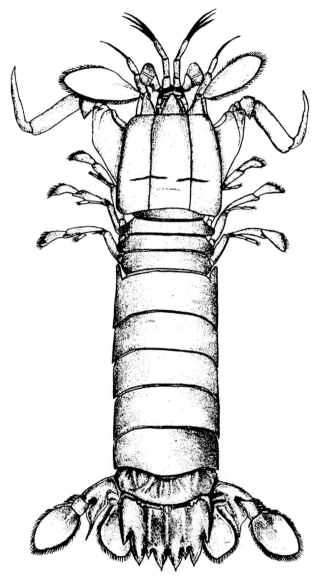

FIGURE 69. *Hemisquilla braziliensis* (Moreira), male, dorsal view (from Moreira, 1903a).

border of the basal segment of the uropod, no such denticles were detected in either species by me.

*Ontogeny.*—Unkown.

*Type.*—Syntypes are deposited in the Museu Nacional, Rio de Janeiro and the U. S. National Museum.

*Type-locality.*—Rio de Janeiro Bay.

*Distribution.*—Brazil, between Espiritu Santo in the north and Santos in the south, in depths between 37 and 100 m.

TABLE 9

MEASUREMENTS IN MM OF SELECTED SPECIMENS OF *Hemisquilla braziliensis*

| Sex | ♂ | ♀ | ♂ | ♂ | ♂ |
|---|---|---|---|---|---|
| TL | 77.8 | 128.5 | 136.0 | 146.0 | 155.0 |
| CL | 15.3 | 26.0 | 25.8 | 27.8 | 28.8 |
| Cornea Width | 5.6 | 6.7 | 6.6 | 6.9 | 6.5 |
| Cornea Length | 5.2 | 6.0 | 6.2 | 6.6 | 6.4 |
| Fifth Abdominal Somite Width | 16.7 | 30.6 | 29.6 | 31.6 | 33.6 |
| Propodus Length | 11.4 | 19.8 | 20.6 | 21.7 | 23.3 |
| Antennular Peduncle Length | 15.8 | 23.5 | 24.3 | 24.4 | 26.1 |

## Genus *Eurysquilla* Manning, 1963

*Eurysquilla* Manning, 1963a, p. 314; 1966, p. 378; 1968, p. 139 [key].

*Definition.*—Surface of body smooth; eyes with cornea bilobed, elongate, or subglobular; rostral plate subtriangular, with or without apical spine; antennal protopod with at least 1 ventral papilla; carapace narrowed anteriorly, anterolateral angles rounded or subacute, not spined; cervical groove not distinct across dorsum of carapace but indicated along lateral plates; broad marginal carinae, not anteriorly recurved, present posteriorly on each lateral plate; exposed thoracic somites with at most 1 longitudinal carina laterally; eighth thoracic somite usually with prominent ventral projection; raptorial claw slender, dactylus with 7 or more teeth; upper margin of propodus pectinate throughout its length; propodus with 3 movable spines at base, middle smallest; carpus with or without dorsal ridge; merus slender, ventrally channeled throughout its length for reception of propodus; mandibular palp present or absent; 3 or more epipods present; endopod of walking legs composed of 1 or 2 segments; abdomen flattened, very loosely articulated; first 3 somites without sharp carinae, fourth and fifth somites occasionally carinate; sixth somite with 6 or more posterior spines, submedians not sharply carinate; telson broad, dorsal surface orna-

248

mented with sharp median carina, terminating in spine, and 1 or more entire or interrupted carinae lateral to median carina; posterior margin of telson with 6 sharp teeth, submedians with movable apices; submedian denticles absent; 1 or 2 sharp intermediate and 1 sharp lateral denticle present; basal prolongation of uropod terminating in 2 spines, inner longer, with or without row of slender spines or 1 spine or lobe on inner margin.

*Type-species.*—*Lysiosquilla plumata* Bigelow, 1901, p. 156, by original designation.

*Gender.*—Feminine.

*Number of species.*—Six, of which two are described here as new. Four species occur in the western Atlantic.

*Nomenclature.*—Manning (1963a) separated this genus from the nominal genus *Pseudosquilla* Dana. Prior to that, Schmitt (1940), Manning (1961) and Serène (1962) had pointed out the heterogeneity of the species then included in *Pseudosquilla.* Of the four species originally placed in *Eurysquilla,* three had been described in *Lysiosquilla* and one in *Pseudosquilla.*

*Remarks.*—The species now splaced here are all very similar in general facies. The loosely articulated abdomen, broad telson with three pairs of marginal teeth, submedians with movable apices, and slender raptorial claw in which the dactylus is armed with more than seven teeth are similar in all six species. On the basis of the structure of the basal prolongation of the uropod, two groups of species can be recognized. The first, including *E. plumata, E. sewelli,* and *E. veleronis,* is characterized by a very slender basal prolongation, the inner margin of which is armed with at most one spine. The second group, including *E. maiaguesensis* and the two new species described below, has a broad basal prolongation with a row of slender spinules on its inner margin.

Although all of the species agree in most major features, several differences in important characters have been found. The following is restricted to American species, for these features could not be investigated in *E. sewelli:*

1. Papillae on antennal protopod: These papillae are absent in *E. maiaguesensis,* but are definitely present in *E. veleronis, E. plumata,* and the two new species described below.

2. Mandibular palp: Present in all American species but *E. maiaguesensis.*

3. Epipods: Five epipods are present in *E. veleronis, E. chacei,* n. sp., and *E. plumata* (one specimen has 4-5). The holotype of *E. holthuisi,* n. sp., has but three, and specimens of *E. maiaguesensis* have either three or four.

4. Segments on endopod of walking legs: This is made up of two seg-

**249**

ments in all but *E. plumata,* in which the two segments, if present, are poorly defined.

In related genera, these features seem to be constant throughout the genus, but in *Eurysquilla* they can be used only as specific characters.

The shape of the eye and rostral plate also varies a great deal throughout the genus. The cornea may be globular, or bilobed and elongate, and the rostral plate may be long and slender or short and broad, with or without an apical spine. These features also seem to be more consistent within other genera, *Pseudosquillopsis* and *Parasquilla,* for example.

Some species live in moderately deep water, to over 400 m. In the western Atlantic, *E. plumata* lives in shallower water than any of the three other species; it is also the most widely distributed. Of the remainder of the species in the genus, one, *E. veleronis* (Schmitt, 1940), occurs in the eastern Pacific, and the other *E. sewelli* (Chopra, 1939), has been recorded in the Gulf of Aden. *E. veleronis* is the eastern Pacific analogue of *E. plumata.*

*Affinities.*—This genus appears to be more closely related to *Manningia* Serène, *Coronidopsis* Hansen, and *Eurysquilloides* Manning than to any of the other genera in the Gonodactylidae. With these three genera *Eurysquilla* shares a loosely articulated body, broad telson, and telson structure. It differs from *Manningia* and *Coronidopsis* chiefly in having a slender raptorial claw with the dactylus armed with more than four teeth. The propodus of the raptorial claw in *Eurysquilloides* is also slender, but that genus has the propodus only partially pectinate and also has more carinae on the abdomen. These four genera, *Coronidopsis, Manningia, Eurysquilla* and *Eurysquilloides* are more closely related to each other than to *Pseudosquilla, Pseudosquillopsis,* and *Parasquilla,* but are nearer these latter three than any other genera in the Gonodactylidae.

#### KEY TO WESTERN ATLANTIC SPECIES OF *Eurysquilla*

1. Basal prolongation of uropod with at most 1 spine or lobe on inner margin ................................... *plumata,* p. 251.
1. Basal prolongation with row of 5 or more slender spines on inner margin .................................................. 2
2. Cornea not noticeably bilobed; telson with 2 sharp intermediate denticles ................................. *maiaguesensis,* p. 254.
2. Cornea strongly bilobed; telson with 1 intermediate denticle with ventral spinule ............................................ 3
3. Lateral processes of sixth and seventh thoracic somites rounded; anal pore flanked laterally by tubercle ........... *chacei,* n. sp., p. 257.
3. Lateral processes of sixth and seventh thoracic somites sharp posterolaterally; no tubercles lateral to anal pore .... *holthuisi,* n. sp., p. 259.

*Eurysquilla plumata* (Bigelow, 1901)

Figure 70

*Lysiosquilla plumata* Bigelow, 1901, p. 156, text-figs. 6-9.—Kemp, 1913, p. 203 [listed].—Chopra, 1939, p. 172 [discussion].—Schmitt, 1940, p. 171 [discussion].—Chace, 1958, p. 141 [discussion].—Manning, 1963a, p. 314 [listed].

*Pseudosquilla plumata:* Manning, 1959, p. 18; 1961, p. 3 [table; analogue of *P. veleronis*].—Serène, 1962, pp. 6-7 [discussion].

*Eurysquilla plumata:* Manning, 1966, p. 378, figs. 8c-e.

*Previous records.*—FLORIDA (Manning, 1959); PUERTO RICO (Bigelow, 1901); BRAZIL (Manning, 1966).

*Material.*—FLORIDA: 1 ♂, 19.4; ½ mi SSW of Alligator Reef Light, Monroe Co.; 13 September 1959; W. A. Starck, II; USNM 111110.— 1 ♀, 23.5; SE Ship Channel, Key West, Monroe Co.; 30 December 1953; W. Hess; UMML 32.1153.—1 ♀, 25.4; Tortugas, Monroe Co.; 21 June 1931; W. L. Schmitt; USNM 66017.—PUERTO RICO: 1 damaged ♂, ca. 15.0; Mayaguez Harbor; FISH HAWK Sta. 6062 (P. R. Sta. 134); holotype; USNM 64823.—BRAZIL: 1 ♂, 22.2; 1 ♀, 21.0; off Salvador; 12° 56.5′S, 38°31.5′W; 20 m; CALYPSO Sta. 59; 24 November 1961; MNHNP.—1 ♂, 18.2; Abrolhos Ids.; 18°10′S, 38°50′W; 39m; CALYPSO Sta. 88; 29 November 1961; USNM 111055.

*Diagnosis.*—Cornea subglobular, placed very obliquely on stalk; dactylus of raptorial claw with 9 teeth; first 4 abdominal somites without carinae; fifth somite with blunt, unarmed intermediate carinae; telson with median carina and, on either side, a tuberculate carina, converging posteriorly with median carina, and 1 sharp, undivided, unspined carina terminating anterior to and inside intermediate marginal tooth; intermediate denticles rounded, outer with a dorsal tubercle and a ventral spinule; lateral denticle with dorsal tubercle and ventral spinule; basal prolongation of uropod slender, with low lobe on inner margin of inner spine.

*Description.*—Eye with cornea subglobular, broader than and set obliquely on stalk; eyes not extending to end of second segment of antennular peduncle; ocular scales fused into broad plate, indented along midline, terminating on each side in acute lobe.

Antennular processes produced into sharp, forwardly curved spines.

Antennal protopod with at least 1 ventral papilla.

Rostral plate short, triangular, narrowed anteriorly, without apical spine.

Carapace without anterolateral spines or longitudinal carinae but with broad marginal carina present at posterior end of each lateral plate.

Raptorial claw with dactylus notched at base of outer margin, inner margin with 8-9 slender teeth.

Mandibular palp and 4-5 epipods present (1 specimen with 4 epipods on 1 side).

Fifth thoracic somite lacking distinct lateral process; lateral processes

251

of sixth and seventh somites rounded; walking legs with endopods slender, obscurely divided into 2 segments.

First 4 abdominal somites without longitudinal carinae but with antero-lateral depressions; fifth somite with longitudinal swelling, noticeably carinate in some specimens, above lateral margin, posterolateral angle with strong spine; sixth somite with 6 posterior spines, submedians not carinate, and ventral spine on each side in front of articulation of uropod.

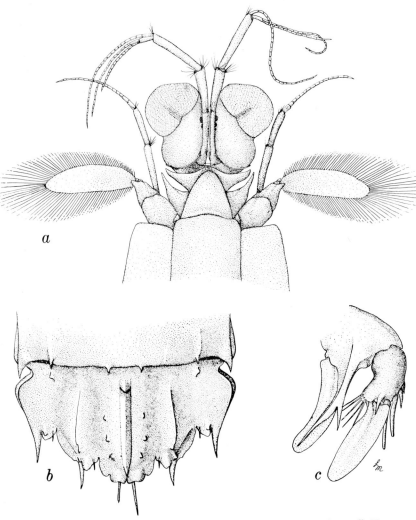

FIGURE 70. *Eurysquilla plumata* (Bigelow), female, TL 25.4, off Tortugas, Florida. *a*, anterior portion of body; *b*, telson; *c*, uropod, ventral view (setae omitted in *c*).

252

Telson almost twice as broad as long, with sharp median carina, feebly notched at base, terminating in sharp spine, and, on either side of median carina, 1 carina, armed dorsally with 2, 3 or 4 tubercles, converging toward posterior apex of median carina, and 1 undivided carina, neither spined nor interrupted, terminating anterior to bases of intermediate marginal teeth; anterior margin of telson with sharp tubercle on either side under intermediate spine of sixth somite; 2 rounded intermediate denticles, outer with ventral spinule, and 1 lateral denticle, with ventral spinule, present; inner margin of intermediate and lateral teeth each with prominent, rounded, upturned lobe.

Uropod with exopod articulated to basal segment almost at right angle; proximal segment of exopod with 5-7 slender, movable spines, last extending almost to midlength of distal segment; inner spine of basal prolongation of uropod the longer, with inconspicuous lobe on inner margin.

*Color.*—In preservative, background pale yellow; eyestalks with prominent black spot on inner surface; base of rostral plate lined with brown; carapace with band of brown chromatophores at level of cervical groove, another behind it; thoracic and abdominal somites with dorsal diffuse patch of brown chromatophores; first and fifth abdominal somites with posterolateral patches of dark chromatophores, arranged in black circle on fifth somite; telson with dark lateral patches and V-shaped patch at distal end of median carina; distal portions of uropod dark.

*Size.*—Males, TL ca. 15.0-22.2 mm; females, TL 21.0-25.3 mm. Measurements of female, TL 25.4: carapace length, 4.2; cornea width, 1.3; eye length, 1.8; rostral plate length, 1.1, width, 1.1; telson length, 2.5, width, 4.7.

*Discussion.*—The slender basal prolongation of the uropod armed with only an inconspicuous lobe on its inner margin will immediately separate this species from its western Atlantic relatives. *E. plumata* agrees most closely with *E. veleronis* and *E. sewelli* and can be distinguished from both of these by the subglobular cornea; in *E. veleronis* the cornea is elongate and in *E. sewelli* it is bilobed. The basal prolongation of these three species is also similar, but in *E. veleronis* there is a strong spine on the inner margin and in *E. sewelli* the inner margin is smooth.

*E. veleronis* is the eastern Pacific analogue of *E. plumata*.

*Remarks.*—The dark posterolateral circles of patches of the fifth abdominal somite will also distinguish this species from the other western Atlantic species. *E. veleronis* has a similar color pattern.

Several of the characteristic features of this species may be difficult to distinguish in view of the small size of the specimens. The endopod of the walking legs appears to be unisegmental in most of the specimens examined, but an indistinct suture was detected in two specimens. The longitudinal carinae of the fifth abdominal somite, which pass through the black

253

posterolateral circles, may appear as indistinct swellings. All but one specimen had five epipods; one had four epipods on one side, five on the other. The ventral papilla of the antennal protopod could be detected with certainty in only one specimen. It seems likely that the small size and soft condition of these specimens contributed to this apparent variation.

The specimen from Alligator Reef Light was found in the net when a specimen of *Astichopus multifidus* was dipped up. It may have been living in association with the holothurian, but it also may have been hiding under it.

*Ontogeny.*—Unknown.

*Type.*—U. S. National Museum.

*Type-locality.*—Mayaguez Harbor, Puerto Rico.

*Distribution.*—Florida Keys, Puerto Rico and Brazil, in shallow water to 55 m.

## *Eurysquilla maiaguesensis* (Bigelow, 1901)
## Figure 71

*Lysiosquilla maiguesensis* Bigelow, 1901, p. 158, text-figs. 10-13.—Kemp, 1913, p. 203 [listed].—Chopra, 1939, p. 172 [discussion].—Schmitt, 1940, p. 171 [discussion; transferred to *Pseudosquilla*].—Chace, 1958, p. 141 [discussion].—Manning, 1963a, p. 314 [listed].
*Pseudosquilla maiaguesensis:* Manning, 1961, p. 38 [discussion].—Serène, 1962, pp. 6, 7 [discussion].

*Previous records.*—PUERTO RICO (Bigelow, 1901).

*Material.*—PUERTO RICO: 1 ♂, 22.5; Mayaguez Harbor; FISH HAWK Sta. 6066 (P. R. Sta. 138); lectotype; USNM 64824.—1 ♂, 18.5; 1 brk. ♀; data as in lectotype; paralectotypes; USNM 110939.—1 ♂, 19.7; N of Puerto Rico; 18°31′30″N, 66°14′55″W; 220 m; Johnson-Smithsonian Deep Sea Exped. Sta. 105; USNM 119137.—BARBADOS: 1 ♀, 19.7; J. B. Lewis; USNM 111111.

*Diagnosis.*—Eyes with cornea broad, not bilobed, placed obliquely on stalk; dactylus of raptorial claw with 8-10 teeth; first 5 abdominal somites without sharp carinae; telson with median carina and 2 sharp carinae on dorsal surface, all with posterior spines; both intermediate denticles sharp; basal prolongation of uropod broad, with row of 16-20 spines on inner margin and no rounded lobe between apical spines.

*Description.*—Eye with broad cornea, not noticeably bilobed, set very obliquely on stalk; eyes extending beyond second segment of antennular peduncle; ocular scales fused into broad plate, tapering laterally to acute spines, directed anteriorly.

Antennal protopod without papillae.

254

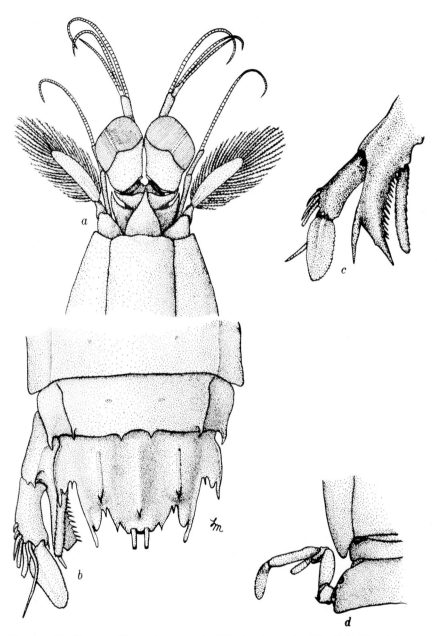

FIGURE 71. *Eurysquilla maiaguesensis* (Bigelow), male, TL 19.7, N of Puerto Rico. *a,* anterior portion of body; *b,* last abdominal somite, telson, and uropod; *c,* uropod, ventral view; *d,* lateral processes of fifth and sixth thoracic somites (setae omitted in *c-e*).

Rostral plate elongate, narrowed anteriorly, apex sharp.

Carapace narrowed anteriorly, with poorly-marked marginal carinae on posterior portions of lateral plates.

Mandibular palp absent; 4 epipods present.

Fifth thoracic somite without distinct lateral process; lateral processes of next 2 somites rounded anteriorly, truncate posteriorly; walking legs with prominent spine on posterior margin of basal segment; ventral surface of eighth thoracic somite with long, slender median projection; endopods of walking legs ovate, composed of 2 segments.

First 4 abdominal somites without carinae or spines; fifth somite with low, longitudinal, lateral swelling, not strongly carinate; fifth somite unarmed posterolaterally; sixth somite with 6 sharp posterior spines, submedians not carinate, intermediates strongly so.

Telson with sharp median carina terminating in slender spine; dorsal surface with pair of sharp carinae, posteriorly spined, at level of intermediate denticle; 2 sharp intermediate and 1 sharp lateral denticles present; inner margins of intermediate and lateral teeth each with broad, curved, upturned lobe, oriented almost vertically; ventral surface of telson smooth except for anal pore.

Uropod with exopod articulated normally, not set at right angles to basal segment; dorsal surface of basal segment carinate; outer margin of proximal segment of exopod with 4-6 slender, movable spines, last recurved, extending to or beyond end of distal segment; basal prolongation of uropod broad, inner spine longer, with inconspicuous lobe on outer margin and row of 16-20 slender, fixed spines on inner margin.

*Color.*—Carapace with scattered dark chromatophores near level of cervical groove and more distinct patches posterolaterally; dorsal surface of thoracic and abdominal somites with scattered brown chromatophores, abdominal somites with dark posterior line; body with 2 dark lateral stripes extending from carapace to telson, coverging distally on telson.

*Size.*—Males, TL 18.5-22.5 mm; only intact female, TL 19.7 mm. Other measurement of male lectotype, TL 22.5: carapace length, 3.7; rostral plate length, 0.9, width, 0.7; telson length, 2.1, width, 3.4.

*Discussion.*—*E. maiaguesensis* differs from all other species in the genus in lacking the mandibular palp, in having two sharp intermediate denticles on the telson, and in the shape of the basal prolongation of the uropod, which terminates in two spines, with no intervening lobe, and with many slender spines on inner margin. This species agrees with *E. plumata, E. veleronis,* and *E. sewelli* (Chopra) in the shape of the ocular scales which in these species are fused along the midline into a broad plate which tapers to acute lateral points. In *E. chacei* n. sp., and *E. holthuis* n. sp., the scales are separate and truncate.

*Remarks.*—*E. maiaguesensis* lacks the dark posterolateral spots characteristic of *E. veleronis* and *E. plumata,* and it occurs in deeper water than either of these species.

The small specimen from Barbados lacks an apical spine on the rostral plate. It agrees in all other respects with the remainder of the material examined.

*Ontogeny.*—Unknown.

*Type.*—U. S. National Museum. The largest male of Bigelow's three syntypes is here selected as a lectotype. The other two specimens, a male and broken female, are paralectotypes.

*Type-locality.*—Mayaguez Harbor, Puerto Rico.

*Distribution.*—Puerto Rico and Barbados, in depths to at least 315 m.

## Eurysquilla chacei, new species
### Figure 72

*Holotype.*—1 ♂, 49.6; Little Bahama Bank; 27°30′N, 78°52′W; 419 m; COMBAT Sta. 238; 3 February 1957; USNM 113643.

*Diagnosis.*—Eyes with cornea broad, bilobed; fifth abdominal somite with 2 carinae on either side; sixth abdominal somite with 8 posterior spines; telson with sharp median carina, and, on either side, a sharp carina flanked by patches of tubercles; outer intermediate denticle and lateral denticle each with ventral spinule; basal prolongation of uropod terminating in 2 spines with intervening lobe and row of slender spines on inner margin; anal pore flanked on either side by sharp tubercle.

*Description.*—Eye triangular, corneal and peduncular axes subequal; cornea strongly bilobed, set obliquely on stalk; eyes not extending much past first segment of antennular peduncle; ocular scales produced into 2 truncate lobes.

Antennular processes produced into long spines directed anteriorly.

Antennal protopod with 1 mesial and 1 ventral papilla.

Rostral plate elongate-triangular, rounded lateral margins converging on apical spine.

Carapace with distinct marginal carinae on posterior portions of lateral plates.

Raptorial claws of holotype both absent.

Mandibular palp and 5 epipods present.

Lateral process of fifth thoracic somite produced into obtuse lobe, directed obliquely downward; last 3 thoracic somties with carina inside of rounded lateral processes; eighth thoracic somite with prominent, rounded, median ventral keel; basal segment of walking legs with outer spinule; endopod of walking legs elongate, composed of 2 segments.

Abdomen without sharp carinae on first 4 somites but with prominent lateral groove; fifth somite with unarmed carina above and below this

257

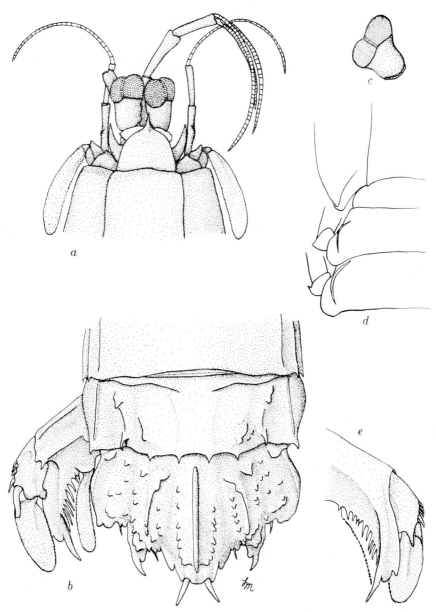

FIGURE 72. *Eurysquilla chacei,* n. sp., male holotype, TL 49.5, Little Bahama Bank. *a,* anterior portion of body; *b,* last abdominal somite, telson, and uropod; *c,* eye; *d,* lateral processes of fifth to seventh thoracic somites; *e,* uropod, ventral view (setae omitted).

258

groove, neither spined, and a small posterolateral spine; sixth abdominal somite with 8 posterior spines, submedians with faint carinae, carinae of intermediates and marginals strong; supplementary spines and few dorsal tubercles present between submedian and intermediate spines; sixth somite with ventral spine in front of articulation of uropod.

Telson with sharp median carina, terminating in strong spine, row of tubercles converging under terminal spine; paired dorsal carinae of telson sharp, not strongly spined posteriorly; area lateral to paired carinae roughly tuberculate, tubercles arranged in irregular longitudinal rows; outer intermediate denticle and lateral denticle each armed with ventral spinule; inner margin of intermediate and lateral teeth of telson with up-raised, broadly rounded lobe; anal pore flanked by sharp tubercle on either side.

Exopod of uropod with 5-6 graded, movable spines on outer margin, last curved, extending about to midlength of distal segment; basal prolongation slender, channeled ventrally, terminating in 2 blunt spines, inner larger, with intervening lobe; inner margin of inner spine with 10 slender fixed teeth, several bifurcate.

*Color.*—No distinct pattern is visible in the holotype.

*Size.*—Male holotype, only known specimen, TL 49.5 mm. Other measurements: carapace length, 10.1; cornea width and eye length, 2.4; rostral plate length, 2.9, width, 2.6; telson length, 6.2, width 9.5.

*Discussion.*—The presence of eight posterior spines on the sixth abdominal somite and a pair of tubercles lateral to the anal pore will separate this species from all others in the genus. Only *E. holthuisi,* n. sp., has a similar basal prolongation of the uropod, with a broad lobe between the apical spines.

*Remarks.*—This is the deepest record for any species in this genus.

*Etymology.*—The species is named for Fenner A. Chace, Jr., in recognition of his work on the systematics of decapod and stomatopod crustaceans.

*Ontogeny.*—Unknown.

*Type.*—U. S. National Museum.

*Type-locality.*—Off Little Bahama Bank.

*Distribution.*—Known only from the type-locality.

### Eurysquilla holthuisi, new species
### Figure 73

*Holotype.*—1 ♂, 34.8; Caribbean Sea, off Panama; 09°04′N, 81°25′W; 273-291 m; OREGON Sta. 3597; 31 May 1962; USNM 113642.

*Diagnosis.*—Eyes with cornea broad, strongly bilobed; dactylus of raptorial claw with 8 teeth; first 5 abdominal somites without sharp carinae; dorsal surface of telson with median carina, and, on either side, 1 sharp carina terminating in blunt tubercle, 2 rounded intermediate denticles, outer with ventral spinule, and 1 lateral denticle with ventral spinule; basal prolongation of uropod slender, inner spine longer, with rounded lobe between terminal spines and 10-12 fixed spines on inner margin.

*Description.*—Eye triangular, corneal axis greater than peduncular, cornea noticeably bilobed; eyes extend past second segment of antennular peduncle; ocular scales separate, truncate.

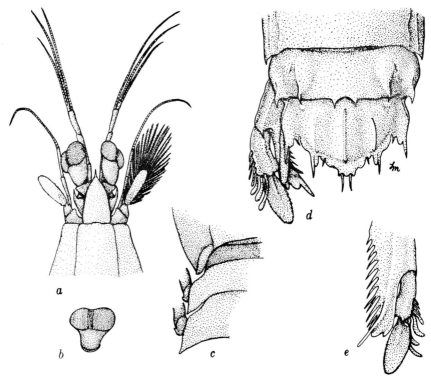

FIGURE 73. *Eurysquilla holthuisi,* n. sp., male holotype, TL 34.8, off Panama. *a,* anterior portion of body; *b,* eye; *c,* lateral processes of fifth to seventh thoracic somites; *d,* last abdominal somite, telson, and uropod; *e,* uropod, ventral view (setae damaged, most omitted).

Antennular processes produced into sharp, anteriorly-directed spines.

Antennal protopod with 1 mesial and 1 ventral papilla.

Rostral plate elongate-triangular, with small but sharp apical spine.

Carapace strongly narrowed anteriorly, with distinct marginal carina along posterior of each lateral plate.

Raptorial claw with notch at base of outer margin of dactylus; carpus lacking prominent dorsal ridge.

Mandibular palp present; 3 epipods present.

Fifth thoracic somite without distinct lateral process but with ventral spine on each side; lateral processes of sixth and seventh thoracic somites with acute, sharp posterolateral angles; basal segments of walking legs with 2 posterior spines, outer more prominent; endopod of walking legs ovate, composed of 2 segments; ventral process of eighth thoracic somite elongate, with apex rounded.

First 4 abdominal somites without longitudinal carinae but with shallow lateral grooves; fifth somite with marked lateral swelling, faintly carinate, and with small spinule at posterolateral angles; sixth somite with 6 sharp posterior spines, submedians faintly carinate, intermediates arising anterior to margin; sixth somite with strong spine in front of articulation of uropod.

Telson with sharp median carina terminating in posterior spine, and, on either side, 1 sharp carina terminating in blunt tooth; inner margin of intermediate and lateral tooth each with a broad, rounded, upturned lobe, as in *E. maiaguesensis;* intermediate denticles rounded, outer smaller, with sharp ventral spinule; lateral denticle with ventral spinule; ventral surface of telson smooth lateral to anal pore.

Uropods with exopod not articulated at right angle to basal segment; proximal segment of exopod with 7-8 movable, graded, spatulate spines, last recurved, not extending to end of distal segment; basal prolongation slender, channeled ventrally, with large inner spine, smaller outer spine and intervening rounded lobe; inner margin of basal prolongation with 10-12 fixed spines, several bifurcate.

*Color.*—Not particularly distinctive in preservative; most body segments outlined in faded orange, with patches of orange along lateral portions; telson with carinae outlined in orange; no dark chromatophores as in *E. plumata* and *E. maiaguesensis.*

*Size.*—Male holotype, only known specimen, TL 34.8 mm. Other measurements: carapace length, 7.1; cornea width, 1.9; eye length, 1.7; rostral plate length, 2.2, width, 1.4; telson length, 3.4, width, 5.9.

*Discussion.*—The bilobed eyes, very elongate rostral plate, and sharp posterolateral angles of the sixth and seventh thoracic somites will separate this species from all other species of *Eurysquilla*. Only *E. holthuisi,* n. sp., and *E. chacei,* n. sp., have a rounded lobe between the terminal spines of the basal prolongation of the uropod; *E. plumata* can be immediaetly distinguished from this species by its slender uropod which lacks additional spines on the inner margin.

*Etymology.*—The species is named for L. B. Holthuis in recognition of his studies on systematics of decapod and stomatopod crustaceans.

*Ontogeny.*—Unknown.

*Type.*—U. S. National Museum.

*Type-locality.*—Caribbean Sea, off Panama.

*Distribution.*—Known only from the type-locality.

## Genus *Pseudosquilla* Dana, 1852

Pseudosquille Eydoux & Souleyet, 1842, p. 263 [originally published in the vernacular].
*Pseudosquilla* Dana, 1852, p. 615.—Serène, 1962, p. 9.—Manning, 1963a, p. 311.—Holthuis & Manning, 1964, p. 137.—Manning, 1968, p. 139 [key].

*Definition.*—Body compact, semicylindrical, surface smooth; eyes with cornea cylindrical or distally flattened, rarely bilobed; rostral plate ovate, anterolateral angles rounded, apical spine present or absent; antennal protopod without papillae but with keeled dorsal process; carapace narrowed anteriorly, rounded anterolaterally and posterolaterally, without carinae or spines; cervical groove not distinct across dorsum of carapace; exposed thoacic somites without carinae; eighth thoracic somite with a feeble median ventral keel; epipods present on first 5 thoracic appendages; mandibular palp three-segmented; raptorial claw slender, dactylus armed with 3 teeth; propodus pectinate proximally only, with 3 movable spines at base, proximal the largest; carpus without sharp dorsal ridge; merus grooved inferiorly throughout its length for reception of propodus; ischiomeral articulation terminal; endopod of walking legs slender, made up of 1 segment only; abdomen without sharp carinae on first 5 somites, anterolateral plates articulated; sixth abdominal usually with 3 pairs of spines, submedians, intermediates, and laterals; telson slender, with sharp median carina and 3 or 4 pairs of dorsal carinae, posterior margin with 3 pairs of teeth, submedians with bases appressed and with movable apices; submedian denticles absent, 2 intermediate and 1 lateral denticle present; basal segment of uropod with distal, dorsal spine; basal prolongation produced into 2 spines, outer usually the longer.

*Type-species.*—*Squilla ciliata* Fabricius, 1787, in accordance with a petition to the ICZN (Holthuis & Manning, 1964) now pending.

*Gender.*—Feminine.

*Number of species.*—Six, of which two occur in the western Atlantic.

*Nomenclature.*—The rather complex nomenclature of this genus has been discussed at some length by Serène (1962), Manning (1963a), and Holthuis & Manning (1964). The following genera have been erected in recent years for species previously placed in *Pseudosquilla: Parasquilla* Manning, 1961; *Pseudosquillopsis* Serène, 1962; *Manningia* Serène, 1962; and *Eurysquilla* Manning, 1963.

*Remarks.*—The members of this genus are readily recognized by their

262

slender, compact shape, the presence of three teeth on the raptorial claw, and the presence of the channeled dorsal process on the antennal protopod. The latter feature is unique in this genus, and its presence will immediately separate species of *Pseudosquilla* from similar species in other genera, especially *Pseudosquillopsis*.

*Pseudosquilla* is circumtropical in distribution, although no single species occurs in all oceans. The most widely distributed species are *P. ciliata* and *P. oculata* which occur in all tropical areas except the eastern Pacific; these species are also the most common in the genus. The species inhabit relatively shallow water, and in general, are typically found on reefs, although *P. ciliata* is as abundant on shallow grass flats as it is on the reefs.

Several of the indices defined in the introduction (above) have been found to be of some value in supplementing the qualitative features traditionally used to separate species of *Pseudosquilla*. These indices include the Corneal Index (CI), the Eye Length-Width Index (ELWI), the Propodal Index (PI), the Propodus Length-Depth Index (PLDI), and the Abdominal Width-Carapace Length Index (AWCLI). Preliminary use of these indices in the material of *P. oculata* from different geographic areas seems to indicate the existence of subspecific (or infra-subspecific) groups. At present the samples are far too small for the indices to be used with confidence. Indices for Atlantic specimens of *P. oculata* are included below to aid future work with the indices when more material is available.

Color is an important taxonomic character in this genus. All of the known species can be distinguished on the basis of color pattern alone. This feature is more important to the field worker than to the museum taxonomist, although in some cases color pattern in *Pseudosquilla* is discernible in specimens that have been preserved for over 50 years. Of the six species now placed in this genus, only one, *P. ciliata*, lacks some sort of distinct eyespot on the carapace.

Ingle (1963) indicated that the gross structure of the male pleopod could be used to distinguish species of *Pseudosquilla* [*sensu lato*] from those of *Lysiosquilla* [*sensu lato*]. His analysis was made before most of the currently accepted genera had been recognized. Possibly the structure of the pleopod is indicative of general affinities of groups of genera. As Ingle pointed out, the utility of this feature is lessened by the fact that unrelated species may have similar structures whereas in related species the structure may be entirely different.

*Affinities.*—Comments on the affinities of this genus have been given in the introduction to the family, above.

KEY TO WESTERN ATLANTIC SPECIES OF *Pseudosquilla*

1. Rostral plate lacking apical spine; cornea cylindrical. . *ciliata,* p. 264.
1. Rostral plate with apical spine; cornea broadened distally . . . . . . . . .
. . . . . . . . . . . . . . . . . . . . . . . . . . . . . . . . . . . . . . . . . . . . . . *oculata,* p. 271.

Pseudosquilla ciliata (Fabricius, 1787)
Figure 74

*Restricted synonymy:*
*Squilla ciliata* Fabricius, 1787, p. 333 [not Atlantic].—Manning, 1963a, p. 311 [listed].
*Squilla stylifera* Lamarck, 1818, p. 189 [no locality].—Gibbes, 1850, p. 200 [p. 36 on separate; listed]; 1850a, p. 25 [listed].—von Martens, 1872, p. 146.—Mellis, 1875, p. 204.
*Squilla quadrispinosa* Eydoux & Souleyet, 1842, p. 262, pl. v, fig. 1 [Hawaii: based on an abormality].
*Alimerichthus cylindricus* Guérin-Méneville, in Sagra, 1855, pl. 3, fig. 12; in Sagra, 1857, p. lxv, pl. 3, fig. 12 [pelagic larval stage].—Gurney, 1946, p. 161.—Manning, 1963a, p. 310 [discussion].—Holthuis & Manning, 1964, p. 138.
non *Pseudosquilla monodactyla:* Miers, 1880, p. 110, pl. III, figs. 1-2 [postlarval *P. oculata*].
*Pseudosquilla ciliata:* Brooks, 1886, p. 53, pl. 15, fig. 10.—Bigelow, 1893, pp. 101, 102; 1894, p. 499.—Faxon, 1896, p. 165.—Nobili, 1898, p. 2.—Rankin, 1898, p. 253.—Rathbun, 1899, p. 628.—Rankin, 1900, p. 545.—Young, 1900, p. 501 [compiled].—Bigelow, 1901, p. 154, text-figs. 3-4.—Verrill, 1901, p. 20.—Rathbun, 1919, p. 347.—Schmitt, 1924, p. 81.—Boone, 1927, p. 6.—Pratt, 1935, p. 445.—Lunz, 1937, p. 6.—Monod, 1939, p. 567.—Holthuis, 1941, p. 35.—Coventry, 1944, p. 544.—Andrews, Bigelow & Morgan, 1945, p. 340.—Gurney, 1946, p. 152.—Dennell, 1950, p. 63.—Morrison & Morrison, 1952, p. 396.—Chace, 1954, p. 449.—Lemos de Castro, 1955, p. 26, text-figs. 20-21, pl. 7, fig. 39, pl. 16, fig. 51.—Manning, 1959, p. 18; 1961, p. 39, pl. 11, figs. 1-2; 1963b, p. 468 [listed].—Elofsson, 1965, p. 5 [anatomy].—Bullis & Thompson, 1965, p. 13 [listed].—Manning, 1966, p. 380, text-figs. 8a-b; 1967b, p. 104.
*Pseuderichthus communis* Hansen, 1895, p. 86, pl. 8, figs. 5-5b [pelagic larva].—? Calman, 1917, p. 142.—Gurney, 1946, p. 169.—Manning, 1963a, p. 310 [discussion].
*Pseudosquilla ciliata* var. *occidentalis* Borradaile, 1899, p. 402.—Verrill, 1923, p. 192, pl. 50, figs. 1-2, pl. 51, figs. 1-1b, pl. 54, fig. 2.—Boone, 1930, pp. 11, 24, pl. 2.
*monodactyla* stage: Bigelow, 1901, p. 156.—Holthuis, 1941, p. 35.
*Pseudosquilla monodactyla:* Nobili, 1898, p. 2.—Rathbun, 1919, p. 348.—Schmitt, 1924, p. 81 [postlarva].
*Pseudosquilla:* Milne & Milne, 1961, p. 422, figs. 10-11 [biology].

*Previous Atlantic records.*—BERMUDA: Bigelow, 1894; Rankin, 1900; Verrill, 1901, 1923; Gurney, 1946; Dennell, 1950; Morrison & Morrison, 1952.—BAHAMAS: Bigelow, 1893, 1894; Rankin, 1898; Andrews, Bigelow, & Morgan, 1945; Manning, 1959.—Milne & Milne, 1961.—FLORIDA: Bigelow, 1893, 1894; Boone, 1930; Lunz, 1937; Chace, 1954; Manning, 1959, 1967b.—PANAMA: Manning, 1961.—CUBA: von Martens, 1872; Boone, 1927. — JAMAICA: Rathbun, 1899. — PUERTO RICO: Bigelow, 1901.—VIRGIN ISLANDS: St. Thomas (Brooks, 1886; Nobili, 1898; Bigelow, 1901).—LESSER ANTILLES: St. Eustatius (Rathbun, 1919; Holthuis, 1941).—Guadeloupe (Monod, 1939).—Martinique (Faxon, 1896).—COLOMBIA: Holthuis, 1941; Coventry, 1944; Manning, 1961.—ARUBA: Rathbun, 1919; Holthuis,

1941.—CURAÇAO: Rathbun, 1919; Schmitt, 1924; Holthuis, 1941.—BRAZIL: Lemos de Castro, 1955; Manning, 1966.—ST. HELENA: Mellis, 1875.—WEST AFRICA: Longhurst, 1958.

Bullis & Thompson (1965) record this species at six OREGON stations.

In addition to these records for adults and monodactyla stages, larvae attributed to this species have been recorded in the Atlantic as follows: ATLANTIC: Guérin-Méneville, in Sagra, 1857; Claus, 1871.—off MASSACHUSETTS: Verrill, 1923.—SARGASSO SEA: Hansen, 1895.—NORTH EQUATORIAL CURRENT: Hansen, 1895.—CUBA: Guérin-Méneville, in Sagra, 1855.—off BRAZIL: Calman, 1917.

Kemp's 1913 record from South Carolina was probably based on von Marten's (1872) mention of Gibbes' (1850) material in the Charleston collection. Gibbes specified no locality for his material.

Boone's (1930) record for the Florida Keys may have erroneously included two specimens of *Squilla discors* Manning that were found from the same locality in the collection of the Vanderbilt Marine Museum. Boone mentioned 10 specimens of *P. ciliata,* but only eight could be located by me (Manning, 1967b).

*Material.*—BERMUDA: 2 ♂; AMNH 8200.—1 ♀; USNM 5136.—1 ♀; Hungry Bay; USNM 25453.—1 ♂, 2 ♀; same; YPM 4424.—1 ♂, 3 ♀; same; YPM 4432.—1 ♀; Ely's Harbor; YPM 4422.—2 ♂, 1 ♀; Long Bird Id.; YPM.—BAHAMAS: 1 ♀; UMML 32.1219.—1 ♂; Bimini; USNM 88687.—1 ♀; Andros Id.; AMNH 9176.—5 ♂, 3 ♀ (4 lots); Clifton Bay, Lyford Cay; USNM.—2 ♂, 1 ♀; off Whale Cay; MCZ.—1 ♀; off Ambergris Cay; MCZ.—1 ♀; Clarence Harbor; USNM 76027.—FLORIDA: 6 ♂, 7 ♀ (7 lots); off Palm Beach; USNM.—1 ♂; Lake Worth; UMML 32.1149.—1 ♀; same; AMNH 9792.— 1 ♂; Straits of Florida, off Hollywood; GERDA Sta. 423; UMML.—1 ♂, 4 ♀ (3 lots); Miami; USNM.— 1 ♂; Biscayne Bay; FISH HAWK Sta. 7482; USNM.—12 ♂, 14 ♀, 1 monodactyla; same; UMML 32.1156, 1210, 1214, 1215, 1216, 1217, 1218, 1220, 1222, 1223, 1224, 1225, 1572.—2 ♂ (2 lots); Bear Cut; USNM.—2 ♂, 1 ♀; same; RMNH.—1 ♂, 1 ♀; off Key Biscayne; USNM 45631.—1 ♂; channel W of Soldier Key; USNM.—1 ♂, 2 ♀; Ragged Keys; USNM.—1 ♂, 1 ♀; Sands Key; UMML 32.430.—1 ♂, 1 ♀; Key Largo; USNM 14105.—2 ♂; same; USNM 14106.—3 ♂; Harry Harris Park, S Key Largo; USNM.—1 ♂; ENE of Crocker Reef; USNM.—5 ♂, 3 ♀ (3 lots); off Alligator Reef Light; USNM.—1 ♀; Lower Matecumbe Key; UMML.—2 ♂, 3 ♀ (3 lots); same; USNM.—2 ♂, 2 ♀; same; USNM 14107.—1 ♂, 1 monodactyla; Indian Key; USNM 14103.—1 ♂; Key Vacas; USNM 14104.—1 ♂; Hawk Channel; FISH HAWK Sta. 7464; USNM.—1 ♂, 1 ♀; same; FISH HAWK Sta. 7469; USNM.—1 ♂, 1 ♀ (2 lots); Little Duck Key; USNM.—1 ♂; N of Knights Key channel; FISH HAWK Sta. 7410; USNM.—1 ♂; Knights Key; USNM.—2 ♀; Summerland Keys; USNM.—2 ♂, 3 ♀; No Name Key; USNM.—1 ♂, 1 ♀; Newfound

Harbor Key; USNM.—6 ♂, 2 ♀; Lignum Vitae Channel; USNM.—1 monodactyla; Grassy Key Lake; Fɪsʜ Hᴀᴡᴋ Sta. 7431; USNM.—1 ♀; (?) Broad Creek; USNM 76031.—1 ♀; (?) off Lose Key; ANSP 4423.— 1 ♂; 2 mi. off Boca Chica; ANSP 4432.—2 ♂; off Key West; Fɪsʜ Hᴀᴡᴋ Sta. 7277; USNM.—1 ♂, 2 ♀ (2 lots); Key West; USNM.—5 ♂, 2 ♀ (6 lots); Key West; USNM 14043, 14108, 15630, 18004, 21494, 46059. —3 ♂, 4 ♀; same; UMML 32.1221.—1 ♀; S of Key West; ANSP 4378.— 2 ♂; off Sand Key, Key West; YPM 4445.—1 ♂; Tortugas; USNM.—1 ♂, 1 ♀; Dry Tortugas; USNM 81412.—1 ♂; moat, Fort Jefferson, Dry Tortugas; USNM.—1 ♀; Long Key seining beach; USNM.—1 ♀; Garden Key; ANSP.—MEXICO: 1 ♂, 2 ♀ (dry); Gulf of Campeche; MCZ.—2 ♂; Cayo Norte, Banco Chinchorro, Yucatan; USNM.—1 ♂; near Allen Point Light, Quintana Roo; Bredin Sta. 89-60; USNM.—BRITISH HONDURAS: 1 ♂; Turneffe Id.; USNM.—1 ♂, 1 monodactyla; E shore Half Moon Cay, Lighthouse Reef; USNM.—NICARAGUA: 1 ♂, 2 ♀; Oʀᴇɢᴏɴ Sta. 3603; USNM.—OLD PROVIDENCE ID.: 2 ♂; USNM 78318.— 1 ♂, 3 ♀; Aʟʙᴀᴛʀᴏss; USNM 9134.—PANAMA: 1 ♀; Oʀᴇɢᴏɴ Sta. 3555; USNM.—1 ♂, 1 ♀; Caledonia Bay; AHF Sta. A2-39; USNM.—1 ♀; same; AHF Sta. A50-39; USNM 106349.—1 ♂; same; AHF Sta. A57-39; AHF. —1 ♂, 2 ♀; Fox Bay, Colon; USNM 106052.—1 ♂; Portobello; USNM 44185.—1 ♀; (?) Playa de Dama; USNM 76051.—CUBA: 1 ♂; Esperanza; USNM 48520.—1 ♂; same; USNM 48561.—1 ♀; between Cape Antonio and Cape Cajon; USNM 48557.—1 ♀; reef, (?) Cayo Lavesos; USNM 48526.—JAMAICA: 1 ♂; Aʟʙᴀᴛʀᴏss; USNM 7693.—1 ♂; Montego Bay; USNM.—1 ♂; Northeast Cay, Morant Cays; USNM.—1 ♀; same; SMIJ.—1 ♂; Rocky Cay, Morant Cays; SMIJ.—DOMINICAN REPUBLIC: 3 ♂, 1 ♀; Sɪʟᴠᴇʀ Bᴀʏ Sta. 5149; USNM.—1 ♀; El Cayo; AMNH 8867.—1 ♂; same; AMNH 8868.—1 brk. ♂; N end El Cayo; AMNH 8865.—PUERTO RICO: 1 ♂; N of Puerto Rico; Johnson-Smithsonian Deep Sea Exped. Sta. 16; USNM.—1 ♀; same; Johnson-Smithsonian Deep Sea Exped. Sta. 26; USNM.—1 monodactyla; San Juan Harbor; Fɪsʜ Hᴀᴡᴋ P. R. Sta. 125 (6053); USNM 64819.—1 ♂; Mayaguez; Fɪsʜ Hᴀᴡᴋ; USNM 64821.—1 ♂; same Fɪsʜ Hᴀᴡᴋ; USNM 64822.— 3 ♀; Rat Id., Joyuda; USNM.—1 ♀ Puerto Real; Fɪsʜ Hᴀᴡᴋ; USNM 64775.—1 ♀; Boqueron Bay; Fɪsʜ Hᴀᴡᴋ; USNM 64810.—1 ♂; off Cabo Rojo; Fɪsʜ Hᴀᴡᴋ P. R. Sta. 144 (6072); USNM 64814.—1 ♂; El Morillo, Cabo Rojo; IMB.—1 ♀; Boqueron, Cabo Rojo; IMB.—1 ♀; off Magueyes Id., Parguera; USNM.—1 ♀; same; IMB.—1 ♀; Parguera; AHF. —1 ♂; Cayo Margarita, Parguera; IMB.—1 ♂, 1 ♀; S of Margarita Reef, Parguera; IMB.—1 ♀; Guanica Harbor; AMNH 2042.—1 ♂; same; AMNH 2045.—1 ♀; Punta Corcabao, Guanica; IMB.—1 ♀; Cayo Romero, Guanica; IMB.—1 ♂; off Guanica Harbor; ANSP.—1 ♂; Bahia de Yegua; UMML 32.1213.—1 ♂; 4 mi. E of Tallaboa; AMNH 2048.—2 monodactylas; between Ratones and Caribe Id., off Tallaboa Bay; AMNH 2043.—1 ♂; off Boca Prieta, Punta Guaniquilla; Fɪsʜ Hᴀᴡᴋ P. R. Sta.

147 (6075); USNM 64815.—1 ♂; Lighthouse Reef, Playa de Ponce; FISH HAWK; USNM 64818.—2 ♀; Ponce; FISH HAWK; USNM 64774.— 1 ♂; same; FISH HAWK; USNM 64812.—1 ♀; off Humacao, Hucares; FISH HAWK P. R. Sta. 171 (6099); USNM 64813.—2 ♂; Fayardo; FISH HAWK; USNM 64776.—1 ♀, 1 mondactyla; off Vieques Id.; FISH HAWK P. R. Sta. 164 (6092); USNM 64808.—1 ♂; off Culebra Id.; FISH HAWK P. R. Sta. 165 (6093); USNM 64809.—1 ♂; Ensenada Honda, Culebra; FISH HAWK; USNM 64807.—3 monodactylas; same; FISH HAWK; USNM 64816.— 1 ♀; same ; FISH HAWK; USNM 64820.—1 ♀; (?) Piedra Prieta Reef; AMNH 8866.—VIRGIN ISLANDS: 3 ♂; no other data; UZM.—1 ♀; St. Thomas; ALBATROSS; USNM 7659.—1 ♂, 2 ♀; St. Thomas; FISH HAWK P. R. Sta. 151 (6079); USNM 64817.—1 ♀; between St. Thomas and St. John; UZM.—2 ♀ (2 lots); St. John; UMML.—1 ♀; Tortola; UMML.—1 ♀; Virgin Gorda; Bredin Stats. 37, 38, 39-58; USNM.— 1 ♀; Christiansted, St. Croix; UZM.—1 dry ♂; Salt River, St. Croix; USNM 104812.— 1 ♂, 1 ♀; Salt River Reefs, St. Croix; USNM 71846.— LESSER ANTILLES: 1 ♂, 1 ♀; Oranjestad Harbor, St. Eustatius; USNM 42985.—2 ♂; N side Cocoa Point, Barbuda; Bredin Sta. 98-59; USNM.— 2 ♀; E side Cocoa Point; Bredin Sta. 108-58; USNM.—1 ♂; Tank Bay, English Harbor, Antigua; Bredin Sta. 74-56; USNM.—1 monodactyla; Freeman's Bay, English Harbor; Bredin Sta. 79-58; USNM.—1 ♀; N of Black's Point, Falmouth Bay, Antigua; Bredin Sta. 109-59; USNM.— 1 brk. ♂; La Feuille, Guadeloupe; USNM 89590.—1 ♀; St. Lucia; ORE-GON Sta. 5058; USNM.—1 ♂, 1 ♀; same; OREGON Sta. 5059; USNM.— COLOMBIA: 1 ♀; OREGON Sta. 3537; USNM.—1 ♂; OREGON Sta. 4906; USNM.—1 brk. ♀; 11 mi. SW of Cape la Vela; AHF Sta. A12a-39; USNM.—1 ♂, 3 ♀; 2 mi. off Bahia Honda; AHF Sta. A15-39; AHF.— VENEZUELA: 1 ♀; Higuerote Miranda; UCVMB 5011.—1 ♂; 4 mi. N of Tortuga Id.; AHF Sta. A44-39; USNM.—BONAIRE: 1 ♀; Klein Bonaire; RMNH.—CURAÇAO: 2 ♂, 2 ♀; Curaçao; ALBATROSS; USNM 21495.—1 ♀; Caracas Bay; USNM 57548.—2 ♂; lagoon, Curaçao; USNM 42986.—1 ♀; Spanish Water; USNM 57547.—BRAZIL: Maranhao State; OREGON Sta. 4227; USNM.—1 ♂; same; OREGON Sta. 4229; USNM.— 2 ♀; same; OREGON Sta. 4231; USNM.—1 ♂, 1 ♀; Piaui State; OREGON Sta. 4243; USNM.—1 ♂; Paraiba State; ALBATROSS Sta. 2758; USNM 23390.—1 ♂; off Parahyba; CALYPSO Sta. 1; USNM.—2 ♂; same; CALYPSO Sta. 2; MNHNP.—1 ♀; off Recife; CALYPSO Sta. 23; MNHNP. —SAINT HELENA: 1 ♂, 4 monodactylas; Sugarloaf; UZM.—3 ♀, 2 monodactylas; Jamestown; UZM.—1 ♂; Flagstaff Bay; UZM.—2 ♂, 2 ♀, 1 monodactyla (in 2 lots); Lemon Valley; UZM.

*Diagnosis.*—Carapace without eyespots; cornea cylindrical, not noticeably expanded; rostral plate without apical spine; telson with median carina and three pairs of dorsal carinae.

*Color.*—Background color yellow-brown, bright green, or almost white,

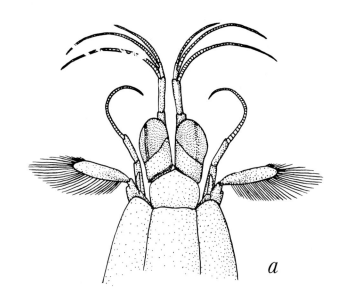

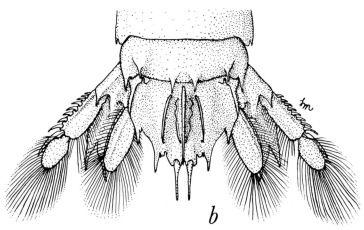

FIGURE 74. *Pseudosquilla ciliata* (Fabricius), male, TL 31.5, CALYPSO Sta. 2. *a,* anterior portion of body; *b,* last abdominal somite, telson, and uropods (from Manning, 1966).

often with two pairs of dark spots on carapace along gastric grooves; sixth thoracic and first abdominal somites each with a pair of dark lateral spots; telson with a pair of dark lateral spots anteriorly as well as a dark spot at base of each intermediate tooth; spines of basal prolongation and movable spines of uropod banded yellow or pink and white; appendages bright yellow, raptorial dactylus pink. Juveniles are often dark laterally with a light median stripe, although specimens up to 50 mm TL may show this pattern. The genital apertures of the female are usually very dark, almost black.

*Size.*—Males, TL 16.5-80.0 mm; females, TL 16.3-89.0 mm; monodactyla stages TL 17.8-23.7 mm.

*Discussion.*—The smaller eyes, unarmed rostral plate, and telson ornamented with a median carina and three pairs of carinae will serve to distinguish this species from *P. oculata,* the only other species of *Pseudosquilla* in the Atlantic.

*Remarks.*—There seems to be some habitat distinctness between *P. ciliata* and *P. oculata* in the western Atlantic. Although *P. ciliata* may occur along with the latter species on the reefs, it is more often found on shallow grass banks; it is often encountered swimming freely rather than in hiding. Townsley (1953) noted that in Hawaii both species were commonly found in the same habitat, on reefs, and that they could be mistaken for one another in the field. This is not the case in Florida specimens of the two species, for the bright color of *P. oculata* makes it difficult to confuse it with *P. ciliata* in the field; moreover, *P. oculata* is cryptic in habit and is rarely or never encountered swimming freely.

Specimens of *P. ciliata* have been examined from the stomach contents of the following fishes: bonefish, *Albula vulpes;* mutton snapper, *Lutjanus analis;* Nassau grouper, *Epinephelus striatus;* southern stingray, *Dasyatis americanus;* and bonnethead shark, *Sphyrna tiburo.* The bonefish typically feeds on shallow grass banks.

Serène (1951) found two morphologically distinct groups of specimens of *P. ciliata* from Indo-China. He designated these as "forme claire," with a broad telson, one spined intermediate denticle, and evenly rounded distal portion on the endopod of the uropod. The "forme foncée" differed in having an elongate telson, no spine on the intermediate denticle, and a distally-tapered endopod on the uropod. All of the material examined by me corresponds to Serène's "forme claire."

*Ontogeny.*—Knowledge of the larval stages of *P. ciliata* is restricted to scattered observations on late pelagic larvae and on post-larvae. Early larvae have not been described, and neither the number of stages nor the duration of larval life is known. An account of the early larval development is in preparation (in cooperation with A. J. Provenzano, Jr.). Late pelagic larvae been held through metamorphosis to the juvenile in the

269

laboratory in experiments conducted at the Institute of Marine Sciences, University of Miami, and a description of the changes from late larva to juvenile is also in preparation.

The names of larval forms listed in the synonymy were compiled with the aid of Gurney (1946) who was correct in the identification of Guérin-Méneville's *Alimerichthus cylindricus.*

Hansen (1895) was the first to show that *Squilla monodactyla* A. Milne-Edwards was the postlarva of a *Pseudosquilla,* and Hansen identified Milne-Edwards' species with *P. oculata* (Brullé). He also distinguished the pelagic larval forms of *P. ciliata* and *P. oculata,* and named the larva of the former *Pseuderichthus communis* and that of the latter *P. distinguendus. P. communis* is smaller (TL 21-24 mm) and has fewer (8) uropod spines than *P. distinguendus* (TL over 24 mm; 9-11 uropod spines). He also pointed out that the postlarva of *P. oculata* was larger than that of *P. ciliata.*

Bigelow (1931) gave distinguishing features of the monodactyla stages of three species from the Pacific, *P. ciliata, P. ornata,* and *P. oculata.* The monodactyla stages of *P. oxyrhyncha* and the eastern Pacific *P. adiastalta* are not known.

The main differences between the monodactyla stage and the juvenile are: (1) the raptorial claw of the monodactyla is unarmed; (2) the lateral dorsal carinae of the telson are not present; (3) the telson is armed with submedian denticles; and (4) the intermediate spines of the sixth abdominal somite are absent. The following notes were made from Atlantic postlarvae.

The eyes are almost cylindrical, with the cornea somewhat broader than in the adults. The rostral plate is broader than long, cordiform, but not as tapered as in *P. oculata.* The fifth abdominal somite is armed with a posterolateral spine, and the sixth abdominal somite lacks intermediate spines. The posterior cleft of the telson is not pronounced. The proximal segment of the uropod is less than 2 times as long as the distal. The outer margin of the uropod is armed with 8-9 spines. The outer spine of the basal prolongation is slightly longer than the inner, and the inner margin of the outer spine is not noticeably angled proximally.

The monodactyla of *P. ciliata* can be distinguished from that of *P. oculata* by the following features: (1) the eyes are slenderer; (2) the rostral plate is broader; (3) the cleft of the telson is not pronounced; (4) the outer margin of the uropod is armed with fewer spines, 8-9; and (5) the inner margin of the outer spine of the uropod is not produced into a distinct angle. The smaller size of *P. ciliata,* of course, will help to distinguish the monodactylas of the two species.

Bigelow (1901) reported monodactylas from Puerto Rico that measured 16-19 mm TL and juveniles with a size range of 16-17.5 mm TL. In the present material the monodactylas show a wide range of overlap with juveniles. Monodactylas measured TL 17.8-23.7 mm (Bigelow's specimens

270

could not all be remeasured) and juveniles as small as 16.3 mm were examined.

*Type.*—Not traced, probably not extant.

*Type-locality.*—Oceano Indico.

*Distribution.*—In the Indo-West Pacific this species is known from South Africa to Hawaii and Japan (Holthuis, 1941a). In the western Atlantic it is known from many localities from Bermuda, the Bahamas, Florida and the Gulf of Campeche to northern Brazil. It is also known from St. Helena and West Africa. It occurs from the littoral zone to 110 m, with the greatest abundance in shallower waters.

*Pseudosquilla oculata* (Brullé, 1836-44)

Figures 75-76

*Restricted synonymy:*
*Squilla oculata* Brullé, 1836-1844, p. 18, pl. unique, fig. 3.
*Squilla monodactyla* A. Milne-Edwards, 1878, p. 232 [postlarva].
*Pseudosquilla oculata:* Miers, 1880, p. 110, pl. III, figs. 3-4.—Koelbel, 1892, p. 109.—Rathbun, 1900, p. 155 [part].—Monod, 1925, p. 92 [listed]; 1932 p. 540 [p. 80 on separate; listed].—Dartevelle, 1951, p. 1032 [listed]. —Monod, 1951, p. 142.—Andrade Ramos, 1951, p. 143 [listed].—Lemos de Castro, 1955, p. 29, figs. 22-23, pl. 8, fig. 40, pl. 16, fig. 52.—Manning, 1959, p. 18; 1961, p. 38 [key]; 1963a, p. 311 [listed].
*Pseudosquilla monodactyla:* Miers, 1880, p. 110, pl. III, figs. 1-2.—Hansen, 1895, p. 85.
*Pseuderichthus distinguendus* Hansen, 1895, pp. 84, 86.—Jurich, 1904, p. 36.—Gurney, 1946, p. 169 [and synonymy].
not *Pseudosquilla monodactyla:* Rathbun, 1919, p. 348.—Schmitt, 1924, p. 81 [= *P. ciliata* (Fabricius)].
not *Pseudosquilla oculata:* Schmitt, 1940, p. 173, text-fig. 15 [= *P. adiastalta* Manning, 1964].

*Previous Atlantic records.*—EASTERN ATLANTIC: CANARY IS-LANDS (Brullé, 1836-44; Koelbel, 1892).—MADEIRA (Miers, 1880). —CAPE VERDE ISLANDS (Monod, 1925, 1951).—GULF OF GUINEA (Monod, 1925). WESTERN ATLANTIC: FLORIDA (Manning, 1959).—BRAZIL: Recife (Lemos de Castro, 1955); Maceio (Rathbun, 1900).

Larval stages and postlarvae have been recorded from the following localities in the Atlantic: Cape Verde Islands (Milne-Edwards); South Equatorial Current (Hansen; Jurich). Miers gave no locality data for his material.

*Material.*—MADEIRA: 1 ♂, 52.2; Camara de Lobos (fishing village east of Funchal); 17 July 1952; MMF 3501.—1 ♂, *ca.* 58.5; Funchal Fish Market; 12 November 1954; MMF 4721.—1 ♂, 68.0; same; 5 May 1955; MMF 5602.—1 ♂, 66.1; same; 30 January 1959; MMF 15344.—ST. HELENA: 1 ♂, 66.6; off West Point; disgorged by St. Helena jack,

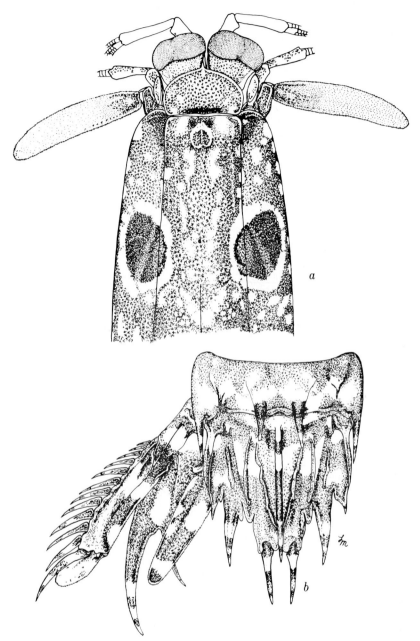

FIGURE 75. *Pseudosquilla oculata* (Brullé), male, TL 84.2, S. of Alligator Reef Light, Florida. *a*, anterior portion of body; *b*, last abdominal somite, telson, and uropod (setae omitted).

272

*Epinephelus ascensionis;* 25 July 1963; H. Wade; USNM 111122.—1 ♀, 66.3; James Bay; August 1963; A. Loveridge; USNM 111119.—BAHAMAS: 1 ♀, 45.7; Hogsty Reef, Northwest Cay; CBC Sta. 577; 28 May 1962; ANSP.—1 ♂, 35.5; same; CBC Sta. 580; 29 May 1962; ANSP.—1 ♀, 68.0; Little Bahama Bank, Grand Bahama Id., off Settlement Point; CBC Sta. 558; 28 July 1961; ANSP.—CAY SAL BANK: 1 ♂, 32.5; off Elbow Key; H. Brown; 9 June 1960; UMML 32.1876.—FLORIDA: 1 ♂, 79.5; ½ mi. off Matecumbe Reef, Monroe Co.; 24 August 1958; W. A. Starck, II; USNM 107766.—2 ♂, 72.5-75.3; 1 ♀; 42.5; 300 yds SW of Alligator Reef Light, Monroe Co.; 5-6 m, rock ledge; 7 January 1962; W. A. Starck, II, *et al.;* USNM 111121.—3 ♂, 34.9-105.0; 1 ♀, 31.4; same; 30 April 1961; UMML 32.2081.—2 ♂, 30.5-105.5; 1 ♀, 124.5; ½ mi SW Alligator Reef Light, Monroe Co.; 20 ft.; 21 July 1961; H. Feddern, *et al.;* USNM 110414.—3 ♂, 56.3-84.2; 2 ♀, 71.9-79.5; S of Alligator Reef Light, Monroe Co.; 5-6 m, inside ledge; 4 January 1963; R. Schroeder; USNM 110415.—MEXICO: 1 ♀, 52.5; reef about 1 mi. NW of Cayo Norte, Banco Chinchorro, Yucatan; 2-3 m; 22 June 1961; W. A. Starck, II; UMML.—1 ♂, 80.6; same; 23 June 1961; USNM 110416.— NICARAGUA: 2 ♂, 73.5-92.8; OREGON Sta. 4836; USNM 120335.—CUBA: 1 ♀, 33.5; Cayo Cristo, 4 mi. N of Isabella; 2-3 July 1918; B. Brown; AMNH 3192.—1 monodactyla, 29.0; northwestern Cuba; Tomas Barrera Expedition; USNM 48512.—DOMINICAN REPUBLIC: 1 ♂, 26.2; beyond sand spit, Piedra Prieta reef; 1 m, rocky; 13 July 1933; J. C. Armstrong; AMNH 8870.—1 ♀, 59.8; near sand pit, Piedra Prieta reef; 1 m, rocky; 14 July 1933; J. C. Armstrong; AMNH 8869.—PUERTO RICO: 29.5; coral reef, Ballena Point; 12 June 1915; R. W. Miner and II. Mueller; AMNH 2044.—VIRGIN ISLANDS: 2 ♀, 39.1-86.1; east side of Ram Head, St. John; shore; 15 August 1962; J. Randall, L. Morera; USNM 110413.—1 ♂, 79.3; 2 ♀, 71.2-78.1; Beehive Point, Lameshur Bay, St. John; 2-6 m; 14 June 1961; J. Randall, R. Schroeder; UMML.—1 ♂, 43.3, 1 monodactyla, 33.3; Road Harbor, Tortola; Bredin Sta. 117-56; 17 April 1956; USNM 111120.—BRAZIL: 1 ♂, 38.0; 1 ♀, 31.7; coral reef off Alagoas, Maceio; 3-4 August 1899; A. W. Greeley, Branner-Agassiz Expedition; USNM 25820.

*Diagnosis.*—Cornea expanded but not bilobed; rostral plate with small apical spine; carapace with pair of dark spots surrounded by light ring; telson with 4 longitudinal carinae, accessory medians, dorsal submedians, intermediates, and marginals, on either side of median carina; basal prolongation of uropod with outer spine the stronger.

*Color.*—Background dark; carapace with pair of large dark eyespots, each surrounded with white ring; anterior portions of lateral plates of carapace very dark with scattered large white spots; carpus, merus, and propodus of raptorial claws with large white spots, arranged in 2 rows on propodus; carpus with prominent dark spot on mesial surface; dactylus pink, macu-

273

lated on inferior border; thoracic and abdominal somites with 1 or more transverse bands of white spots; last 3 thoracic somites each with pair of dark spots on ventral surface, which may join to form bar on eighth somite of male; telson with transverse light band on anterior third, posterior teeth speckled, tips red; uropods maculated or banded, tips of spines red; basal segment of uropod with prominent dark, ventral spot, setae of uropods pink. The color pattern is shown in figures 75 and 76.

The dactyli of the Madeiran specimens were not maculated, and the bodies were uniformly gray.

*Size.*—Males, TL 30.5-105.5 mm; females, TL 29.5-124.5 mm; mono-dactyla stages, TL 29.0-33.3 mm. Measurements of and indices for selected specimens are given in Tables 10-12.

*Discussion.*—Although the specimens of *P. oculata* from both sides of the Atlantic are very similar morphologically, several of the indices that were calculated show differences between the two populations. Western Atlantic specimens have a larger eye with the cornea more expanded in relation to the stalk. The mean ELWI is 109 in western Atlantic forms, 121 in the eastern Atlantic specimens. The Corneal Index of the two largest Madeiran specimens is 490, whereas it ranges between 405 to 447 in American specimens of similar size. The raptorial propodus is shorter and deeper in the St. Helena and Madeira specimens. The mean PLDI for specimens from these localities is 424 as opposed to 468 for the American. The PLDI for the two largest Madeiran specimens is 423-424, whereas for American specimens of similar size it ranges from 467-532.

TABLE 10

Indices for Selected Specimens of *Pseudosquilla oculata* (Brullé) from the Western Atlantic.

| Sex | TL | CL | CI | ELWI | AWCLI | PLDI | PI |
|---|---|---|---|---|---|---|---|
| ♂ | 30.5 | 5.6 | 311 | 139 | 768 | 429 | 093 |
| ♀ | 31.4 | 6.6 | 314 | 114 | 727 | 444 | 093 |
| ♂ | 34.9 | 7.3 | 348 | 129 | 727 | 476 | 090 |
| ♀ | 42.5 | 9.3 | 358 | 119 | 677 | 524 | 085 |
| ♀ | 52.5 | 12.9 | 403 | 094 | 651 | 469 | 095 |
| ♀ | 71.2 | 15.7 | 405 | 108 | 662 | 532 | 089 |
| ♂ | 72.5 | 16.1 | 447 | 111 | 689 | 467 | 096 |
| ♂ | 74.7 | 17.7 | 443 | 108 | 678 | 452 | 101 |
| ♂ | 75.3 | 16.3 | 429 | 111 | 675 | 500 | 091 |
| ♀ | 78.1 | 19.1 | 415 | 097 | 639 | 460 | 097 |
| ♂ | 79.3 | 17.7 | 454 | 108 | 665 | 440 | 093 |
| ♂ | 80.6 | 19.2 | 427 | 100 | 651 | 415 | 099 |
| ♂ | 105.0 | 26.5 | 465 | 096 | 634 | 433 | 104 |
| ♂ | 105.5 | 23.7 | 465 | 106 | 675 | 454 | 092 |
| ♀ | 124.5 | 28.4 | 473 | 095 | 658 | 529 | 087 |
| Range | | | | 094-139 | 634-768 | 415-532 | 085-104 |
| Mean | | | | 109 | 678 | 468 | 094 |

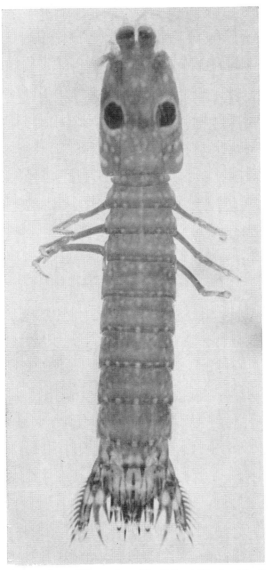

FIGURE 76. *Pseudosquilla oculata* (Brullé), Alligator Reef, Florida.

Samples from both areas are relatively small. In the absence of marked differences in morphology or color pattern and in view of the sample size it is felt that the differences observed cannot be used at present to show subspecific distinctness of the forms from the two areas. With larger samples, subspecific differences might well be indicated. Indices for all eastern Atlantic specimens are given in Table 11.

The eastern Pacific species formerly referred to *P. oculata* by Schmitt (1940) and others is a distinct species, *P. adiastalta* Manning (1964). It differs from *P. oculata* in lacking distinct eye-spots on the carapace; it also has smaller eyes and a stouter raptorial claw than its eastern American analogue.

Variation in morphological features in Indo-West Pacific specimens of *P. oculata,* with particular reference to features that can be expressed as indices, will be published in a review of the Indo-West Pacific species of *Pseudosquilla* now in progress.

*Remarks.*—The western Atlantic representatives of this widely distributed species appear to be restricted to the reef habitat, where they may be taken together with *P. ciliata* (see remarks under *P. ciliata*). The bright color of *P. oculata* will immediately distinguish the two species in the field.

The largest female examined (TL 124.5) is by far the largest specimen reported for this species. Miers (1880) reported a specimen three and one-quarter inches (ca. 83 mm) long.

Townsley (1953) reported that specimens from Hawaii had intermediate spines on the fifth abdominal somite. These spines were not seen by me in either Atlantic or Pacific specimens.

*Ontogeny.*—Only the late pelagic larval stages and the postlarva of this species are known. The pelagic larvae were named *Pseuderichthus distinguendus* and identified with *P. oculata* by Hansen (1895). Hansen distinguished *P. distinguendus* from *P. communis,* the larva of *Pseudosquilla ciliata,* primarily on the basis of size, TL over 24 mm in the former, 21-24 mm in the latter. He also showed that the species described as *Squilla*

TABLE 11

INDICES FOR ALL SPECIMENS OF *Pseudosquilla oculata* (Brullé) FROM MADEIRA AND ST. HELENA

| Loc | Sex | TL | CL | CI | ELWI | AWCLI | PLDI | PI |
|-----|-----|------|------|-----|---------|---------|---------|---------|
| M | ♂ | 52.2 | 12.6 | 450 | 114 | 698 | 433 | 097 |
| M | ♂ | 58.5 | 12.9 | 461 | 129 | 729 | 450 | 096 |
| M | ♂ | 66.1 | 14.7 | 490 | 123 | 682 | 424 | 105 |
| SH | ♀ | 66.3 | 15.1 | 444 | 115 | 715 | 416 | 096 |
| SH | ♂ | 66.6 | 14.9 | 438 | 115 | 691 | 395 | 102 |
| M | ♂ | 68.0 | 14.7 | 490 | 130 | 707 | 423 | 099 |
| Range | | | | | 114-130 | 682-729 | 395-450 | 096-105 |
| Mean | | | | | 121 | 704 | 424 | 099 |

276

TABLE 12
MEASUREMENTS (IN MM) OF SELECTED ATLANTIC SPECIMENS OF
*Pseudosquilla oculata* (Brullé)

| Source | Florida | | Yucatan | Virgin Ids. | | Madeira | Bahamas |
|---|---|---|---|---|---|---|---|
| Sex | ♀ | ♂ | ♂ | ♀ | ♀ | ♂ | ♀ |
| TL | 124.5 | 105.5 | 80.6 | 78.1 | 71.2 | 68.0 | 45.7 |
| CL | 28.4 | 23.7 | 19.2 | 19.1 | 15.7 | 14.7 | 10.8 |
| Cornea width | 6.0 | 5.1 | 4.5 | 4.6 | 3.5 | 3.0 | 2.6 |
| Eye length | 5.7 | 15.4 | 4.5 | 4.5 | 3.8 | 3.9 | 3.3 |
| Propodus length | 32.8 | 25.4 | 19.3 | 19.7 | 17.7 | 14.8 | 12.1 |
| Propodus depth | 6.2 | 5.6 | 4.6 | 4.3 | 3.3 | 3.5 | 2.5 |
| Width of 5th abdominal somite | 18.7 | 16.0 | 12.5 | 12.2 | 10.4 | 10.4 | 7.6 |

*monodactyla* by A Milne-Edwards, 1878, was the postlarva (monodactyla stage) of *P. oculata,* and *P. monodactyla:* Miers, 1880, was identified as the postlarva of the same species.

Bigelow (1931), who outlined the major features of the monodactyla stages of *P. oculata, P. ornata,* and *P. ciliata* from the Pacific, has shown that the monodactyla of *P. ciliata* is smaller (TL 17.5-24.0 mm) and has fewer movable spines on the uropod (9) than that of *P. oculata* (TL 24.0-34.0 mm; 10-11 spines). The following notes on a monodactyla stage of *P. oculata* are based on the smaller specimen examined by me.

The eyes are large, with the cornea noticeably broadened. The rostral plate is slightly longer than broad, cordiform in outline. There are postero-lateral spines on the fourth and fifth abdominal somites, and the sixth somite has submedian and lateral spines but no intermediates. The median cleft of the telson is very pronounced. The outer margin of the uropod is armed with 11 teeth, and the proximal segment of the exopod of the uropod is over three times as long as the distal segment. The outer spine of the basal prolongation is slightly longer than the inner, and the inner margin of the outer spine is noticeably angulated rather than smoothly rounded.

The monodactylas examined by me measured 29.0-33.3 mm TL. Milne-Edwards specimen was 27 mm long, and Miers specimen was about 25 mm long. The major differences in morphological features between the monodactylas of *P. oculata* and *P. ciliata* were discussed under the latter species.

Gurney (1946) has provided a synonymy of both *P. communis* and *P. distinguendus,* as well as most other larval forms, and his work should be consulted for Pacific records.

*Type.*—A dry specimen, Muséum National d'Histoire Naturelle, Paris.

*Type-locality.*—Canary Islands.

*Distribution.*—Indo-West Pacific, from Mauritius to the Hawaiian Islands (Holthuis, 1941a) and Atlantic. In the eastern Atlantic it has been reported from Madeira, the Cape Verde Islands, the Canary Islands, and the Gulf of Guinea. In the western Atlantic it is known from the Bahamas, Florida, Yucatan, off Nicaragua, Cuba, Dominican Republic, Puerto Rico, Virgin Islands, and Recife and Maceio, Brazil. It has not been recorded previously from St. Helena.

## Genus *Parasquilla* Manning, 1961

*Parasquilla* Manning, 1961, p. 7; 1962, p. 80.—Serène, 1962, p. 7.—Manning, 1963a, p. 312; 1968, p. 139 [key].
*Pseudosquilla:* Figueiredo, 1962, p. 5.—Ingle, 1963, p. 11 [reference to *Pseudosquilla ferussaci* only].
*Pseudosquillopsis (Faughnia)* Serène, 1962, p. 17.—Manning, 1963a, p. 311.
not *Pseudosquillopsis (Pseudosquillopsis)* Serène, 1962, p. 16.

*Definition.*—Body compact, depressed, surface minutely pitted; eyes with cornea bilobed, outer margin of eye longer than inner; rostral plate trapezoidal, broader than long, apex unarmed; antennal protopod with 1 ventral papilla; carapace narrowed anteriorly, anterolateral angles acute or truncate, unarmed, posterolateral margins rounded; maginal carinae prominent posterolaterally; cervical groove present across dorsum of carapace; exposed thoracic somites with longitudinal carinae; eighth thoracic somite with a prominent subtriangular or rounded median, longitudinal, ventral keel; epipods present on first 5 thoracic appendages; mandibular palp three-segmented; raptorial claw stout, dactylus armed with 3 teeth; propodus fully pectinate, with 3 movable teeth at base, proximal much the largest; carpus with well-defined dorsal ridge divided into 2 teeth or lobes; merus grooved inferiorly throughout its length for reception of propodus; ischiomeral articulation terminal; endopod of walking legs slender, two-segmented; abdomen flattened, with carinae on first 5 somites, submedians absent in subgenus *Faughnia;* anterolateral plates of abdomen articulated; sixth abdominal somite with 6 carinae, submedians and laterals spined, intermediates with or without spines; telson slender, with sharp median carina and at most one pair of dorsal carinae; posterior margin with 3 pairs of teeth, submedians with movable apices; submedian teeth with bases separate, minute submedian denticles present; 2 intermediate and 1 lateral denticle present; basal segment of uropod with distal, dorsal spine; basal prolongation produced into 3 spines, outer longest.

*Type-species.*—*Parasquilla meridionalis* Manning, 1961, by original designation.

*Gender.*—Feminine.

*Number of species.*—Four, of which two occur in the western Atlantic.

*Nomenclature.*—Manning (1961) separated this genus from the nominal *Pseudosquilla* Dana, 1852, and originally included two species, *Squilla ferussaci* Roux, 1828, and *Parasquilla meridionalis.* In 1962 Serène de-

278

scribed *Pseudosquillopsis,* with two subgenera, one of which, *Faughnia,* was later (1963a) transferred to *Parasquilla* by Manning.

*Remarks.*— The members of this genus are rather rare. They can be recognized by the combination of *Pseudosquilla*-like characters, including the slender telson with three pairs of marginal teeth and the raptorial claw armed with three teeth, with *Squilla*-like features, such as abdominal carination.

In the Atlantic, all of the species of *Parasquilla* occur in moderate depths, in excess of 45 m, and would not ordinarily be encountered with any species of *Pseudosquilla,* with which they might be confused.

This genus now contains four species. *P. (Faughnia) haani* (Holthuis, 1959) has been reported from Japan, Formosa, and the Gulf of Siam (Serène, 1962); it is the only Indo-West Pacific representative of the genus. *P. (Parasquilla)* contains three species, all Atlantic, one of which, *P. ferussaci* (Roux, 1828) is known from the Mediterranean and adjacent Atlantic.

*Affinities.*—*Parasquilla* shows close affinities with *Pseudosquilla* and *Pseudosquillopsis.* It agrees with them in general body shape, slender outline of the telson, and possession of three teeth on the raptorial dactylus. With the latter it agrees in having a bilobed cornea set obliquely on the stalk, in having the raptorial propodus fully pectinate, and in having a similar armature on the basal prolongation of the uropod. It differs from both in having a rugose body, pentagonal rostral plate, a carinate abdomen, and submedian denticles on the telson.

The compact body, presence of only three teeth on the raptorial claw and slender telson will distinguish *Parasquilla* (as well as *Pseudosquilla* and *Pseudosquillopsis*) from *Eurysquilla, Eurysquilloides, Manningia,* and *Coronidopsis,* all of which are apparently more distantly related to the *Pseudosquilla* line.

KEY TO WESTERN ATLANTIC SPECIES OF *Parasquilla*

1. Rostral plate with anterolateral spines; lateral processes of sixth and seventh thoracic somites posteriorly acute ........ *coccinea,* p. 279.
1. Anterolateral angles of rostral plate blunt; lateral processes of sixth and seventh thoracic somites rounded ........ *meridionalis,* p. 283.

*Parasquilla (Parasquilla) coccinea* Manning, 1962
Figure 77

*Parasquilla coccinea* Manning, 1962, p. 181, fig. 1; 1963a, p. 312 [listed].— Bullis & Thompson, 1965, p. 13 [listed].

*Previous records.*—CAMPECHE BANK: northern GULF OF MEXICO; off EAST FLORIDA (Manning, 1962; Bullis & Thompson, 1965).

*Material.*—BAHAMAS: 1 ♂, 105.0; Santaren Channel; SILVER BAY Sta.

2468; USNM 119156.—FLORIDA: 1 ♂, 109.2; Straits of Florida; SILVER BAY Sta. 2390; holotype; USNM 107586.—1 ♂, 111.0; same; paratype; UMML 32.1758.—1 ♂, 85.9; S of Cape St. George; SILVER BAY Sta. 100; paratype; USNM 107587.—1 ♂, 95.3; NW of Cape San Blas; OREGON Sta. 944; paratype; USNM 96405.—OFF MISSISSIPPI DELTA: 1 ♂, 111.9; OREGON Sta. 83; paratype; USNM 91117.—MEXICO: 1 ♀, 47.5; NE to NW of Cayo Arenas, Campeche Bank; found in stomach of red snapper; 101-137 m; 1958; paratype; UMML 32.1170.

*Diagnosis.*—Anterolateral angles of rostral plate armed with 1 or 2 spines; posterolateral angles of sixth and seventh thoracic somites acute, directed posteriorly; oblique dorsal carinae of telson scarcely visible.

*Description.*—Lateral margins of rostral plate strongly concave, divergent; anterolateral angles armed with 1 or 2 spines directed obliquely outward; anterior margins sloping to obtuse apex.

Dorsal ridge of carpus of raptorial claw with 2 acute lobes, spined in smaller specimens.

Lateral processes of sixth and seventh thoracic somites spined posteriorly in small specimens, acute in larger ones.

Abdominal carinae spined as follows: submedian, 6; intermediate, 5; lateral, (5) 6; marginal, (1) 2-5.

Dorsal surface of telson with an obscure, oblique longitudinal prominence on either side of median carina; 7-11, usually 9-10, movable submedian denticles present.

Outer margin of penultimate segment of uropodal exopod with 7-11 graded, movable spines, last extending past midlength of distal segment.

*Color.*—Fresh specimens are brilliantly colored. Setae and spinules of appendages pink to crimson; body off-white, lateral portions dark, metallic; median portions of thoracic and abdominal somites outlined in gold; highlights of carinae and spines of body gold; median carina and spines of telson and uropods outlined in gold. In preservative all color fades but the light background and contrasting dark, metallic lateral portions.

*Size.*—Males, TL 85.9-111.9 mm; only known female, TL 47.5 mm. Other measurements of a male, TL 109.2: carapace length, 23.2; cornea width, 4.2; rostral plate length, 3.1, width, 5.5; telson length, 15.2, width, 17.0.

*Discussion.*—P. coccinea mainly differs from its western Atlantic relative, *P. meridionalis,* in having the anterolateral angles of the rostral plate spined and the lateral processes of the sixth and seventh thoracic somites produced into acute processes. Both *P. coccinea* and *P. meridionalis* differ from the East Atlantic *P. ferussaci* in lacking spines on the intermediate carinae of the sixth abdominal somite.

280

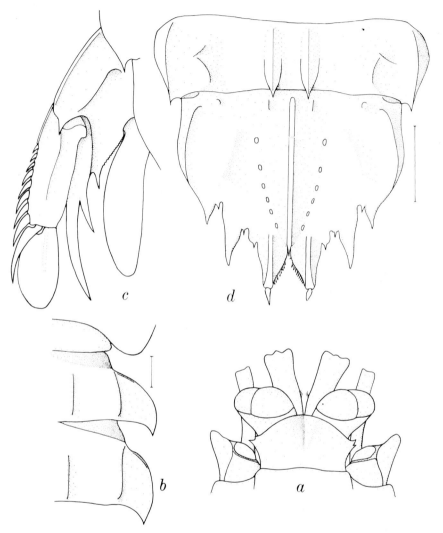

FIGURE 77. *Parasquilla coccinea* Manning, male holotype, TL 109.2, Straits of Florida. *a,* anterior portion of body; *b,* lateral processes of fifth and sixth thoracic somites; *c,* uropod; *d,* last abdominal somite and telson (setae omitted; from Manning, 1962).

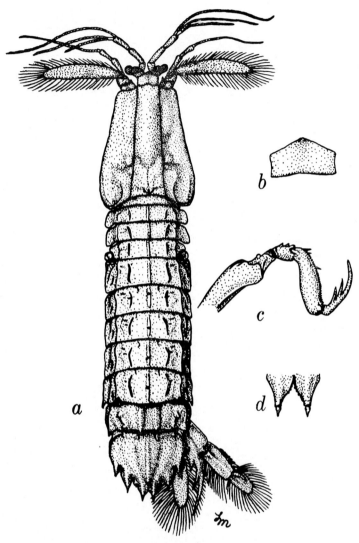

FIGURE 78. *Parasquilla meridionalis* Manning. Female holotype, TL 92.2, OREGON Sta. 2249: *a,* dorsal view; *c,* raptorial claw; *d,* submedian teeth of telson. Male paratype, TL 91.0, OREGON Sta. 2267: *b,* rostral plate (redrawn from Manning, 1961).

*Ontogeny.*—Unknown.

*Type.*—U.S. National Museum.

*Type-locality.*—24°42'S, 80°44'W; Straits of Florida.

*Distribution.*—Southeast Florida, Santaren Channel, Northern Gulf of Mexico, and Gulf of Campeche; 82-382 m.

### *Parasquilla (Parasquilla) meridionalis* Manning, 1961
### Figure 78

*Parasquilla meridionalis* Manning, 1961, p. 8, pl. 2; 1963a, p. 312 [listed].— Bullis & Thompson, 1965, p. 13 [listed].

*Previous records.*—BRITISH GUIANA; SURINAM; BRAZIL (Manning, 1961, Bullis & Thompson, 1965).

*Material.*—BRITISH GUIANA: 1 ♀, 92.2; OREGON Sta. 2249; holotype; USNM 105989.—SURINAM: 1 ♂, 91.0; OREGON Sta. 2267; paratype; USNM 105990.—BRAZIL: 1 ♂, 99.4; Amapa State; OREGON Sta. 2052; paratype; UMML 32.1171.—1 ♂, 54.7; same; OREGON Sta. 4209; USNM 119157.

*Diagnosis.*—Anterolateral portions of rostral plate rounded; posterolateral margins of sixth and seventh thoracic somites rounded or subacute posteriorly, unarmed; no oblique swellings or dorsal carinae on telson.

*Description.*—Rostral plate with lateral margins concave, anterolateral angles rounded; anterior margins converge on obtuse apex.

Dorsal ridge of carpus of raptorial claw with 2 subacute processes, neither spined.

Lateral processes of sixth thoracic somite rounded anteriorly, subtruncate posteriorly; lateral processes of seventh somite rounded anterolaterally and posterolaterally.

Abdominal carinae spined as follows: submedian, 6; intermediate, 5; lateral, 6; marginal, (3) 4-5.

Dorsal surface of telson smooth, without oblique longitudinal prominences or carinae on either side of median carina; 4-9 minute, movable submedian denticles present.

Outer margin of penultimate segment of uropodal exopod with 7-10, usually 9, graded, movable spines, last extending past midlength of distal segment.

*Color.*—In preservative, dorsal surface of body light brown, lateral portions of body dark, metallic brown; posterior margin of telson orange-red; uropod with movable spines, distal segment of exopod and endopod red; notches of abdominal carinae red.

*Size.*—Males, TL 54.7-99.4 mm; only known female, TL 92.2 mm. Other measurements of a male, TL 91.9: carapace length, 22.1; cornea width, 4.1; rostral plate length, 2.6, width, 5.7; telson length, 15.0, width, 15.8.

*Remarks.*—The differences between this species, *P. coccinea,* and *P. ferussaci* have been summarized in the discussion of *P. coccinea.*

*Ontogeny.*—Unknown.

*Type.*—U.S. National Museum.

*Type-locality.*—07°40'N, 57°34'W; British Guiana.

*Distribution.*—Off British Guiana, Surinam, and Amapa State, Brazil; 46-92 m.

## Genus *Odontodactylus* Bigelow, 1893

*Odontodactylus* Bigelow, 1893, p. 100.—Manning, 1967a, p. 1 [other references; review of Indo-West Pacific species]; 1968, p. 139 [key].

*Definition.*—Surface of body smooth; cornea subglobular, set very obliquely on stalk; rostral plate without anterior spine; antennal protopod without papillae; carapace rounded anterolaterally and posterolaterally, without carinae; cervical groove not distinct across dorsum of carapace, position indicated on gastric grooves only; 5 epipods present; mandibular palp present; raptorial claw short, stout, dactylus inflated at base; inner margin of dactylus with teeth; propodus without proximal movable spine, superior margin not pectinate but distally serrate; ischiomeral articulation distal to terminus of merus; inferior surface of merus not ventrally channeled throughout its length for reception of propodus; endopod of walking legs elongate, two-segmented; exposed thoracic somites without sharp carinae; abdomen, semicylindrical, without sharp carinae on first 5 somites; sixth somite with 6 or more carinae but only 6 terminating in spines; telson with sharp median carina and numerous longitudinal carinae on dorsal surface; 3 pairs of marginal teeth, submedians with movable apices; submedian denticles numerous; 2 fixed intermediate and 1 fixed lateral denticle present; basal prolongation of uropod with outer spine the longer.

*Type-species.*—*Cancer scyllarus* Linnaeus, 1758, p. 633, by subsequent selection by Bigelow, 1931, p. 144.

*Gender.*—Masculine.

*Number of species.*—Five, of which one occurs in the western Atlantic. Manning (1967a) reviewed the Indo-West Pacific species.

*Nomenclature.*—*Odontodactylus* was first proposed by Bigelow (1893) as a subgenus of *Gonodactylus;* it was recognized as a full genus by Bigelow in 1894. *Odontodactylus* is no. 731 on the Official List of Generic Names in Zoology.

Holthuis & Manning (1964) discussed the nomenclatural difficulties

284

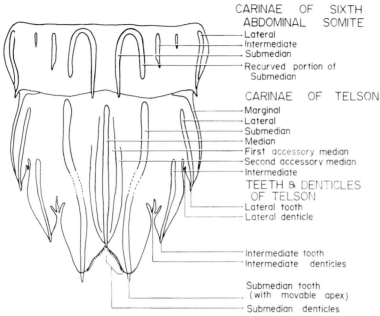

CARINAE OF SIXTH ABDOMINAL SOMITE
- Lateral
- Intermediate
- Submedian
- Recurved portion of Submedian

CARINAE OF TELSON
- Marginal
- Lateral
- Submedian
- Median
- First accessory median
- Second accessory median
- Intermediate

TEETH & DENTICLES OF TELSON
- Lateral tooth
- Lateral denticle

- Intermediate tooth
- Intermediate denticles

- Submedian tooth (with movable apex)
- Submedian denticles

FIGURE 79. Terms used in descriptive account of *Odontodactylus*.

involved in the designation of the type-species for *Gonodactylus* and *Odontodactylus*. The International Commission was requested to take action to preserve these genera as currently known.

*Remarks.*—Manning (1967a) has given a more complete account of the genus, including a key to the five species.

The terminology of the carinae of the telson, which has been discussed in detail by Manning (1967a), is shown in figure 79.

*Affinities.*—*Odontodactylus* has always been considered to be related to *Gonodactylus,* based on the similarity of the articulation of the ischium and merus of the raptorial claw. In both of these genera this articulation is slightly in advance of the proximal end of the merus. Also, members of both genera have the dactylus of the raptorial claw dilated basally. The genus also resembles *Hemisquilla* in some features, particularly the structure of the eye, the presence of a flattened plate on the basal segment of the antenna, and the broad, flattened shape of th uropodal exopod.

*Odontodactylus brevirostris* (Miers, 1884)
Figures 80-81

*Restricted synonymy.*—
*Gonodactylus brevirostris* Miers, 1884, p. 567, pl. 52, fig. C.

285

*Gonodactylus Hansenii* Pocock, 1893, p. 477, pl. 20 B, figs. 3-3b.
*Gonodactylus Havanensis* Bigelow, 1893, p. 101.
*Odontodactylus havanensis* Bigelow, 1894, p. 497, pl. 20, text-figs. 1-2.—
    Kemp, 1913, p. 204 [listed].—Rathbun, 1919, p. 346.—Lunz, 1937, p. 5,
    text-fig. 2.—Balss, 1938, p. 132 [discussion].—Holthuis, 1941, p. 37.—
    Chace, 1954, p. 449.—Manning, 1959, p. 17; 1961, p. 7 [discussion].—
    Bullis & Thompson, 1965, p. 13 [listed].
*Odontodactylus latirostris* Borradaile, 1907, p. 212, pl. 22, figs. 3-3a.
*Odontodactylus southwelli* Kemp, 1911, p. 94.
*?Odontodactylus:* Fish, 1925, p. 154 [discussion; larva].
*Odontodactylus nigricaudatus* Chace, 1942, p. 88, pl. 28; 1954, p. 449.—
    Manning, 1959, p. 15 [discussion].
*Odonodactylus brevirostris:* Manning, 1967a, p. 22, figs. 7-9 [detailed ac-
    count, with Indo-West Pacific references; erroneous spelling].

*Previous Atlantic Records.*—BAHAMAS: New Providence (Rathbun,
1919.—FLORIDA: Key West (Lunz, 1937; Chace, 1954); Boca Grande
(Manning, 1959).—LOUISIANA: Bullis & Thompson, 1965.—MEXICO:
Gulf of Campeche (Rathbun, 1919; Chace, 1942, 1954).—CUBA: Bigelow, 1893, 1894; Rathbun, 1919.—CURAÇAO: Rathbun, 1919; Holthuis, 1941.

*Material.*—BAHAMAS: 1 ♂, 37.8; Grand Bahama Bank; 29 August 1954;
U. S. Fish and Wildlife Service; USNM 120262.—1 ♀, 32.5; same; UMML
32.1750.—1 juv., ca. 13.0; New Providence; surface, electric light; W.
Nye, ALBATROSS; USNM 11550.—FLORIDA: 1 ♀, 54.3; off Sand Key,
Key West; 10 March 1926; PAWNEE; YPM 4444.—1 ♂, 29.0; Dry
Tortugas; 10 February 1931; M. B. Bishop; YPM 4446.—1 ♂, 75.5; SW
of Boca Grande; November 1956; J. Stephens; CHML.—MEXICO: 1
dry ♀, 61.0; Campeche Banks; from red snapper; USNM 44692.—1 ♂,
75.2; Yucatan; holotype of *O. nigricaudatus;* MCZ 12087.—2 ♂, 65.9-
70.2; 4 ♀, 51.0-63.4; same; paratypes of *O. nigricaudatus;* MCZ 12362.—
1 ♂, 78.1; 1 ♀, 59.7; same; USNM 120259.—CUBA: 1 ♂, 20.0; off
Havana; ALBATROSS Sta. 2323; holotype of *G. havanensis;* USNM 17997.
—GRENADA: 1 ♀, 30.3; 309 m; BLAKE; MCZ 7934.—MONTSERRAT:
1 ♀, 32.5; H.M.S. ACHILLES; BMNH 1935.7.135.—CURAÇAO: 1 ♂,
51.2; Piscadera Bay; 20 March 1905; Prof. Boeke; RMNH 90.

*Diagnosis.*—Cornea subglobular; rostral plate usually rounded anteriorly,
over twice as broad as long; raptorial dactylus with 6 to 8 teeth; first 5
abdominal somites with 2 lateral, longitudinal ridges, not sharply carinate;
3 pairs of spined carinae on sixth somite, submedians recurved posteriorly;
telson with median carina, and, on either side, an accessory median carina,
a dorsal submedian carina (poorly connected to carina of submedian
tooth), an intermediate carina and a lateral carina converging distally, but
not always fusing with, the marginal carina.

*Description.*—Cornea subglobular, set obliquely on short stalk, greatest
breadth of cornea contained between 1.9 and 4 times in carapace length;

286

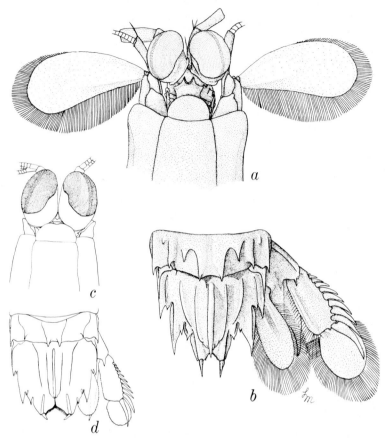

FIGURE 80. *Odontodactylus brevirostris* (Miers). Male, TL 37.8, Grand Bahama Bank: *a,* anterior portion of body; *b,* last abdominal somite, telson, and uropod. Juvenile, TL ca. 13.0, New Providence: *c,* anterior portion of body; *d,* last abdominal somite, telson, and uropod (setae omitted in *d*).

ocular scales separate, area between scales deeply excavate; eyes extending past end of second segment of antennular peduncle.

Antennular peduncle short, stout, about three-fourths median length of carapace; peduncle with a basal, plate-like, bilobed dorsal process and overhangs basal articulation.

Antennal scales broadly rounded, shorter than median length of carapace; basal segment of antenna with ovate, plate-like process on mesial margin; most of margin lined with setae.

Rostral plate evenly rounded anteriorly, not extending to ocular scales, greatest width more than twice median length.

287

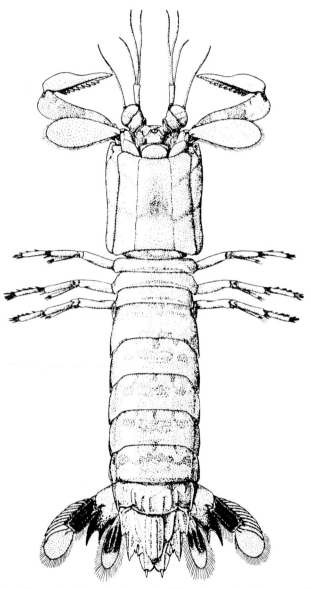

FIGURE 81. *Odontodactylus brevirostris* (Miers), male holotype of *O. nigricaudatus* Chace, TL 75.2, Yucatan, dorsal view (from Chace, 1942).

288

Carapace without carinae or spines, broadly rounded anterolaterally and posterolaterally.

Raptorial dactylus armed with 6 to 8 teeth on inner margin.

Thoracic somites smooth, lateral processes of last 3 broadly rounded, eighth somite with a median, acute, triangular prominence on ventral surface.

First 5 abdominal somites without sharp carinae, but with 2 longitudinal lateral prominences; third, fourth, and fifth somites with posterolateral spines; sixth somite with submedian, intermediate, and lateral carinae spined posteriorly; median tubercle absent; submedian carinae recurved posteriorly; proximal denticle mesial to lateral carina.

Telson with sharp median carina terminating in posterior spine, and, on either side, an accessory median carina, a dorsal submedian carina, the carinae of the submedian, intermediate and lateral teeth, and a marginal carina; median carina with two prominences under terminal spine; accessory medians converge under median carina; dorsal submedians approach and may meet carinae of submedian teeth; carina of intermediate tooth not extending anteriorly past base of tooth; carina of lateral tooth posteriorly converging toward, but not always joining, distal portion of marginal carina; intermediate denticles, especially inner, carinate dorsally.

Outer margin of penultimate segment of uropodal exopod with 7-11 graded, movable spines, last not reaching apex of distal segment; distal segment equal to, or shorter than, penultimate.

*Color.*—In preservative, eyestalks reddish; posterolateral margins of abdominal somites reddish; uropods with a dark patch across endopod of uropod, basal prolongation, and proximal segment of exopod. Chace (1942) noted that his specimens were light green, with dark markings, as noted above.

*Size.*—Males, TL 20.0-75.5 mm; females, TL 30.3-63.5 mm; postlarva, TL 13.0 mm. Measurements and indices for selected specimens are given in Table 13.

*Discussion.*—Chace (1942) was the first to point out differences between the adults of *O. havanensis* (as *O. nigricaudatus*) from the Gulf of Mexico and *O. hansenii* from the Indo-West Pacific. Kemp (1913) had compared *O. brevirostris* with *O. havanensis*, basing his remarks on Bigelow's 1894 account, but Kemp was dealing only with juveniles. Chace noted the following differences:

1. The Campeche specimens had smaller eyes. However, Bigelow (1931) noted that in his specimens of *O. hansenii* (TL 16-62 mm) the proportion of maximum corneal diameter to carapace length varied between 1:2.5 to 1:3.9. In present specimens (Table 12) this proportion varies from 1:1.9 to 1:3.3 in specimens between TL 20.0 and 54.3 mm.

TABLE 13

| Sex | ♂ | ♂ | ♀ | ♀ | ♀ | ♂ |
|---|---|---|---|---|---|---|
| Total Length | 20.0 | 29.0 | 30.3 | 54.3 | 61.0 | 75.2 |
| Carapace Length | 4.2 | 6.0 | 6.9 | 12.1 | 14.0 | 16.4 |
| Cornea Length | 2.2 | 2.5 | 2.8 | 3.7 | . . . | 4.5 |
| Cornea Width | 1.7 | 2.3 | 2.3 | 3.5 | 3.4 | 4.1 |
| Fifth Abdominal Somite Width | 4.6 | 6.3 | 7.2 | 12.7 | . . . | 16.3 |
| Telson Length | 3.1 | 3.9 | 4.5 | 7.6 | 8.9 | . . . |
| Telson Width | 3.7 | 5.2 | 5.8 | 9.7 | 11.6 | 13.3 |
| CI | 247 | 261 | 300 | 346 | 412 | 400 |
| AWCLI | 1095 | 1050 | 1044 | 1049 | . . . | 994 |
| Uropod P/D | 1500 | . . . | . . . | . . . | . . . | 1033 |

The ranges of the corneal indices are almost identical in specimens from the Pacific and the Atlantic.

2. The submedian carinae of the telson were interrupted in the Campeche specimens. This feature is variable, and intergrades between specimens in which the submedians are completely interrupted to those in which they are almost complete can be found in both Atlantic and Indo-West Pacific specimens.

3. The intermediate carinae of the telson did not turn distally to join the marginal. This is also a variable feature, for the intermediate may lie parallel to the marginal carina and its apex does not always turn toward or fuse with the marginal carina.

4. The distal segment of the uropod was longer in the Campeche specimen than in those from the Indo-West Pacific. This character changes with age. In Bigelow's type of *O. havanensis* (TL 20.0 mm) the distal segment is 2/3 the length of the proximal. In Chace's large specimens (TL 51.0-78.1 mm) the segments are almost subequal.

A series of Atlantic specimens has been compared with specimens of *O. brevirostris* from the Indo-West Pacific area and no differences could be found. As Manning (1967a) pointed out in a review of the Indo-West Pacific species, *O. brevirostris* is the oldest available name for the species formerly known as *O. havanensis, O. hansenii, O. latirostris, O. southwelli,* and *O. nigricaudatus.*

Remarks.—Bigelow (1894) did not illustrate the lateral carinae of the telson for the young type of *O. havanensis;* the lateral carinae are present and do not turn toward or fuse with the marginal carinae.

Manning (1967a) discussed specific characters in *Odontodactylus* and pointed out that the position of the ocular scales, the number of teeth on the claw, the setation of the antennal scale, and the carination of the last abdominal somite and telson were valuable characters that did not change with age. Manning also pointed out the differences between postlarvae and juveniles, and noted that two sizes of postlarvae were known in both *O. brevirostris* and *O. scyllarus* (Linnaeus). As in *Pseudosquilla ciliata* (Fabricius), there seems to be overlap in sizes between postlarvae and juveniles.

The synonymy given here is complete for only the Atlantic records. References from other areas are limited to the original description and one or two other pertinent references. Complete synonymies for all species have been presented in a revision of the genus (Manning, 1967a).

*Ontogeny.*—Bigelow (1894) ascribed a larval form to this genus and erected the name *Odonterichthus* for it. The evidence that the odonterichthus larva is the larva of this genus is slender, for no larva has been held through the molt to postlarva. Calman (1917) has identified two specimens of an odonterichthus larva from the south Atlantic, off the Brazilian coast, but no adults have been taken in this area. Fish (1925) discussed larvae attributed to this genus.

*Type.*—British Museum (Natural History).

*Type-locality.*—Providence Is., S of the Seychelles Ids., western Indian Ocean.

*Distribution.*—In the western Atlantic this species is known from scattered localities between the Bahamas, northern Gulf of Mexico, through the Caribbean to Curaçao, in depths to 309 m. In the Indo-West Pacific the species has a wide distribution, from the western Indian Ocean to Hawaii and Japan.

### Genus *Gonodactylus* Berthold, 1827

*Gonodactylus* Berthold, 1827, p. 271 [part].—Holthuis & Manning, 1964, p. 139 [petition to ICZN to designate type-species].—Manning, 1968, p. 139 [key].

*Definition.*—Size moderate, TL 110 mm or less; body smooth, subcylindrical, compact; eye small, cornea rounded, set obliquely on stalk; rostral plate with broad basal portion and long apical spine; antennal protopod with at most 1 papilla, basal segment of antenna with long, channeled anterointernal process; carapace not conspicuously narrowed anteriorly, without carinae or spines, cervical groove absent; lateral plates of carapace extending anteriorly beyond base of rostral plate, broadly rounded anteriorly; thoracic somites without dorsal carinae, rounded laterally; eighth thoracic somite with median ventral groove; 5 epipods present; mandibular palp present; raptorial claw stout, dactylus unarmed, basally inflated; propodus neither pectinate nor armed proximally, inferior margin with distal dilation; carpus without dorsal ridge; merus stout, channeled inferiorly on distal two-thirds only; ischiomeral articulation not terminal, at about proximal third of merus; endopods of walking legs slender, two-segmented; abdomen almost tubular, without dorsal carinae on first 5 somites; antero-

291

lateral plates articulated, not carinate; sixth abdominal somite with 3 pairs of broad carinae, apical spines usually obsolete in adults; telson broad, carinae variously swollen, dorsal surface usually with median and anterior submedian carinae, marginal teeth with dorsal carinae; posterior margin of telson with at most 3 pairs of teeth, laterals often reduced, occasionally absent; submedian marginal teeth with bases separate, movable apices present or absent; series of small submedian denticles, 1 or 2 intermediate, if 2, mesial largest, both occasionally with apical spine, and 1 lateral denticle present; proximal segment of uropodal exopod with lateral spines, segment projecting beyond articulation of distal segment; endopod elongate, with complete or reduced complement of setae; basal prolongation produced into 2 spines, outer usually stouter, always longer.

*Type-species.*—*Squilla chiragra* Fabricius, 1781, in accordance with a petition to the ICZN, now pending (Holthuis & Manning, 1964).

*Gender.*—Masculine.

*Number of species.*—Approximately 24, of which eight occur in the western Atlantic.

The number of Indo-West Pacific species is likely to increase with future study. Many lettered varieties and nominal varieties have been described in the past [see Kemp (1913) for discussion and references] and most of these have been relegated to the synonymy of *G. chiragra*.

*Nomenclature.*—The generic name *Gonodactylus* has long been attributed to Latreille, 1825. That work, although dated 1825, was not published until 1828. Berthold was the first to use the name in 1827. He included two species, *Squilla chiragra* Fabricius and *Cancer scyllarus* Linnaeus; the latter was designated as the type-species by H. Milne-Edwards in Cuvier, 1837. Inasmuch as *Cancer scyllarus* is also the type-species of *Odontodactylus* Bigelow, Holthuis & Manning (1964) have petitioned the ICZN to set aside Milne-Edwards' type-designation and designate *Squilla chiragra* as the type-species of *Gonodactylus*.

In 1893 Bigelow described *Odontodactylus* as a subgenus of *Gonodactylus* and in 1894 recognized the former as a genus; it has been recognized by all subsequent students of the group.

Kemp (1913) recognized four groups of species in *Gonodactylus*. Group I corresponded to *Gonodactylus s.s.*, in the sense adopted here. Group II included *Mesacturus* Miers and Group III included *Protosquilla* Brooks. Group IV has recently been named *Hoplosquilla* by Holthuis (1964). All four of these genera are recognized by Holthuis & Manning (in press) and Manning (1968).

*Remarks.*—The American species of *Gonodactylus* have long posed a problem to students of the group. Hansen (1926) was inclined to lump all variants from both eastern Pacific and western Atlantic into one species,

292

*G. oerstedii.* Schmitt (1924, 1924a) described two Atlantic varieties of *G. oerstedii,* var. *spinulosus* and var. *curacaoensis.* In a discussion of telson morphology of eastern Pacific *Gonodactylus,* Schmitt (1940) pointed out a distinction between "Atlantic" and "Pacific" types of telsons. He also noted that although these types of telsons were largely restricted to the oceans for which they were named, some specimens of each type could be found in samples from both oceans.

To some extent the present study bears out Schmitt's observations on the distinctness of these telson types. The large number of specimens of American *Gonodactylus* now available for study clearly shows that distinct species with each telson type occur on each side of the isthmus of Panama. Pacific specimens of *Gonodactylus,* whether of Atlantic or Pacific type, usually have more angulated rostral plates than Atlantic specimens and most Pacific specimens can be distinguished from Atlantic specimens on the basis of the shape of the rostral plate alone. All American species of *Gonodactylus* differ from Indo-West Pacific species of the genus in possessing a mesial accessory carina on the intermediate carina of the telson. This carina may not be well-formed in specimens under TL 15-20 mm, but it is present in all adults.

Inasmuch as the geographical designation of Atlantic and Pacific telsons is misleading, these telson types will be referred to herein as Oerstedii-type and Bredini-type. The distinctions between these two basic telson shapes are as follows:

OERSTEDII-TYPE TELSON (Fig. 89):

1. The intermediate marginal teeth are distinctly separated from the submedian teeth.

2. The longitudinal axes of the submedian and intermediate teeth, if extended posteriorly, are subparallel.

3. The intermediate denticle is recessed, *i.e.,* it is situated anterior to the apex of the intermediate tooth, usually by the width of the denticle or more. Usually both intermediate denticles are provided with sharp, spined apices.

4. The movable apices of the submedian teeth, unless damaged, are present in specimens of all sizes.

The western Atlantic species included in this category are: *G. curacaoensis, G. minutus,* n. sp., *G. oerstedii, G. austrinus,* n. sp., *G. spinulosus,* and *G. torus,* n. sp.

BREDINI-TYPE TELSON (Figs. 87-88):

1. The intermediate marginal teeth are not widely separated from the submedians; in some cases the apex is appressed to the margin of the submedian.

2. The longitudinal axes of the submedian and intermediate teeth, if extended posteriorly, are convergent.

3. The intermediate denticle is not anteriorly recessed, but is usually

293

situated at or posterior to the level of the apex of the intermediate tooth.

4. The movable apices of the submedian teeth may be present in small specimens, TL less than 20 mm, but are always absent from adults.

The western Atlantic species included in this category are: *G. bredini*, n. sp., and *G. lacunatus* Manning.

Of the species placed in the Oerstedii-group, *G. minutus* and *G. torus* are dwarf species. The first clue that these represented distinct taxa was the presence of males with well-developed copulatory tubes and secondary sexual characteristics at TL 25-30 mm. Dwarf species of *Gonodactylus* have also been recorded in the Indo-West Pacific, where Serène (1954) recognized *G. viridis* as a variety of *G. chiragra*.

In his analysis of *G. chiragra* and other Pacific species of *Gonodactylus*, Kemp (1913) commented on the fact that variation was more noticeable in small specimens (TL 60 mm or less) than in large ones (TL 70 mm or more). Three possible explanations were set forth by Kemp: (1) small specimens showing wide variation die before reaching maturity, (2) small specimens may represent dwarf races, maturing at small sizes, capable of reproducing their morphological characteristics, and (3) small specimens lose their characteristics at subsequent molts and approach a typical form as they grow. Kemp felt that the second hypothesis was improbable and accepted the third as the most likely. The existence of dwarf or small species, the presence of which has been demonstrated in both Atlantic and Pacific Oceans, should help to solve the riddle of variation of Pacific species as it has in the present study.

*G. minutus* and *G. spinulosus* differ from the other western Atlantic species of the Oerstedii-group in having dorsal spinules on the telson. None of the species of the Bredini-group have such ornamentation. Three of the eastern Pacific species of *Gonodactylus*, *G. festae* Nobili, *G. bahiahondensis* Schmitt and *G. stanschi* Schmitt, also are characterized by the presence of dorsal spinules and tubercles on the telson.

*G. curacaoensis* Schmitt differs from other species of the Oerstedii-group in having sharp dorsal carinae on the telson, a feature shared by the Indo-West Pacific *G. smithii* Pocock and *G. acutirostris* de Man.

The terminology employed in the descriptions of the telson morphology of *Gonodactylus* is the same as that used for other members of the Gonodactylidae, especially *Odontodactylus*. These terms are illustrated in Fig. 82. The terminology differs from that used by Kemp (1913) and Schmitt (1940, text-fig. 25) in that (1) their submedian carinae are herein called the accessory medians, (2) their intermediate carinae are herein referred to as the anterior or dorsal submedians, and (3) their lateral carinae, which extend on to the intermediate tooth, are herein called the intermediate carinae.

The median carina and the accessory medians which converge under its apex form the anchor, which is present except in adult males in which the tumidity of the median carina often completely obscures the accessory medians.

In *Gonodactylus* the anterior submedian carina of the telson is never connected to the carina of the submedian tooth.

### CHARACTERS IN *Gonodactylus*

In the analysis of the large series of specimens reported herein, a number of features were investigated in an effort to discover more useful characters for recognition of the species. Most of these investigations proved to be fruitless, so that in this genus we are still faced with the apparent fact that the most important distinguishing features are subjective aspects of telson morphology, features based on relative shape. Although most of the characters investigated appear to be useless, it seems to be worthwhile to give a short account of them here as a guide to future work. When more specimens are available for study, detailed statistical analysis of some of these features may prove to be of some value.

*A. Shape of abdominal pleura.*—In lateral view, the posterior margin of the abdominal pleura may be evenly curved from top to bottom, as in most species, or it may be sinuous, with the dorsal portion concave, the ventral portion convex, as in *G. oerstedii*. This feature seems to hold up well for *G. oerstedii,* but in most of the other species it is variable. In no species however, is the inflection as marked as in *G. oerstedii*.

*B. Shape of lateral processes of thoracic somites.*—The anterolateral portion of the lateral process of the sixth thoracic somite appears broadly rounded in some species, angled, indented, or flattened in others. It is of little value as a specific character in American species, for it can be completely different in two specimens from the same lot which are otherwise identical.

*C. Male pleopod.*—The endopod of the second pleopod of the male is modified into an accessory sexual organ. Although this appendage is of specific shape in many other crustaceans, in *Gonodactylus* the shape, proportions, and general setation are identical in all western Atlantic species. The shape of the male pleopod may well be a generic character in stomatopods but it is useless as a specific character.

*D. Mouthparts.*—No differences could be found in shape or ornamentation of the mouthparts and maxillipeds. Study of this feature was halted after examining a series of appendages of *G. oerstedii* and *G. bredini* without detecting any noticeable differences. This feature was not examined in other species.

*E. Antennule segmentation.*—In all species, the number of segments of the antennules increases with increasing size. At any given size, the number of segments was found to be the same in *G. oerstedii, G. bredini,* and *G. lacunatus.*

*F. Relative width of the body.*—The Abdominal Width/Carapace Length Index (AWCLI) is used herein as a measure of the relative width of the body. As in other stomatopods, telescoping of the body in preservative negates use of Total Length (TL) in setting up Indices; Carapace Length (CL) should be used rather than TL.

The AWCLI is defined herein as the Abdominal width, measured at the fifth somite, divided by the carapace length, x 100. Data on AWCLI for all western Atlantic species are summarized in tabular form under the description of each species. The Index varies with age in each species, with the body becoming relatively narrower with age. In none of the western Atlantic species is the Index as different as it is for the Indo-West Pacific *G. chiragra* and *G. platysoma,* but use of the Index helped to distinguish *G. torus,* which is by far the slenderest of the western Atlantic species. Not enough specimens are available for study at each size to permit recognition of sexual differences in the indices.

*G. Posterolateral angles of fifth abdominal somite.*—In most species examined this angle is rounded, unarmed. In *G. curacaoensis* a posterolateral spine is always present, and some small specimens of *G. torus* also have a small spinule or at least an angular projection.

*H. Dorsal spinulation of telson.*—The presence of dorsal spines and tubercles on the telson is a good specific character. The pattern of spinulation may also be of some importance. This character seems to hold up as well for Indo-West Pacific species of the genus, including *G. demanii* Henderson and its allies.

*I. Marginal teeth of the telson.*—The shape of the marginal teeth of the telson has proved to be of some importance. In *G. oerstedii* and *G. curacaoensis* the teeth are normally sharp, particularly the submedians and intermediates. In *G. curacaoensis* the laterals are also sharp, with well-formed apices. In *G. bredini* the marginal teeth of the telson are rounded off. The relative convergence of the marginal teeth must also be examined, for convergence of longitudinal axes of teeth or the lack of it can be of importance.

*J. Presence of movable apices of submedian marginal teeth.*—The presence or absence of these teeth in adults is an important character. In *G. oerstedii* and *G. curacaoensis,* for example, these teeth are present at all sizes. Adults of *G. bredini* lack these movable apices, as do most specimens of that species, but a few juveniles appear to have them.

*K. Presence and position of intermediate denticles of telson.*—The presence and position of these denticles is an important character. In the Oerstedii-group there are two denticles, both usually with sharp apices, and they are recessed into the gap between the submedian and intermediate teeth. In the Bredini-group there is but one rounded denticle that may even be situated posterior to the apex of the intermediate marginal tooth.

296

*L. Shape of uropodal endopod.*—The endopod shape is generally somewhat variable, but it is characteristic for *G. oerstedii* and *G. curacaoensis.* In *G. oerstedii* the endopod is tapered distally, with the inner margin sinuous, convex proximally and concave distally. In *G. curacaoensis* the endopod is broad, oval, with the inner margin evenly convex. The setation of the uropod seems to be of more importance in the Indo-West Pacific *G. demanii* and its allies; in all American species the uropod has a normal complement of setae.

*M. Spination, setation and shape of walking legs.*—Attempts were made to analyze variation in these features for *G. bredini, G. curacaoensis, G. lacunatus, G. minutus, G. oerstedii,* and *G. spinulosus.* The spination of the exopod and the setation and relative lengths of the segments of the endopod were examined in all of these species and were found to overlap broadly. No diagnostic characters were found in these features.

Color and color pattern may well prove to be of some value in future studies on members of this genus; pattern, if diagnostic, will probably be masked by color polymorphism in living animals. In *G. oerstedii* color pattern is dimorphic. In preservative, males always have a pair of dark bars on the merus of the claw whereas females have bars composed of numerous

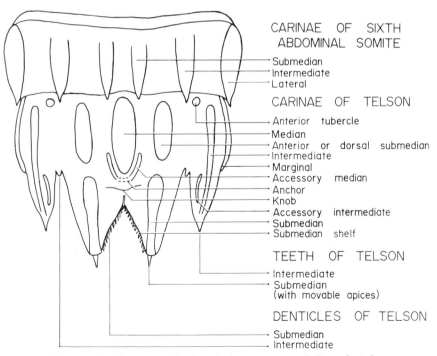

FIGURE 82. Terms used in descriptive accounts of *Gonodactylus.*

black chromatophores. Neither of these patterns is well-marked in fresh specimens which are extremely variable in overall color. Preserved specimens of *G. spinulosus* always have a median patch of black chromatophores on the sixth thoracic and first abdominal somites; live specimens also have these "ideographs" but they are masked completely by the marbled color of living specimens. Dingle (1964) has commented upon the color polymorphism of a population of *Gonodactylus* from Bermuda. Further studies of speciation and systematics in American *Gonodactylus* will have to be accompanied by field studies on habitat and color variation.

For further comments on members of the genus and problems associated with their study the reader is referred to Lanchester (1903), Kemp (1913), Hansen (1926), and Bigelow (1931). Their comments are invaluable to any study of *Gonodactylus*.

Inasmuch as the key to species presented below is constructed with seemingly minor differences in telson morphology, it is suggested that the reader examine the illustrations, particularly those of *G. bredini* and *G. oerstedii*, before attempting to work with the key.

KEY TO WESTERN ATLANTIC SPECIES OF *Gonodactylus*

1. Dorsal surface of telson with spinules ...................... 2
1. Dorsal surface of telson without spinules ..................... 3
2. Larger intermediate denticles of telson with dorsal tubercle; knob armed with 2-4 tubercles .................... *spinulosus*, p. 299.
2. Larger intermediate denticles of telson without dorsal tubercle; knob usually unarmed, rarely with 1-2 tubercles...*minutus*, n. sp., p. 304.
3. Median and anterior submedian carinae each with strong posterior spine; all marginal teeth of telson sharp; fifth abdominal somite with posterolateral spine ...................... *curacaoensis*, p. 307.
3. Median and anterior submedian carinae of telson each with at most an apical tubercle; lateral tooth of telson not produced into a spine; fifth abdominal somite unarmed posterolaterally ................... 4
4. Telson of Bredini-type (fig. 88), intermediate marginal teeth not widely separate from submedians, apex of intermediate denticle at level of or posterior to apex of intermediate tooth ...................... 5
4. Telson of Oerstedii-type (fig. 89), intermediate marginal teeth widely separate from submedians, apex of intermediate denticle anterior to apex of intermediate tooth ............................. 6
5. Carinae of submedian teeth of telson longitudinally sulcate, sulci carinate ................................ *lacunatus*, p. 311.
5. Carinae of submedian teeth of telson at most pitted, never longitudinally sulcate ........................... *bredini*, n. sp., p. 315.
6. Endopod of uropod tapering distally, with inner margin sinuous, convex proximally, concave distally ................ *oerstedii*, p. 325.

298

6. Endopod of uropod oval, inner margin convex or sinuous, not markedly concave distally .................................... 7

7. Posterior margin of first to four abdominal somites concave dorsally, convex ventrally; apices of intermediate teeth sharp ............. ................................. *torus,* n. sp., p. 335.

7. Posterior margin of first to fourth abdominal somites evenly concave, lacking strong double inflection; apices of intermediate teeth blunt, obtuse .............................. *austrinus,* n. sp., p. 338.

## *Gonodactylus spinulosus* Schmitt, 1924
## Figure 83

*Gonodactylus oerstedii* var. *spinulosus* Schmitt, 1924a, p. 96, pl. 5, fig. 5.—Bigelow, 1931, p. 123 [part].
*Gonodactylus oerstedii:* Hansen, 1926, p. 27 [part].
not *Gonodactylus oerstedii* var. *spinulosus:* Andrade Ramos, 1951, p. 143 [listed].—Lemos de Castro, 1955, p. 45 [= *G. minutus,* n. sp.].
*Gonodactylus spinulosus:* Manning, 1961, p. 41, pl. 11, figs. 3-4.
not *Gonodactylus spinulosus:* Manning, 1966, p. 372, text-fig. 6b [= *G. minutus,* n. sp.].

*Previous records.*—PUERTO RICO: Bigelow, 1931.—VIRGIN ISLANDS: Hansen, 1926.—BARBADOS: Schmitt, 1924a.—CURAÇAO: Manning, 1961.

*Material.*—BAHAMAS: 1 ♀, 54.4; W shore of Wood Cay, Little Bahama Bank; CBC Sta. 503; 28 June 1959; R. Schroeder, S. Gross; ANSP.—1 ♀, 23.2; SE of Settlement Point, Grand Bahama Island; CBC Sta. 553; 27 July 1961; ANSP.—2 ♂, 17.3-32.4; 1 ♀, 25.2; shoreline on E side of Northwest Cay, Hogsty Reef; CBC Sta. 576; 27 May 1962; ANSP.—1 ♂, 35.8; S. Bimini; 1-4m; 30 June 1959; W. R. Courtenay, Jr., *et al.;* USNM 124637.—1 ♀, 37.2; Bimini; 2 November 1948; A. S. Pearse; USNM 124630.—1 ♀, 14.6; Andros Id.; 1912; P. Bartsch; USNM 45576.—2 ♀, 17.7-25.4; along E shore of Oyster Cay, Exuma Chain; 0-2 + m; 21 August 1963; H. A. Feddern, *et al.;* USNM 124641.—1 ♂, 21.6; 5 ♀, 25.1-34.0; 1 juv., 9.8; E shore of small cay NW of Little Major's Spot Cay, Exuma Chain; 0-2 m; 25 August 1963; H. A. Feddern, *et al.;* USNM 124642.—1 ♀, 22.8; rocky point 2 mi S of Cockburr Town, San Salvador; 19 March 1937; Smithsonian-Hartford Exped. Sta. 9; USNM 124622.—FLORIDA: 1 ♂, 42.9; Long Reef, E of Elliott Key; 13 December 1959; A. J. Provenzano, Jr.; USNM 124638.—1 ♀, 35.4; ¼ mi. SSW of Alligator Reef Light; 5 m; 26 August 1961; W. A. Starck, II, *et al.;* USNM 124639.—1 ♂, 26.1; 200 yds. SW of Alligator Reef Light; rocky ledge, ca. 5-6 m; 7 January 1962; W. A. Starck, II, *et al.;* USNM 119320.—1 ♀, 26.8; from rocks below lighthouse dock, E side of Loggerhead Key, Tortugas; 24 July 1930; W. L. Schmitt; USNM 124620.—MEXICO: 1 ♀, 11.6; central part of Nicchehabin Reef, Ascension Bay, Quintana Roo; coral pieces behind reef; Bredin Sta. 72-60; 14 April 1960; USNM

124635.—1 ♀, 13.9; same; Bredin Sta. 67-60; 13 April 1960; USNM 124634.—1 ♂, 33.3; 1 ♀, 37.3; Mujeres Id., Quintana Roo; Smithsonian-Bredin Sta. 9-60; 31 March 1960; USNM.—OLD PROVIDENCE ID.: 2 ♂, 18.2-30.0; shore, reef, and tide pool; 6 August 1938; W. L. Schmitt (Presidential Cruise); USNM 78319.—WEST INDIES: 1 ♂, 29.7; 1 ♀, 24.5; no specific data; C. L. Thomas; paralectotypes of *G. oerstedii* Hansen; UZM.—1 brk. ♀; no specific data; Ørsted; paralectotypes of G. *oerstedii* Hansen; UZM.—HAITI: 41.3; near Dames Point, Cape Haitien; shore; 22 March 1937; Smithsonian-Hartford Exped. Sta. 15; USNM 124624.— 1 ♂, 29.0; 2 ♀, 10.0-15.8; reef to E side of Tierra Baja Road, Tortuga Id.; 21 March 1937; Smithsonian-Hartford Exped. Sta. 12; USNM 124623.— PUERTO RICO: W end of San Juan Id., near Fort San Geronimo; 27 March 1937; Smithsonian-Hartford Exped. Sta. 16; USNM 124625.—1 ♂, 35.8; Playa de Ponce Reef; FISH HAWK; 1 February 1899; USNM 64772. —1 ♂, 12.5; 1 ♀, 20.4; Arayo; FISH HAWK; 3 February 1899; USNM 64802.—VIRGIN ISLANDS: 6 ♂, 9.0-45.3; 6 ♀, 11.9-43.6; Danish West Indies; 15 December 1911; Meng; UZM.—1 ♂, 17.0; E shore, seaward side of coal dock, St. Thomas Harbor; shore and tide pool; 4 April 1937; Smithsonian-Hartford Exped. Sta. 23; USNM 124626.—1 ♂, 23.5; reef and beach, Smith Bay, St. Thomas; 25 April 1937; Smithsonian-Hartford Exped. Sta. 68; USNM 124629.—1 ♂, 24.2; Coral Harbor, base of Coral Bay, St. John; SJHK 58-31; 20 December 1958; L. P. Thomas; USNM 124636.—4 ♂, 16.8-46.1; 4 ♀, 15.4-49.2; E shore Europa Bay, St. John; 16 February 1959; H. Kumpf, *et al.;* UMML.—2 ♀ (1 brk.); from stomach of red hind, *Epinephelus guttatus*, taken off Whistling Cay, St. John; 8 May 1959; J. Randall; UMML.—4 ♂, 17.4-34.7; 7 ♀, 19.8-32.6; E side of Ram Head, St. John; shore; 15 August 1962; J. Randall, L. Morera; USNM 124640.—1 ♂, 23.0; 1 ♀, 27.0; same; BMNH.—1 ♂, 35.6; Long Point, St. Croix; 12 August 1906; Th. Mortensen; UZM.—1 ♂, 20.8; 1 ♀, 22.0; same; 12 February 1906; Th. Mortensen; UZM.—1 ♀, 32.8; reef, St. Croix; 8 April 1937; Smithsonian-Hartford Exped. Sta. 32; USNM 124627. —2 ♀, 13.4-15.3; off Buck Id., N of St. Croix; 7 m; 19 February 1906; Th. Mortensen; UZM.—LESSER ANTILLES: 1 ♀, 25.2; Point Planche Bay, Great Bay, St. Martin; 26 June 1949; P. W. Hummelinck 1125A; RMNH.—2 ♀, 15.3-19.7; reefs off Martello Tower, S coast, Barbuda; 7 April 1956; Bredin Sta. 92-56; USNM 124633.—1 ♂, 39.6; Gallows Bay, N of Oranjestad, St. Eustatius; 28 February 1957; L. B. Holthuis 1133; RMNH.—1 ♀, 29.6; English Harbor, Antigua; Univ. Iowa Barbados-Antigua Exped., 1918; USNM 68943.—1 ♀, 26.7; Rat Ids., Guadeloupe; 30-31 March 1956; Bredin Sta. 69-56; USNM 124632.—1 ♂, 24.6; 2 ♀, 15.8-52.7; Portsmouth, Dominica; 19 April 1959; Bredin Sta. 75-59; USNM 106064.—1 ♀, 14.7; Pigeon Id., St. Lucia; 22 March 1956; Bredin Sta. 47-56; USNM 124631.—1 ♀, 32.3; off the Castle, E side, Barbados; 2-7 m; Univ. Iowa Barbados-Antigua Exped., 1918; holotype; USNM 68945.—1 ♂, 30.6; data as in holotype; paratype; USNM 57984.

—1 ♀, 23.0; Barbados; 4 June 1918; Univ. Iowa Barbados-Antigua Exped.; USNM 57985.—1 ♀, 31.2; tidepools, Pelican Id., Barbados; Univ. Iowa Barbados-Antigua Exped., 1918; USNM 124621.—1 ♂, 27.8; Okra Reef, Barbados; 13 May 1918; Univ. Iowa Barbados-Antigua Exped.; USNM 68941.—1 brk. juv., 11.5; off Needham's Point, Barbados; Univ. Iowa Barbados-Antigua Exped., 1918; USNM 57983.—1 abdomen; Carlide Bay, N and NE end of Pelican Id., Barbados; 19 April 1937; Smithsonian-Hartford Exped. Sta. 56; USNM 124628.—ARUBA: 1 ♂, 18.2; 1 juv., 10.0; Malmok, Aresji; 14 August 1955; P. W. Hummelinck 1301; RMNH.—CURAÇAO: 1 ♀, 14.4; near Willemstad; 26 December 1956; L. B. Holthuis 1028; RMNH.—2 ♂ (1 brk.), 30.3; 3 ♀, 16.8-49.1; strand van Marie Pompoen, by Willemstad; 26 December 1958; L. B. Holthuis 1028; RMNH.—1 ♂, 14.0; Piscadera Bay; 20 November 1956; L. B. Holthuis 1002; RMNH.—6 ♂, 15.4-27.0; 4 ♀, 17.5-25.5; 1 fragment; same; 12 November through December 1956; L. B. Holthuis 1002; RMNH.—1 ♂, 22.8; same; November-December 1956; L. B. Holthuis 1002; RMNH.—1 ♀, 30.2; bay NW of Piscadera Bay; 25 January 1957; L. B. Holthuis 1068; RMNH.—1 ♂, 37.5; Boca Lagoon; 13 November 1948; P. W. Hummelinck 1020A; RMNH.—1 ♀, 28.7; N side entrance St. Joris Bay; 20 April 1955; P. W. Hummelinck 1354; RMNH.—1 ♀, 25.6; same; (?) 20 February 1955; P. W. Hummelinck 1354; RMNH.—1 ♀, 31.8; Fuik Bay; 13 January 1957; L. B. Holthuis 1051; RMNH.—1 ♂, 31.5; Vista Alegre; AHF Sta. A46-39; 23 April 1939; USNM 106350.— 2 ♂, 30.6-34.6; same; AHF.—BONAIRE: 1 ♀, 11.7; near Hoop, S of Kralendijk; 10 September 1948; P. W. Hummelinck 1058C; RMNH.—

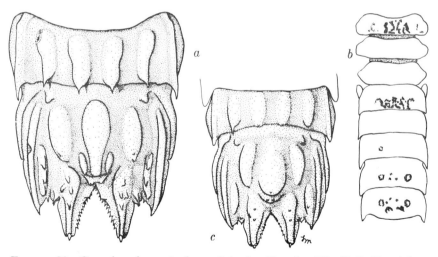

FIGURE 83. *Gonodactylus spinulosus* Schmitt. Female, TL 52.7, Dominica: *a*, sixth abdominal somite and telson. Male, TL 35.8, S. Bimini: *b*, chromatophore pattern on body; *c*, sixth abdominal somite and telson.

1 ♀, 29.7; Point Vierkunt, S of Kralendijk; 10 March 1957; L. B. Holthuis 1146; RMNH.—1 ♀, 24.4; NE coast; 31 August 1948; P. W. Hummelinck; RMNH.—1 ♀, 14.1; Boca Washikemba; 7 April 1955; P. W. Hummelinck 1375B; RMNH.—1 ♀, 14.5; Paloe Lechi, S of Salinja; 4 September 1948; P. W. Hummelinck 1056C; RMNH.—1 ♂, 25.5; E coast at landing place, Klein Bonaire; 13 September 1948; P. W. Hummelinck 1049B; RMNH.—2 ♂, both ca, 14.5; NE coast Klein Bonaire; 3 August 1948; P. W. Hummelinck; RMNH.—1 ♀, 23.0; W point, Klein Bonaire; 28 March 1955; P. W. Hummelinck 1367; RMNH.—1 ♀, 22.9; Klein Bonaire; 11 March 1957; L. B. Holthuis 1147; RMNH.—1 juv., ca. 8.0; same; RMNH.

*Diagnosis.*—Lower portion of posterior margin of pleura of first to fourth abdominal somites straight or slightly convex; pleura, in lateral view, with posterior margin evenly curved or concave above, slightly convex below; sixth abdominal somite with 6 carinae, variously swollen, each usually with apical spinule. A summary of abdominal width-carapace length indices for specimens of various sizes follows:

| CL in mm | n | AWCLI mean | AWCLI range |
|---|---|---|---|
| 3.0 | 6 | 884 | 829-923 |
| 4.0 | 6 | 844 | 818-872 |
| 5.0 | 11 | 832 | 774-869 |
| 6.0 | 13 | 824 | 785-873 |
| 7.0 | 4 | 827 | 792-853 |
| 8.0 | 3 | 826 | 782-864 |
| 9.0 | . | . . . | . . . . . . |
| 10.0 | 2 | 823 | 814-832 |
| 11.0 | 1 | 796 | . . . . . . |

Telson broader than long, of Oerstedii-type; median carina flash-shaped in all but large males, very inflated in males, apical tubercle usually present; carinae of telson with dorsal spinules as summarized in Table 14; accessory median carinae armed posteriorly, fusing with median carina to form anchor; knob always armed with 2 or more tubercles; anterior submedian carinae short, inflated, usually with apical tubercle and 1 or more ventral subapical tubercles; carinae of submedian teeth sinuous, armed dorsally with 2 or more tubercles, usually 3-6; submedian teeth with inconspicuous shelf on inner margin, larger specimens with angled prominence on anterior third of shelf; oblique swelling extending posteriorly from knob to base of each submedian carina low, inconspicuous; submedian teeth subparallel, not convergent, movable apices always present; intermediate teeth with sharp apices, widely separate from submedians, longitudinal axes parallel with submedians; intermediate carinae unarmed dorsally, accessory intermediates always with dorsal spinules; submedian denticles small, numerous, distinct throughout length of teeth; 2 sharp, anteriorly-recessed intermediate denticles present, both with spined apices, mesial denticle with 1 or more dorsal tubercles.

Endopod of uropod slender, tapered, usually convex proximally, concave distally; endopod shorter, blunter in juveniles.

*Color.*—In preservative, body mottled blue or green, dactyli blue; sixth thoracic and first abdominal somites with dark, rectangular patch of chromatophores ("Chinese ideographs" of Schmitt) as shown in fig. 83b; remainder of abdominal somites often with paired submedian black circles, with or without dark markings between them; carinae, spines, and tubercles with bluish cast.

*Size.*—Males TL 9.0-48.0 mm; females, TL 10.0-54.4 mm; juveniles, 8-0-11.5 mm. Other measurements of a male, TL 34.6: carapace length, 7.9; fifth abdominal somite width, 6.5; telson length, 5.2, width, 5.6.

TABLE 14

COMPARISON BETWEEN PATTERNS OF DORSAL SPINULATION OF TELSON IN *G. minutus*, N. SP., AND *G. spinulosus* Schmitt

| CARINA | *G. minutus* range | | | *G. minutus* usual | | | *G. spinulosus* range | | | *G. spinulosus* usual | | |
|---|---|---|---|---|---|---|---|---|---|---|---|---|
| | R | M | L | R | M | L | R | M | L | R | M | L |
| Median carina | | 1 | | | 1 | | | 1 | | | 1 | |
| Anchor | 0-1 | | 0-1 | 1 | | 1 | 1-3 | | 1-3 | 1 | | 1 |
| Knob | | 0-2 | | | 0 | | | 2-4 | | | 2 | |
| Ant. submedian | 0-1 | | 0-1 | 1 | | 1 | 1-2 | | 1-2 | 2 | | 2 |
| Submedian | 0-3 | | 0-3 | 2 | | 2 | 2-7 | | 2-10 | 3-6 | | 3-6 |
| Accessory intermediate | 0-2 | | 0-2 | 0-1 | | 0-1 | 2-4 | | 2-5 | 2-3 | | 2-3 |

*Discussion.*—G. *spinulosus* can be differentiated from all other western Atlantic species by the dorsal spinules on the telson and the characteristic patches of black chromatophores that are found on the sixth thoracic and first abdominal somites. *G. minutus*, n. sp., a Brazilian species, resembles *G. spinulosus* in having dorsal spinules, but in the Brazilian species the spinules are less numerous (see Table 14) and similar dark patches of pigment are never present on the body.

The eastern Pacific species of *Gonodactylus* discussed by Schmitt (1940) lack the dark color on the sixth thoracic and first abdominal somites and futher differ in having spiniform anterolateral angles on the rostral plate. Of the eastern Pacific species, *G. festae* Nobili most closely resembles *G. spinulosus* and may be considered as its analogue.

Remarks.—G. spinulosus is predominantly a reef-dweller, where it often occurs in association with both G. oerstedii and G. curacaoensis. G. spinulosus can be distinguished from those species in the field by the dark patches mentioned above which are always present but which may be masked by overall color. These patches were aptly named "Chinese ideographs" by Schmitt (1924a) in his original description of the species.

The color pattern and spinulation of the telson can be distinguished with no difficulty in specimens between 12-15 mm TL.

Marked sexual differences have not been noted, other than the normal inflation of the telson carinae in large males. Some of the larger females also tend to have at least the median carina inflated to some extent.

Ontogeny.—Unknown.

Type.—U. S. National Museum.

Type-locality.—Barbados.

Distribution.—Western Atlantic, from the Bahamas and southern Florida through the Antilles to Curaçao and Bonaire, including Mexico, Old Providence Island, Haiti, Puerto Rico, the Virgin Islands, St. Martin, Barbuda, St. Eustatius, Antigua, Guadeloupe, Dominica, St. Lucia, Barbados, and Aruba. Sublittoral to 10 m, usually in coral reefs, Porites, or rocks.

**Gonodactylus minutus,** new species
Figure 84

Gonodactylus chiragra var. minutus Brooks, 1886, p. 22 [nomen nudum].
Gonolactylus oerstedii var. spinulosus: Bigelow, 1931, p. 123 [part].—
  Andrade Ramos, 1951, p. 143 [listed].—(?) Lemos de Castro, 1955, p.
  45 [not G. o. var. spinulosus Schmitt, 1924].
Gonodactylus spinulosus: Manning, 1966, p. 372, text-fig. 6b.

Previous records.—BRAZIL: Fernando de Noronha, Rocas Id., off Recife (Manning, 1966); off Cape St. Roque (Bigelow, 1931); Ilha Trinidade (Lemos de Castro, 1955). Brooks also mentioned Cape St. Roque for his specimens but the CHALLENGER Station referred to was at Fernando de Noronha.

Holotype.—1 ♀, 26.4; off Recife, Pernambuco, Brazil; 07°29'S, 35°30'W; 45 m; CALYPSO Sta. 1; 16 November 1961; USNM 113247.

Paratypes.—1 ♂, 23.5; 1 ♀, 25.2; Fernando de Noronha; 03°48.6'S, 32° 24.8' W; 52 m; CALYPSO Sta. 17; 18 November 1961; MNHNP—4 ♂, 17.0-26.3; 2 ♀, 14.5-27.8; off Fernando de Noronha; 03°47'S, 32°24.5'W; 13-66 m; CHALLENGER Sta. 113A; 2 September 1873; BMNH.—5 ♂, 12.5-16.0; 5 ♀, 13.4-27.6; 4 juv., less than 11.0; off Recife, Pernambuco; 06° 59'30"S, 34°47'W; 36 m; ALBATROSS Sta. 2758; 16 December 1887; USNM 18490.

Other material.—BRAZIL: 2 ♂, 17.4-22.7; 1 juv., 9.5; Rocas Id.; 03°50'

S, 33°54'W; 47-54 m; Calypso Sta. 7; 17 November 1961; USNM 113248.—7 ♂, 11.5-30.5; 4 ♀, 9.5-25.8; same; 03°51.1'S, 33°51.1'W; 27 m; Calypso Sta. 8; 17 November 1961; MNHNP.—2 ♂, 17.1-18.2; 4 ♀, 16.8-20.2; 3 juv., 8.8-9.5; Fernando de Noronha; 03°49.7'S, 36°26' W; 31 m; Calypso Sta. 19; 18 November 1961; MNHNP.—1 ♀, 20.6; off Recife, Pernambuco; 08°22.5'S, 34°44'W; 38-52 m; Calypso Sta. 25; 21 November 1961; MNHNP.

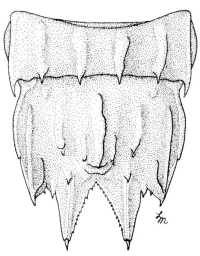

FIGURE 84. *Gonodactylus minutus*, n. sp., female holotype, TL 26.4, Calypso Sta. 1, sixth abdominal somite and telson (from Manning, 1966).

*Diagnosis.*—Lower portion of posterior margin of pleura of first to fourth abdominal somites convex; pleura, in lateral view, concave above, convex below; fifth abdominal somite angled posteriorly, unarmed; sixth abdominal somite with 6 carinae, swollen in males, each usually with apical tubercle or spine. A summary of abdominal width-carapace length indices for specimens of various sizes follows:

| CL in mm | n | AWCLI mean | AWCLI range |
|---|---|---|---|
| 3.0 | 9 | 834 | 758-923 |
| 4.0 | 8 | 819 | 775-846 |
| 5.0 | 11 | 810 | 774-846 |
| 6.0 | 2 | 791 | 789-793 |

Telson broader than long, of Oerstedii-type; median carina flask-shaped, tapering distally; median and anterior submedian carinae very swollen in large males, slender in females and juveniles; carinae of telson armed

305

dorsally, spination summarized in Table 14; carinae of submedian teeth not inflated, sinuous; swollen oblique prominence present on each side between knob and carina of submedian tooth; submedian teeth with apices subparallel, movable apices always present; intermediate marginal teeth with blunt apices, widely separate from submedians, longitudinal axes not convergent with axes of submedian teeth; accessory intermediate carinae present; submedian denticles small, numerous, present throughout length of tooth; 2 intermediate denticles present, recessed, both with apical spinule.

Uropodal endopod with mesial margin sinuous.

*Color.*—Body with scattered dark chromatophores, with some concentration of pigment mesially on sixth thoracic and first abdominal somites, "Chinese ideographs" characteristic of *G. spinulosus* not present. Dactylus orange-pink. Some specimens, particularly males, with background color greenish.

*Size.*—Males, TL 11.5-30.5 mm; females, TL 9.5-27.8 mm; juveniles, TL 8.8-9.5 mm. Other measurements of a male, TL 26.3: carapace length, 4.9; fifth abdominal somite width, 3.8; telson length, 3.3, width, 3.5.

*Discussion.*—*G. minutus* is a dwarf species, with adults not known to exceed 30.5 mm in total length. The only other American species which matures at such a small size is *G. torus,* n. sp., which differs in having a noticeably slenderer body and in lacking supplementary dorsal spinules on the telson.

*G. minutus* and *G. spinulosus* are the only two species of the genus in the western Atlantic in which the dorsal carinae of the telson are provided with supplementary spinules. In this respect they resemble several eastern Pacific species, such as *G. festae* Nobili, but the Atlantic species differ from their Pacific counterparts in having rounded or subtruncate ocular scales, not produced into lateral angles, and in having the anterolateral angles of the rostral plate broadly rounded.

*G. minutus* and *G. spinulosus* are similar but the former has fewer dorsal spines on the telson (Table 14), matures at a much smaller size, and lacks the characteristic patches of dark pigment on the sixth thoracic and first abdominal somite that are always found on *G. spinulosus.*

In my report on the Brazilian stomatopods collected by the CALYPSO (1966) I suggested that Brooks' *G. minutus* might be conspecific with *G. lacunatus,* then described as new. Examination of Brooks' specimens leaves no doubt that they are conspecific with the species described here.

*Etymology.*—The name is from the Latin "minutus," and alludes to the small size of the species.

*Ontogeny.*—Unknown.

*Type.*— U.S. National Museum. A series of paratypes has also been

306

deposited in the Muséum National d'Histoire Naturelle, Paris, and in the British Museum (Natural History).

*Type-locality.*—Off Recife, Pernambuco, Brazil.

*Distribution.*—From Fernando de Noronha to Ilha Trinidade, Brazil, in shallow water to 66 m.

*Gonodactylus curacaoensis* Schmitt, 1924

Figure 85

*Gonodactylus oerstedii* var. *curacaoensis* Schmitt, 1924, p. 80, pl. 8, fig. 6; 1924a, p. 96.
*Gonodactylus curacaoensis:* Manning, 1961, p. 40.—Bullis & Thompson, 1965, p. 13 [listed].

*Previous records.*—BAHAMAS: Cay Sal Bank (Manning, 1961).—BARBADOS: Schmitt, 1924a.—HONDURAS: Bullis & Thompson, 1965.—CURAÇAO: Schmitt, 1924.

*Material.*—BAHAMAS: 1 ♂, 29.0; Bean, Geographical Society of Baltimore, col.; no other data; USNM 33025.—1 ♂, 46.5; S of Great Abaco; Silver Bay Sta. 5130; USNM 119269.—1 ♀, 38.0; E of Sandy Cay, New Providence; CBC Sta. 303; 16 May 1956; ANSP.—1 ♂, 50.2; S shore Grand Bahama Id.; CBC Sta. 504; 1 July 1959; ANSP.—1 ♂, 36.4; same; CBC Sta. 506; 2 July 1959; ANSP.—3 ♂, 32.1-54.0; 1 ♀, 28.0; SE of Settlement Point, Grand Bahama Id.; CBC Sta. 553; 27 July 1961; ANSP.—1 ♂, 41.0; 2 ♀, 26.0-42.3; 1 brk. juv.; same; CBC Sta. 558; 28 July 1961; ANSP.— 2 ♂, 22.0; Hogsty Reef, isolated coral head off westernmost tip of Northwest Cay; CBC Sta. 580; 29 May 1962; ANSP.— 2 ♂, 18.2-23.7; 1 ♀, 49.7; isolated composite coral head off large bay on NW end of Conception Id.; CBC Sta. 589; 2 June 1962; ANSP.—6 ♂, 11.0-30.0; 4 ♀, 15.2-31.6; 2 juv., 9.0-9.2; E side of small cay NW of Little Major's Spot Cay, Exuma Chain; 0-2 m; 25 August 1963; H. A. Feddern, *et al.;* USNM 124271.—2 ♂, 32.8-35.5; 3 ♀, 60.0-63.3; along E shore of Oyster Cay, Exuma Chain; 0-2.5 m; 21 August 1963; H. A. Feddern, *et al.;* USNM 124270.—2 ♀, 39.1-71.8; off Elbow Key, Cay Sal, Bank; 9 June 1960; H. Brown; UMML 32.1875.—1 ♀, 19.2; Water Cays, Cay Sal Bank; 12 m; 22 May 1960; H. Brown; USNM 124262.—FLORIDA: 2 ♀, 50.0-55.4; off Miami Beach; 15 m; 24 April 1940; J. S. Schwengel; ANSP 4526.—2 ♂, 33.1-52.6; Margate Fish Shoal; 20 June 1961; UMML personnel; UMML 32.2077.—1 ♂, 34.7; Crocker Reef, off Plantation Key; 28 September 1958; W. A. Starck, II; USNM 124259.—1 dam. spec.; ¼ mi. SW of Alligator Reef Light; from stomach of 100 mm *Hypoplectrus unicolor;* 23-30 m; 29 October 1960; J. Randall, *et al.;* UMML.—1 ♂, 41.0; off Alligator Reef Light; 30 April 1961; W. A. Starck, II, *et al.;* UMML 32.2082.—5 ♂, 25.0-49.7; 8 ♀, 15.0-28.3; 3 juv., 12.0 or less; same; USNM 124263.—1 ♀, 37.3; 21 July 1961; UMML 32.2078.—3 ♂, 22.5-33.8; 3 ♀, 34.2-44.6; same; 26 August

1961; USNM 124266.—9 ♀, 15.2-34.7; 3 juv., 8.9-10.5; same; 7 January 1962; USNM 119273.—7 ♂,13.2-43.0; 3 ♀, 16.4-17.2; same; 4 January 1963; R. Schroeder; USNM 124268.—1 ♂, 25.5; 1 ♀, 27.5; same; 8 August 1963; W. Davis; USNM 124269.—1 ♂, 58.7; 1 mi. off Lower Matecumbe Key, S of Lignumvitae Pass; 10 July 1961; R. M. Bailey and sons; USNM 124265.—1 juv., 10.0; 2 mi. WSW Rebecca Shoal Light; 24°34′N, 82°37′W; 19 m; 2 October 1948; UMML 32.689.—BRITISH HONDURAS: 2 ♂, 19.7-28.0; 2 ♀, 22.5-40.0; 1 juv., 8.5; 200 yds. S of Cay Bokel, Turneffe Id.; 30 June 1961; W. A. Starck, II; USNM.—HONDURAS: 1 ♀, 15.2; OREGON Sta. 3620; USNM 119268.—HAITI: 1 ♂, 49.6; St. Marc Bay; 22 December 1959; J. Randall, et al.; USNM 124261. —PUERTO RICO: 1 ♂, 21.9; off Vieques Id.; FISH HAWK P. R. Sta. 161 (6089); USNM 64801.—VIRGIN ISLANDS: 1 ♂, 42.1; St. Thomas; A. H. Riise; UZM.—1 ♂, 9.0; 1 ♀, 11.5; Thatch Id., N of St. John; 25-29 m; 12 March 1906; Th. Mortensen; UZM.—3 ♂, 25.4-41.6; 3 ♀, 27.3-27.6; Coral Harbor, Coral Bay, St. John; SJHK 58-31; 20 December 1958; L. P. Thomas; USNM 124260.—1 ♂, 38.2; 1 ♀, 26.5; same; BMNH.—1 dam. ♂; near Concordia Bay, St. John; from stomach of 320 mm Nassau grouper; 6 m; 17 December 1958; J. Randall; UMML 32.1169.—2 dam. spec.; Kiddle Bay, St. John; from grouper stomachs; 3 September 1959; J. Randall; UMML.—4 ♀, 34.1-48.5; reef and sand near Cabritte Horn Point, St. John; 24 May 1961; J. Randall, R. Schroeder; USNM 119272 (2 ♀ to IM).—3 ♂, 34.0-58.7; 4 ♀, 25.1-54.8; Beehive Point, Lameshur Bay, St. John; 2-6 m; 14 June 1961; J. Randall, et al.; USNM 124264.—2 ♂, 22.0-26.0; 6 ♀, 32.0-59.5; E side of Ram Head, St. John; 15 August 1962; J. Randall, L. Morera; USNM 124267. —1 juv., ca. 9.0; 2 km S of Sandy Point, St. Croix; 9 m; 2 February 1906; Th. Mortensen; UZM.—LESSER ANTILLES: 1 ♀, 10.4; N lagoon, Aves Id., W of Dominica; sand, rocks, tidal zone; 12 May 1949; P. W. Hummelinck 1114; RMNH.—1 dam. juv.; off Needham's Point, Barbados; Univ. Iowa Barbados-Antigua Exped., 1918; USNM 57983.—1 ♂, 26.7; Okra Reef, Barbados; 13 May 1918; Univ. Iowa Barbados-Antigua Exped.; USNM 68942.—1 ♂, 11.7; SW of Pelican Id., Barbados; 13 May 1918; C. C. Nutting; USNM 68952.—CURAÇAO: 1 ♀, 47.9; Caracas Bay; in Porites furcata; 5 May 1920; C. J. van der Horst; lectotype; USNM 57527.—1 ♀, 43.1; same; paralectotype; ZMA.—1 ♀, 32.5; Lagoen Bay; 9 January 1949; P. W. Hummelinck; RMNH.—BONAIRE: 1 ♀, 31.4; Strand van Klein Bonaire; 11 March 1957; L. B. Holthuis 1147; RMNH.

Diagnosis.—Lower portion of posterior margin of pleura of first to fourth abdominal somites convex; pleura, in lateral view, with posterior margin concave above, convex below; fifth abdominal somite spined postero-laterally; sixth abdominal somite with 6 slender, sharp carinae, all with posterior spines. A summary of abdominal width-carapace length indices for specimens of various sizes follows:

| CL in mm | n | AWCLI mean | AWCLI range |
|---|---|---|---|
| 3.0 | 8 | 860 | 828-889 |
| 4.0 | 6 | 835 | 805-857 |
| 5.0 | 6 | 807 | 774-844 |
| 6.0 | 9 | 814 | 746-867 |
| 7.0 | 10 | 811 | 757-857 |
| 8.0 | 6 | 797 | 782-808 |
| 9.0 | 3 | 804 | 763-835 |
| 10.0 | 3 | 817 | 760-878 |
| 11.0 | 2 | 831 | 784-877 |
| 12.0 | 3 | 770 | 739-786 |
| 13.0 | 5 | 800 | 776-832 |

Telson longer than broad, of Oerstedii-type; median carina slender, sharp, with long apical spine; accessory median carinae sharp, unarmed, converging under apex of median to form anchor; knob present, well-defined, unarmed, often with median depression; anterior submedian carinae sharp, slender, terminating in apical spine; median and anterior submedians often tumid, unarmed in large males; submedian marginal carinae sharp; indistinct oblique swelling extending from submedians toward knob; submedian teeth with axes subparallel, movable apices always present; intermediate marginal teeth sharp, widely separated from

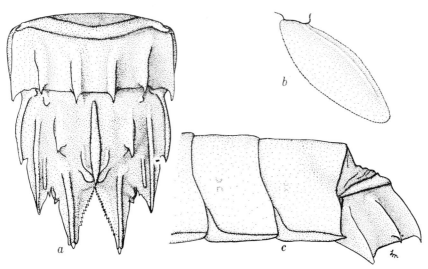

FIGURE 85. *Gonodactylus curacaoensis* Schmitt, female, TL 45.5, St. John, V. I. *a,* sixth abdominal somite and telson; *b,* uropodal endopod (setae omitted); *c,* last three abdominal somites, lateral view.

submedians, longitudinal axes subparallel with axes of submedians; accessory intermediate carinae well-defined, rarely fusing posteriorly with intermediate carinae; lateral tooth distinct, sharp; submedian denticles small, numerous, often obsolete on posterior third of tooth; intermediate denticles armed posteriorly, sharp, deeply recessed anteriorly, larger inner denticle with dorsal tubercle or short carina.

Endopod of uropod broad, inner margin evenly convex.

*Color.*—In preservative, background color orange; inflated portions of chelae crimson or purple; articulations of chela outlined in purple; level of cervical groove indicated by crimson spot on each gastric groove; thoracic and abdominal somites outlined posteriorly in red or purple; setae of uropods pink or crimson.

*Size.*—Males, TL 9.0-58.7 mm; females, TL 10.4-71.8 mm; juveniles, TL 8.5-12.0 mm. Other measurements of female, TL 48.5 mm: carapace length, 9.8; fifth abdominal somite width, 7.6; telson length, 7.9, width, 6.6.

*Discussion.*—*G. curacaoensis* can be distinguished immediately from all other American species of the genus by the possession of three pairs of spined marginal teeth and the presence of spined dorsal submedian and median carinae on the telson, and by the presence of posterolateral spines on the fifth abdominal somite. In some large males the spines of the dorsal carinae of the telson may be obsolete, but in those specimens the lateral teeth of the telson are distinct and the fifth abdominal somite always has posterolateral spines.

The uropodal endopod has a very characteristic shape; none of the other American species have the inner margin of the endopod so broad and evenly curved.

Some specimens of *G. torus,* n. sp., may have posterolateral spines on the fifth abdominal somite but the carinae and marginal teeth of *G. torus* never approach the condition found in *G. curacaoensis,* and the uropodal endopod in *G. torus* is of different shape. Also *G. torus* is a much smaller species.

*Remarks.*—Even the smallest specimens examined are clearly referable to this species; juveniles can be distinguished on the basis of the shape of the uropodal endopod alone.

*Ontogeny.*—Unknown.

*Type.*—A syntype in the collection of the U. S. National Museum, a female, TL 47.9 (USNM 57527) is here selected as the lectotype. A female paralectotype, TL 43.1 mm, is in the Zoological Museum, Amsterdam.

*Type-locality.*—Caracas Bay, Curaçao.

*Distribution.*—Western Atlantic, from the Bahamas, southern Florida,

British Honduras, off Honduras, Haiti, Vieques Id. (Puerto Rico), St. Thomas, Thatch Id., St. John, St. Croix, Aves Id., Barbados, Curaçao, and Bonaire. Sublittoral to 38 m, usually on coral reefs.

## Gonodactylus lacunatus Manning, 1966
## Figure 86

*Gonodactylus chiragra:* (?) Smith, 1869, pp. 31, 41; 1869a, p. 391.— Bigelow, 1893, p. 101 [part]; 1894, p. 495 [part].—Rathbun, 1900, p. 155 [not *G. chiragra* (Fabricius, 1781)].

?*Gonodactylus falcatus:* Moreira, 1901, pp. 1, 69 [not *G. falcatus* (Forskål, 1775)].

*Gonodactylus oerstedii:* (?) Andrade Ramos, 1951, p. 143 [listed].—(?) Lemos de Castro, 1955, p. 42, text-figs. 30-31, pl. 12, fig. 44, pl. 18, fig. 56 [possibly more than one species].—Bullis & Thompson, 1965, p. 13 [part] [not *G. oerstedii* Hansen, 1895].

*Gonodactylus lacunatus* Manning, 1966, p. 374, text-fig. 6c.

*Previous records.*—BRAZIL: Jacuma, Paraiba (Rathbun, 1900); Recife (Manning, 1966); Ilha de Noqueira (Rathbun, 1900); Maceio (Rathbun, 1900; Lemos de Castro, 1955; Manning, 1966); off Belmonte, Bahia (Manning, 1966); Abrolhos Ids. (Smith, 1869, 1869a; Bigelow, 1893, 1894; Manning, 1966); Cabo Frio and Ilha Trinidade (Lemos de Castro, 1955). Specimens from OREGON Sta. 3604 were reported by Bullis & Thompson (1965) as *G. oerstedii.*

*Material.*—NICARAGUA: 1 ♀, 32.9; OREGON Sta. 3604; USNM 119276. —WEST INDIES: 1 ♂, 1 ♀ (both damaged); 1886; Krebs; paralectotype of *G. oerstedii* Hansen; UZM.—VIRGIN ISLANDS: 1 ♀, 33.0; Danish West Indies; 1906; F. Borgesen; UZM.—1 ♂, 28.4; 1 ♀, 36.3; between St. John and Tortola; 40 m; 2 March 1906; Th. Mortensen; UZM.—1 ♀, 23.5; between St. John and Thatch Id.; ca. 27 m; 9 March 1906; Th. Mortensen; UZM.—COLOMBIA: 1 ♀, 30.2; OREGON Sta. 4850; USNM 119277.—BRAZIL: 1 brk. ♂, CL 9.2; no other data; C. F. Hartt; USNM 119274.—1 ♂, 24.0; 1 ♀, 15.3; Cabo Baco Pari, Bahia Formosa, Rio Grande do Norte; 29 January 1964; A. Lemos de Castro; MNRJ.— 1 ♀, 18.0; off Jacuma, Paraiba; 5 m; Branner-Agassiz Expedition; 18 June 1899; A. W. Greeley; USNM 25817.—1 ♂, 20.2; 1 ♀, 14.7; off Recife, Pernambuco; ALBATROSS Sta. 2758; USNM 18490.—1 ♂, 19.4; same; 08°27′S, 34°55.5′W; 27 m; CALYPSO Sta. 28; 21 November 1961; paratype; MNHNP.—1 ♀, 15.0; same; 08°28′S, 34°55.5′W; 22-30 m; CALYPSO Sta. 29; 21 November 1961; paratype; USNM 113249.—1 ♀, 46.5; Tamandaré Bay, Rio Formoso, Pernambuco; in reefs; 8 July 1959; MNRJ. —2 ♂, 25.2-29.1; 1 ♀, 42.7; Ilha de Noqueira, Pernambuco; stone reef; Branner-Agassiz Expedition; 10 July 1899; A. W. Greeley; USNM 25818. — 1 ♀, 28.8; Maceio, Alagoas; 5 m; CALYPSO Sta. 34; 22 November 1961; paratype; MNHNP.—12 ♂, 20.8-41.3; 14 ♀, 21.5-48.0; Maceio, Alagoas; coral reef; Branner-Agassiz Expedition; 25 July and 3-4 August 1899;

A. W. Greeley; USNM 25819.—3 ♂, 22.2-24.2; 1 ♀, 19.2; N of Belmonte, Bahia; 15°37.5'S, 38°44.5'W; 39 m; CALYPSO Sta. 69; 27 November 1961; paratypes; USNM 113250.—2 brk. spec.; Abrolhos Ids.; ALBATROSS; 28 December 1887; USNM 18486.—1 ♂, 46.5; 1 ♀, 16.4; same 18°00'S, 38°18'W; 48 m; CALYPSO Sta. 77; 28 November 1961; paratypes; USNM 113251.—2 ♂, 18.0-19.5; 1 soft ♀, 17.7; 1 postlarva, 8.5; same; 18°09'S, 38°20'W; 33 m; CALYPSO Sta. 79; 28 November 1961; paratypes; MNHNP.—2 ♀, 13.9-16.5; 2 juv., both ca. 9.0; same; 18°09.5'S, 38°30'W; 50 m; CALYPSO Sta. 80; 28 November 1961; paratypes; MNHNP.—1 ♀, 38.7; same; 18°06'S, 38°42'W; 37 m; CALYPSO Sta. 81; 28 November 1961; holotype; MNHNP.—1 ♂, 25.5; 2 ♀, ca. 13.0-23.2; same data; paratypes; MNHNP.—1 ♀, 25.6; same; 18°05'S, 38°46'W; 27 m; CALYPSO Sta. 82; 28 November 1961; paratype; MNHNP.—1 ♂, 14.5; 3 ♀, 11.0-15.0; same; 17°59'S, 38°43.5'W; 17 m; CALYPSO Sta. 83; 28 November 1961; paratypes; MNHNP.—2 post larvae, ca. 8.0; same; stomach contents of fish; CALYPSO Sta. 87; 29 November 1961; MNHNP.—3 ♀, 14.5-28.0; same; 18°18'S, 38°53'W;

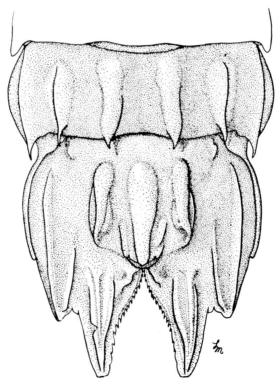

FIGURE 86. *Gonodactylus lacunatus* Manning, female holotype, TL 38.7, CALYPSO Sta. 81, sixth abdominal somite and telson (from Manning, 1966).

312

38 m; CALYPSO Sta. 89; 29 November 1961; paratypes; USNM 113252.—
1 ♀, 30.3; S of Abrolhos Ids.; 18°51'S, 39°08'W; 49 m; CALYPSO Sta.
90; 29 November 1961; paratype; MNHNP.—1 ♂, 21.7; 2 ♀, 18.2-19.6;
off Parahyba(?), Espiritu Santo; 21°22'S, 40°43.5'W; 25 m; CALYPSO
Sta. 98; 1 December 1961; paratypes; MNHNP.—1 ♂, 29.6; Maguinhos,
Cabo Frio, Rio de Janeiro; 9 May 1952; J. Becker; MNRJ.—1 ♂, 40.4;
Praia do Forno, Arrail do Cabo, Cabo Frio, Rio de Janeiro; 7 December
1956; MNRJ.—1 ♀, 58.5; Cabo Frio, Rio de Janeiro; 5 September 1952;
J. Becker; MNRJ.

*Diagnosis.*—Lower portion of posterior margin of pleura of first to fourth
abdominal somites straight or slightly convex; pleura, in lateral view,
particularly of first and second somites, with posterior margin concave
above, convex or straight below; fifth abdominal somite rounded or angled
posterolaterally, unarmed; sixth abdominal somite with 6 slender carinae,
each with posterior spine; carinae of sixth somite occasionally swollen in
adult males. A summary of abdominal width-carapace length indices for
specimens of various sizes follows:

| CL in mm | n | AWCLI mean | AWCLI range |
|---|---|---|---|
| 3.0 | 7 | 819 | 800-867 |
| 4.0 | 4 | 828 | 811-854 |
| 5.0 | 7 | 805 | 791-826 |
| 6.0 | 8 | 796 | 725-833 |
| 7.0 | 3 | 809 | 773-833 |
| 8.0 | 5 | 811 | 783-833 |
| 9.0 | 9 | 796 | 774-796 |
| 10.0 | 1 | 792 | . . . . . . |
| 11.0 | 4 | 789 | 757-809 |
| 12.0 | 1 | 759 | . . . . . . |

Telson as long as broad or slightly broader than long, appearing
elongate, of Bredini-type; median carina flask-shaped, with posterior
median tubercle, noticeably inflated in larger males; accessory median
carinae prominent, elongate, usually unarmed, occasionally swollen in
large males, forming anchor; knob present, unarmed; anterior submedian
carinae short, dimpled or excavate dorsally and posteriorly, rarely with
posterior tubercle; oblique swollen prominences diverging posteriorly from
under knob; carinae of submedian teeth longitudinally sulcate, sulci
marked by sharp carinae; movable apices of submedian teeth of telson
absent, apices of teeth convergent posteriorly; inner margins of submedian
teeth with smooth shelf, not angled anteriorly; intermediate marginal teeth
not widely separate from submedians, longitudinal axes of intermediates

and submedians convergent posteriorly; intermediate carinae each with accessory carina on inner face, neither armed; lateral carina well-marked laterally, point of fusion with intermediate carina not prominent; submedian denticles small, numerous; 1 rounded, unarmed intermediate denticle present, situated at level of intermediate tooth or anterior to it, not deeply recessed.

Endopod of uropod tapering, with inner margin sinuous, convex proximally, concave distally.

*Color.*—Largely faded in the present material; body yellow-brown or light green in preservative; carinae of telson green; walking legs and uropods yellow. One male had traces of mottled pattern on the telson; no specimens exhibited the dark bar on the merus of the claw, although many had a black spot on the proximal portion of the dorsal depression of the merus.

*Size.*—Males, TL 14.5-46.5 mm; females, TL 13.9-58.5 mm; postlarva, 8.5 mm. Other measurements of female, TL 46.5: carapace length, 10.5; fifth abdominal somite width, 8.5; telson length, 6.6, width, 7.0.

*Discussion.*—The longitudinally channelled carinae of the submedian teeth of the telson will immediately distinguish G. *lacunatus* from all other species in the genus. The telson of G. *lacunatus* seems to be intermediate between that found in G. *oerstedii,* with its sharp marginal teeth, intermediates separated from the submedians, and anteriorly-recessed intermediate denticle, and that of G. *bredini,* in which the intermediate denticles are not recessed and the intermediate marginal teeth are not widely separate from the submedians. G. *lacunatus* and G. *bredini* both lack the movable apices of the submedian teeth of the telson.

As in G. *austrinus* and G. *minutus,* both of which occur in the same general area, the submedian teeth of the telson of G. *lacunatus* are posteriorly convergent. Both of those species, however, have movable apices present on the submedian marginal teeth of the telson, and G. *minutus* further differs in having dorsal spinules on the telson.

Some Caribbean specimens of G. *bredini* approach G. *lacunatus* in having the submedian marginal teeth of the telson pitted or with a single long dorsal depression. The depression, when present, is never carinate in G. *bredini.* That species also differs in having noticeably shorter accessory median carinae.

One female from off Nicaragua (OREGON Sta. 3604) is tentatively identified with G. *lacunatus,* although it differs in numerous features. The telson is of the Oerstedii-type with widely-spaced, sharp marginal teeth, well-formed intermediate denticles, both armed, and the movable apices of the submedian marginal teeth are present. The carinate sulci on the submedian teeth are present. In some respects the specimen also resembles G. *torus,* but the body is much broader than in any specimens of that species. More comparative material will be needed to determine the status of the specimen from Nicaragua. Other than that record, the specimen from off Colombia, and the few specimens from the Virgin Islands in the collection from Copenhagen, *G. lacunatus* is known only from Brazilian waters.

*Remarks.*—The specimens recorded by Smith (1869, 1869a) and Lemos de Castro (1955) were not examined; more than one species may be present in their collections.

The CALYPSO station, 81, at which the holotype was collected was erroneously given as 80 by me (1966) in the legend for figure 6.
*Ontogeny.*—Unknown.

*Type.*—The holotype is in the Muséum National d'Histoire Naturelle, Paris. A series of paratypes was deposited in that institution and in the U. S. National Museum.

*Type-locality.*—18°06'S, 38°42'W, Abrolhos Ids., Brazil.

*Distribution.*—Western Atlantic, from scattered localities in the Caribbean Sea, including Nicaragua, Colombia, and the Virgin Islands, and off Brazil. It is most abundant off Brazil. Shallow water to 50 m.

## Gonodactylus bredini, new species
### Figures 87, 88

*Gonodactylus chiragra:* (?) Goode, 1879, p. 7 [biology].—(?)Brooks, 1886, p. 56 [part].—(?)Heilprin, 1888, p. 323; 1889, p. 152 not seen.—(?) Verrill, 1901, p. 20 [not *G. chiragra* (Fabricius, 1781)].
*Gonodactylus falcatus:* (?)Sharp, 1893, p. 105 [part] [not *G. falcatus* (Forskål, 1775)].
*Gonodactylus oerstedii* Hansen, 1895, p. 65 [part; footnote.—(?)Rankin, 1900, p. 545.—Bigelow, 1901, p. 152 [part].—Rathbun, 1919, p. 348 [part].—(?)Verrill, 1923, p. 189, pl. 50, figs. 3-4, pl. 51, figs. 2-2b, pl. 55, fig. 3 (larva), text-fig. 1.—Schmitt, 1924, p. 80 [part]; 1924a, p. 96 [part].—Bigelow, 1931, p. 120 [part].—Lunz, 1935, p. 152, text-fig. 1.— Holthuis, 1941, p. 38 [part].—(?)Gurney, 1946, p. 155, text-fig. 14 [larva].—(?)Dennell, 1950, p. 63.—(?)Pearse & Williams, 1951, p. 144 [listed].—(?)Morrison & Morrison, 1952, p. 396 [discussion].—(?)Chace, 1954, p. 449 [part].—(?)Menzel, 1956, p. 46 [listed].—Springer & Bullis, 1956, p. 22.—Manning, 1959, p. 16 [part]; 1961; p. 43 [part].— (?)Hazlett & Winn, 1962, p. 27 [biology].—(?)Dingle, 1964, p. 236.— Bullis & Thompson, 1965, p. 13 [part].
*Gonodactylys oerstedi:* (?)Reed, 1941, p. 46 [discussion].

*Previous records.*—BERMUDA: ?Goode, 1879; ?Brooks, 1886; ?Heilprinn, 1888, 1889; ?Sharp, 1893; ?Rankin, 1900; ?Verrill, 1901, 1923; ?Gurney, 1946; ?Dennell, 1950; ?Morrison and Morrison, 1952; Manning, 1959; ?Hazlett & Winn, 1962; ?Dingle, 1964.—CAROLINAS: Pearse & Williams, 1951.—NORTH CAROLINA: Cape Fear, Beaufort (Lunz, 1935).—SOUTH CAROLINA: Charleston Harbor (Lunz, 1935).— FLORIDA: Crocker Reef, Tortugas, Fort Myers, Clearwater, SE of

315

Apalachicola (Manning, 1959); Gulf of Mexico W of Florida (Springer & Bullis, 1956; Manning, 1959); Apalachee Bay (Menzel, 1956).— TEXAS: ?Reed, 1941.—WEST INDIES: Hansen, 1895.—PUERTO RICO: Bigelow, 1901, 1931.—LESSER ANTILLES: Antigua (Schmitt, 1924a); Barbados (Schmitt, 1924a).—COLOMBIA: SW of Cape la Vela (Manning, 1961).—VENEZUELA: Cubagua Id. (Manning, 1961). —TRINIDAD: Port of Spain (Manning, 1961).—ARUBA: Punta Basora (Manning, 1961).—CURAÇAO: Schottegat (Rathbun, 1919); Caracas Bay, Spanish Water (Schmitt, 1924).—BONAIRE: Holthuis, 1941.

Bullis & Thompson (1965) recorded specimens (as *G. oerstedii*) from COMBAT Sta. 203, SILVER BAY Stats. 53, 54, 69, 71, 581, 1695, 1704, 1992, and 2142; and OREGON Stats. 275, 1215, 1719, and 2355.

*Holotype.*—1 ♂, 58.7; W side of Baradal, Tobago Cays, Grenadine Islands, Lesser Antilles; under stones and in coral rock at low tide; Smithsonian-Bredin Sta. 21-56; 17 March 1956; USNM 119140.

*Paratypes.*—1 ♂, 43.8; flats E of Vixen Point, Virgin Gorda, Virgin Islands; from grass beds, *Porites, Millepora,* and some *Acropora,* shore to 2 m; Smithsonian-Bredin Sta. 10-58; 28 March 1958; USNM 119144. —1 ♀, 63.0; Prickly Pear Id., Vixen Point, Virgin Gorda, Virgin Islands; beach with *Pocillopora;* Smithsonian-Bredin Sta. 111-56; 15 April 1956; USNM 119142.—1 ♀, 48.0; Gravenor Landing, Barbuda, Lesser Antilles; collecting along shore; Smithsonian-Bredin Sta. 113a-58; 28 April 1958; USNM 119145.—1 ♂, CL 6.9; 3 ♀, 42.0-59.3; N of Black's Point, Falmouth Bay, Antigua, Lesser Antilles; cracked from rocks and coral; Smithsonian-Bredin Sta. 110-59; 30 April 1959; USNM 119141.—4 ♀, 21.7-47.2; Rat Islands, Pointe-à-Pitre, Guadeloupe, Lesser Antilles; exposed reef and submerged weedy rocks; Smithsonian-Bredin Sta. 69-56; 30-31 March 1956; USNM 119143.—2 ♂, 53.2-55.4; 1 ♀, 42.0; Spanish Water, near Newport, Curaçao; L. B. Holthuis 1013; 25 November 1956; RMNH.—4 ♂, 28.8-60.0; 1 ♀, 35.6; Marie Pompoen, near Willemstad, Curaçao; L. B. Holthuis 1028; 26 December 1956; RMNH.—3 ♂, 23.6-47.5; 2 ♀, 50.3-52.7; NW end of Fuik Bay, Curaçao; L. B. Holthuis 1052; 13 January 1957; RMNH.

*Other material.*—BERMUDA: 3 ♂, 4 ♀; no specific locality; USNM.— series of fragments; same; USNM 43044.—1 ♂; same; USNM 33026.— 1 ♀; tidepools at Whalebone Bay; USNM.—2 ♀; Whalebone Bay; USNM. —3 ♂; Hungry Bay; USNM 25452.—1 ♀; SW end of St. George's Id.; USNM.—2 ♂, 1 ♀; Church Bay; USNM.—NORTH CAROLINA: 1 ♀; Black Rocks, off New River; USNM 90097.—2 ♀; Beaufort; FISH HAWK; USNM 51010.—2 ♀; SILVER BAY Sta. 1695; USNM.—2 ♀; SILVER BAY Sta. 1704; USNM.—1 ♂; off Cape Fear; ALBATROSS Sta. 2623; USNM 11261.—SOUTH CAROLINA: 1 ♀; 15 mi. SE of Charleston; USNM 5064.—1 ♂; Blackfish Banks, 12 mi. off Charleston; USNM 5135.—1 ♂; PELICAN Sta. 182-11; USNM.—1 ♀; PELICAN Sta. 194-10; USNM.—

GEORGIA: 1 ♀; PELICAN Sta. 181-13; USNM.—BAHAMAS: 3 ♂; probably Bahamas; USNM 21057.—6 fragmented spec.; Andros; USNM. —3 ♀; S of Fresh Creek, Andros Id.; UMML.—1 brk. ♂; Abaco; ALBATROSS; USNM.—2 ♀; Nassau; USNM.—2 ♂, 4 ♀ (3 lots); Lyford Cay, New Providence; USNM.—2 ♂; SE end of Cat Cay; USNM.—2 ♂; Rabbitt Keys, off South Bimini; UMML.—10 ♂, 18 ♀; Highborne Cay, Exuma Chain; USNM.—1 ♂, 1 ♀; Thomas Cay, Exuma Chain; USNM.—FLORIDA: 1 ♂; off St. Augustine; COMBAT Sta. 203; USNM 101612.—1 ♂, 1 ♀; same; SILVER BAY Sta. 2142; USNM.—1 ♂, 1 ♀; off Cape Canaveral; SILVER BAY Sta. 1992; USNM.—1 ♂; Bear Cut, Key Biscayne, Miami; USNM.—1 ♀; Cape Florida flats; UMML.—1 ♀; N of Pumpkin Key, Card Sound; FISH HAWK Sta. 7493; USNM.—2 ♂, 1 ♀; Key Largo; USNM.— 1 ♂; off Tennessee Reef; USNM.—1 ♂, 1 ♀ (2 lots); off Alligator Reef Light; UMML.—1 ♀; same; USNM.—2 ♂, 2 ♀; Content Key; UMML.— 4 ♂, 2 ♀; same; USNM.—1 ♀; Key West; ALBATROSS; USNM.—30 ♂, 27 ♀, 1 juv. (32 lots); localities around Dry Tortugas; USNM.—1 ♀; WSW of Rebecca Shoal Light; UMML 32.689a.—3 ♂, 2 ♀; off Key West; FISH HAWK Sta. 7277; USNM.—1 ♂; same; FISH HAWK Sta. 7278; USNM. —1 ♂; same; FISH HAWK Sta. 7293; USNM.—1 ♀; N of Key West; SILVER BAY Sta. 71; USNM 101615.—8 ♀; off Cape Sable; FISH HAWK Sta. 7361; USNM.—1 ♂; same; FISH HAWK Sta. 7364; USNM.—2 ♂; same; FISH HAWK Sta. 7370; USNM.—3 ♀; same; FISH HAWK Sta. 7372; USNM. —1 ♂, 1 ♀; same; FISH HAWK Sta. 7375; USNM.—1 ♂, 1 ♀; same; FISH HAWK Sta. 7379; USNM.—1 ♂; same; FISH HAWK Sta. 7387; USNM.— 1 ♀; same; FISH HAWK Sta. 7390; USNM.—1 ♂; same; FISH HAWK Sta. 7391; USNM.—1 ♂, 1 ♀; same; FISH HAWK Sta. 7392; USNM.—2 ♂, 2 ♀; W coast of Florida; USNM.—1 ♂, 1 ♀; off Marco; ALBATROSS Sta. 2413; USNM 9845.—1 ♀; same; OREGON Sta. 1215; USNM 99502.— 1 ♂; W of Naples; SILVER BAY Sta. 69; USNM 101614.—1 ♂; off Charlotte Harbor; 26°04′N, 82°49′W; 39 m; GRAMPUS Sta. 5099; 17 March 1889; USNM 18223.—2 ♂, 2 ♀; same; ALBATROSS Sta. 2411; USNM 21499.— 7 ♂, 6 ♀ (3 lots); Lemon Bay; USNM.—2 ♂; off Venice; ALBATROSS Sta. 2409; USNM 9819.—1 ♀; Little Sarasota Pass; USNM 6661.—2 ♂, 4 ♀; W of St. Petersburg; OREGON Sta. 4089; USNM.—1 ♂, 3 ♀; same; OREGON Sta. 4090; USNM.—1 ♂; 6 mi. N of Anna Maria Id., St. Petersburg; USNM.—2 ♂; 6 mi. W of Anna Maria Id., St. Petersburg; USNM.—1 ♂, 1 ♀; same lot; ret'd to U. S. Fish and Wildlife Service.—1 ♂, 1 ♀; off Madeira Beach; USNM.—2 ♀; off Tarpon Springs; SILVER BAY Sta. 53; USNM 101613.—3 ♂, 2 ♀; same; SILVER BAY Sta. 54; USNM 101616.— 1 ♂, 1 ♀; Anclote Keys; FISH HAWK Sta. 7234; USNM.—1 ♂, 1 ♀; off St. Martin's reef; USNM 13049.—1 ♂; same; FISH HAWK Sta. 7217; USNM.—1 ♀; same; FISH HAWK Sta. 7231; USNM.—1 ♂; S of Cedar Key; USNM 13068.—1 ♂; off Cedar Key; FISH HAWK Sta. 7184; USNM. —1 ♂, 1 ♀; same; FISH HAWK Sta. 7207; USNM.— 2 ♂, 2 ♀; same; FISH HAWK Sta. 7208; USNM.—1 ♀; same; FISH HAWK Sta. 7209; USNM.—

317

1 ♂; same; Oregon Sta. 1719; USNM 101611.—1 ♀; Pepperfish Key; Fish Hawk Sta. 7158; USNM.—1 ♂; same; Fish Hawk Sta. 7160; USNM. —3 ♂; same; Fish Hawk Sta. 7161; USNM.—1 ♂; same; Fish Hawk Sta. 7171; USNM.—1 ♀; same; Fish Hawk Sta. 7181; USNM.—1 ♂, 2 ♀; Deadman's Bay; Fish Hawk Sta. 7151; USNM.—1 ♀; same; Fish Hawk Sta. 7152; USNM.—1 ♂, 1 ♀; same; Fish Hawk Sta. 7153; USNM. —1 ♀; Apalachee Bay; Fish Hawk Sta. 7190; USNM.—1 ♂, 1 ♀; same; Fish Hawk Sta. 7191; USNM.—4 ♂; same; Fish Hawk Sta. 7192; USNM. —1 ♂, 1; same; Fish Hawk Sta. 7195; USNM.—1 ♂; same; Fish Hawk Sta. 7201; USNM.—1 ♂; off Alligator Point, Franklin Co.; USNM 93467. —1 ♀; same; USNM 104268.—1 ♂; off Cape St. George; Silver Bay Sta. 581; USNM.—1 ♂, 1 ♀; same; Albatross Sta. 2373; USNM 9627. —1 ♀; same; Albatross Sta. 2405; USNM 18006.—1 ♂; same; Albatross Sta. 2407; USNM 9813.—2 ♀; same; Oregon Sta. 276; USNM 92383.—1 ♀; same; Pelican Sta. 156-2; USNM.—1 ♀; off Panama City; USNM.—1 ♂; off Fort Walton; USNM.—1 ♂; Pensacola; USNM 23289. —TEXAS: 1 ♀; Heald Bank, Sabine; USNM 97678.—1 ♀; Sabine Pass; USNM 101084.—MEXICO: 1 ♂; Espiritu Santo Bay, Quintana Roo; Bredin Sta. 35-60; USNM.—1 ♂; Nicchehabin Reef, Quintana Roo; Bredin Sta. 82-60; USNM.—1 ♂, 1 ♀; Ascension Bay, Quintana Roo; Bredin Sta. 60-60; USNM.—6 ♂, 7 ♀, 3 juv.; same; Bredin Sta. 77-60; USNM.— 4 ♂, 5 ♀; same; Bredin Sta. 85-60; USNM.—4 ♂, 1 ♀; same; Bredin Sta. 93-60; USNM.—1 ♀; same; Bredin Sta. 95-60; USNM.—1 ♂; Cozumel Id.; Albatross; USNM 9319.—1 ♀; same; Bredin Sta. 106-60; USNM. —1 ♀; S of San Miguel, Cozumel Id.; USNM.—PANAMA: 5 ♂, 5 ♀; Colon Reef, Colon; USNM.—1 ♂; Fox Bay, Colon; USNM 106050.— 2 ♂, 2 ♀; Coco Solo, Canal Zone; USNM.—WEST INDIES: 2 ♂; no specific data; paralectotypes of *G. oerstedii* Hansen; UZM.—1 ♂, 1 ♀; West Indies; paralectotypes of *G. oerstedii* Hansen; UZM.—CUBA: 1 ♀; Havana; paralectotype of *G. oerstedii* Hansen; UZM.—1 ♂, 2 ♀; Cabanas; USNM 48566.—1 ♂; Mariel; USNM 23836.—JAMAICA: 1 ♂; Umbrella Point, nr. Montego Bay; USNM.—1 ♂, 2 ♀; Northeast Cay, Morant Cays; USNM.—4 ♂; Southeast Cay, Morant Cays; USNM.—2 ♂; Rocky Cay, Morant Cays; UMML.—3 ♂, 1 ♀; Northeast Cay, Pedro Cays; USNM.— DOMINICAN REPUBLIC: 1 ♂; no specific locality; USNM 4086.—1 ♂; same; USNM 21501.—PUERTO RICO: 1 ♀; W end San Juan Id.; Smith-sonian-Hartford Exped. Sta. 16; USNM.—1 ♂, 2 ♀; off Boca Prieta, Punta Guaniquilla; Fish Hawk P. R. Sta. 147 [6075]; USNM 64796.—1 ♂; Boqueron Bay; Fish Hawk; USNM 64786.— ♀; off Cabo Rojo; Fish Hawk P. R. Sta. 144 [6072]; USNM 64789.—2 ♂, 1 ♀; off Humacao, Fish Hawk P. R. Sta. 170 [6098]; USNM 64779.—3 ♂; same; Fish Hawk P. R. Sta. 171 [6099]; USNM 64781.—1 ♂; Fajardo; Fish Hawk; USNM 64793.—1 ♀; off Vieques Id.; Fish Hawk; P. R. Sta. 156 [6084]; USNM 64799.—2 ♂, 2 ♀; same; Fish Hawk P. R. Sta. 157 [6085]; USNM 64785.—4 ♂, 8 ♀; same; Fish Hawk P. R. Sta. 163 [6091]; USNM 64783.

318

—15 ♂, 21 ♀; same; FISH HAWK P. R. Sta. 164 [6092]; USNM 64770.—
4 ♀; same; FISH HAWK P. R. Sta. 168 [6096]; USNM 64794.—2 ♀;
Ensenada Honda, Culebra; FISH HAWK; USNM.—2 ♂, 4 ♀; off Culebra;
FISH HAWK P. R. Sta. 158 [6086]; USNM 64792.—3 ♂, 4 ♀; same; FISH
HAWK P. R. Sta. 165 [6093]; USNM 64788.—VIRGIN ISLANDS: 1 ♂,
2 ♀; Danish West Indies; UZM.—1 ♂, 1 ♀; no specific locality; UZM.—
1 ♂; St. Thomas; paralectotype of *G. oerstedii* Hansen; UZM.—3 ♂; same;
paralectotypes of *G. oerstedii* Hansen; UZM.—1 ♀; same; paralectotype of
*G. oerstedii* Hansen; UZM.—9 ♂, 13 ♀, 6 juv. (4 lots); St. Thomas; UZM.
—6 ♂, 11 ♀; same; ALBATROSS; USNM.—1 ♀; Mosquito Bay, St. Thomas;
UZM.—1 ♂; St. James Bay, St. Thomas; UZM.—1 ♂, 4 ♀, 1 juv.; W of
Water Id., St. Thomas; UZM.—2 ♂, 2 ♀, 4 juv.; between Saba and St.
Thomas; UZM.—2 ♀, 2 juv.; W end of Thatch Id., N of St. John; UZM.—
2 ♂, 2 ♀, 1 juv. (2 lots); St. John; UZM.—1 ♂; Coral Harbor, St. John;
Smithsonian-Hartford Sta. 27; USNM.—1 ♂; W side of Coral Bay, St.
John; Smithsonian-Hartford Sta. 28; USNM.—2 ♀; NW of Cruz Bay, St.
John; UZM.—1 ♂, 1 ♀; Fish Bay, St. John; UMML.—1 ♂; Great Salt
Pond Bay, St. John; UMML.—2 ♀; Greater Lameshur Bay, St. John;
UMML.—2 ♂; Leinster Bay, St. John; UZM.—1 ♀; E shore Lesser Lame-
shur Bay, St. John; USNM.—1 ♂; Mary's Creek Point, St. John; USNM.—
1 ♀; Princess Bay, St. John; UMML.—1 ♂; Turner Bay, St. John; RMNH.
—1 ♂, 2 ♀, 1 juv.; Loango; UZM.—4 ♂, 1 ♀, 1 juv.; between Jost van
Dyke and Loango; UZM.—1 ♀; Tortola; UMML.—4 ♂; off Buck Id. N of
St. Croix; UZM.—LESSER ANTILLES: 1 ♂, 2 ♀ (3 lots); Great Bay, St.
Martin; RMNH.—1 ♂; St. Kitts; OREGON Sta. 2355; USNM.—1 ♀;
Charlestown, Nevis; Bredin Sta. 99-56; USNM.—2 ♂, 4 ♀; Charlotte
Point, English Harbor, Antigua; Bredin Sta. 73-56; USNM.—1 ♂; English
Harbor, Antigua; USNM 68955.—2 ♂, 2 ♀; N of Black's Point, Falmouth
Bay, Antigua; Bredin Sta. 110-59; USNM.—1 ♀; Pillars of Hercules,
Antigua; USNM 68946.—1 ♂, 1 ♀; Cochons Id., Pointe-à-Pitre, Guade-
loupe; Bredin Sta. 70-56; USNM.—1 ♂; Woodbridge Bay, Dominica;
Bredin Sta. 58-56; USNM.—1 ♂; Barbados; USNM.—1 ♂; Bathsheba,
Barbados; USNM 68951.—1 juv.; Pelican Id., Barbados; USNM.—5 ♂;
Grenadines; OREGON Sta. 5050; USNM.—COLOMBIA: 2 ♂, 3 ♀; 11 mi.
SW of Cape la Vela; AHF Sta. A12a-39; USNM 106347.—VENE-
ZUELA: 1 ♀; Tortuga Id.; AHF Sta. A22-39; USNM.—3 ♂; Cubagua Id.;
AHF Sta. A25-39; USNM.—3 ♂, 7 ♀; same; AHF Sta. A28-39; USNM.—
1 ♂, 2 ♀; Cubagua Id.; USNM.—TRINIDAD: 2 ♂; Port of Spain; AHF
Sta. A35-39; USNM 106346.—TOBAGO: 1 ♂; UZM.—ARUBA: 1 ♂,
3 ♀; no specific locality; RMNH 53a.—1 ♀; N coast; USNM 42969.—1 ♂,
1 ♀; Paarden Bay; RMNH.—4 ♂, 1 ♀; Pta. Basora; AHF Sta. A16-39;
USNM.—CURAÇAO: 4 ♂, 4 ♀; no specific locality; ALBATROSS; USNM.
—1 ♂, 1 ♀; Caracas Bay; RMNH 153.—1 ♂; same; USNM 57528.—1 ♂;
St. Jan; RMNH.—1 ♀; Schottegat Lagoon; USNM 42970.—1 ♀; Spanish
Water; USNM 57526.—6 ♂, 7 ♀ (2 lots); Brakke Point, Spanish Water;

RMNH.—1 ♂, 2 ♀; Plaja Frankie, Spanish Point; RMNH.—12 ♂, 1 juv. (8 lots); localities around Piscadera Bay; RMNH.—BONAIRE: 1 ♂; Lac Soerebon; RMNH.—1 ♂, 1 ♀; Paloe Lechi, S of Salinja; RMNH.—1 ♀; klein Bonaire; RMNH 143.—1 ♂; W coast of Bonaire; RMNH.

*Diagnosis.*—Lower portion of posterior margin of pleura of first to fourth abdominal somites straight or slightly concave; pleura, in lateral view, particularly of first and second somites, with posterior margin usually evenly concave; fifth abdominal somite rounded or angled posterolaterally, unarmed; sixth abdominal somite with 6 carinae, variously swollen, submedians and intermediates often with apices unarmed in adults. A summary of abdominal width-carapace length indices for specimens of various sizes follows:

| CL in mm | n | AWCLI mean | AWCLI range |
|---|---|---|---|
| 3.0 | 2 | 808 | 756-851 |
| 4.0 | 6 | 851 | 810-891 |
| 5.0 | 5 | 828 | 811-866 |
| 6.0 | 17 | 817 | 750-868 |
| 7.0 | 15 | 804 | 746-845 |
| 8.0 | 24 | 811 | 779-850 |
| 9.0 | 27 | 805 | 774-851 |
| 10.0 | 17 | 793 | 744-825 |
| 11.0 | 12 | 799 | 742-839 |
| 12.0 | 11 | 783 | 730-829 |
| 13.0 | 6 | 798 | 760-869 |
| 14.0 | 7 | 781 | 751-824 |
| 15.0 | 1 | 781 | . . . . . . |
| 16.0 | 2 | 763 | 762-765 |
| 17.0 | 2 | 794 | 760-829 |

Telson usually broader than long, of Bredini-type; median carina often inflated, armed at most with a small apical tubercle; accessory median carinae short, occasionally inflated, fusing posteriorly with median carina to form anchor; knob usually present, inflated, unarmed; anterior submedian carinae short, swollen, usually unarmed, rarely with apical tubercles; carinae of submedian teeth occasionally inflated and/or pitted dorsally, never longitudinally sulcate; submedian teeth usually with shelf on inner surface, shelf lacking anterior angled prominence; swollen oblique carina usually present extending from anterior base of each submedian carina towards knob; movable apices of submedian teeth usually absent; intermediate teeth not widely separate from submedians, longitudinal axes converging posteriorly; intermediate carina with accessory carina on inner face; lateral carina distinct, apex separated from intermediate tooth by

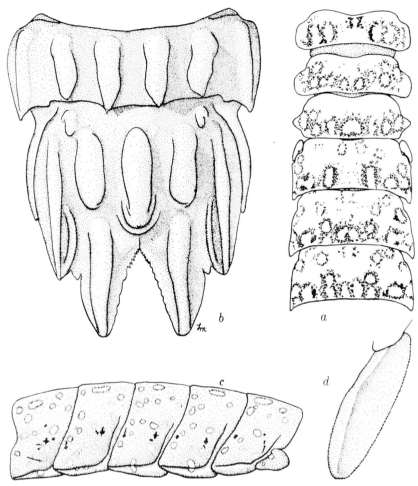

Figure 87. *Gonodactylus bredini*, n. sp. Male paratype, TL 28.9, Curaçao. *a*, chromatophore pattern of body, dorsal view. Male holotype, TL 58.7, Grenadine Ids.: *b*, sixth abdominal somite and telson; *c*, abdomen, lateral view; *d*, uropodal endopod (setae omitted).

obscure notch; submedian denticles small, numerous, most distinct anteriorly; usually 1 unarmed intermediate denticle present, not deeply recessed, situated at level of or posterior to apex of intermediate tooth.

Uropodal endopod usually broad, with inner margin straight, slightly convex, occasionally faintly sinuous; endopod not markedly tapering towards apex.

*Color.*—Variable, body usually light-green; littoral Caribbean specimens

321

with greenish body, often with numerous black circles across each somite (fig. 87), walking legs yellowish, dactyli pink; merus lacking bars of dark pigment in Caribbean form; occasionally some Caribbean specimens with dark rectangular patches on sixth thoracic and first abdominal somites, lateral dark patches present on sixth thoracic somite or absent; some specimens from Florida dark blue or green, with orange walking legs; other specimens from Florida mottled as shown in fig. 88. Dingle (1964) has commented on color polymorphism in Bermudan specimens.

*Size.*—Males, TL 9.0-70.8 mm; females, TL 9.3-74.6 mm. Other measurements of male holotype, TL 58.7: carapace length, 13.5; fifth abdominal somite width, 10.2; telson length, 9.1, width, 8.7.

*Discussion.*—Although this species exhibits a wide range of variation in ornamentation of the telson there has been no difficulty in separating any of the specimens examined from *G. oerstedii. G. bredini,* which may be as common as *G. oerstedii* in the Caribbean, may be distinguished from it as follows: (a) the posterior margin of the abdominal pleura of the first four abdominal somites are evenly concave in *G. bredini,* sinuous in *G. oerstedii;* (b) adults of *G. bredini* as well as most specimens over TL 20 mm lack the movable apices on the submedian teeth of the telson; these are present at all sizes of *G. oerstedii;* (c) the intermediate marginal teeth converge on the submedians in *G. bredini* whereas they are subparallel in *G. oerstedii;* (d) the apex of the intermediate marginal tooth is blunter in *G. bredini* than in *G. oerstedii* and the apex of the intermediate tooth is not widely separate from the outer margin of the submedian in *G. bredini;* (e) in *G. oerstedii* the intermediate denticles are usually armed with small spinules or a small tubercle and they are always recessed anteriorly by at least their width; in *G. bredini* there is usually only one rounded denticle which is not recessed but is often situated posterior to the apex of the intermediate tooth; (f) the uropodal endopod is more sinuous and tapering in *G. oerstedii* than in *G. bredini;* and (g) the color pattern of *G. bredini* differs in that there is no sign of dimorphism; specimens are usually light green or mottled green, and both males and females may have black circles scattered on the body; in Caribbean *G. bredini,* at least I have seen no indications of bars of pigment on the merus of the raptorial claw in either sex; the dactyli of the claws are always pink in *G. bredini,* blue in *G. oerstedii.*

Of the western Atlantic species reported herein, *G. lacunatus* is the only one which may be confused with *G. bredini.* In *G. lacunatus* the longitudinal channels on the submedian carinae are deep and carinate; in *G. bredini* the carinae of the submedian teeth may be slightly pitted but they are never longitudinally channelled.

Specimens of *G. bredini* fall into at least five distinct "populations" from different areas in the western Atlantic. These may well be shown to be distinct species or subspecies but they cannot be adequately characterized

322

on the basis of material now available for study. Future work on *G. bredini* and its American allies will require field observaitons, notes on color in life, and as much information on habitat as possible.

The type-series of *G. bredini* has been restricted to a few specimens taken littorally in the Caribbean proper. In this series and in other littoral Caribbean forms, the body is slender, appearing elongate. The median and anterior submedian carinae of the telson are not usually inflated; the anchor is usually distinct, and the knob is always visible. The intermediate denticles of the telson are always slightly recessed but never armed, and the submedian marginal teeth never have movable apices. The black circles on a green background is a characteristic color pattern of these forms; it is shown in fig. 87.

In specimens from Bermuda the body appears to be stouter and more compact and the telson is broader, but the AWCLI overlaps completely with that of Caribbean specimens. The carinae of the marginal teeth are sharper than in Caribbean specimens, and the marginal teeth are broader and blunter. The intermediate denticle is rounded and appears to be broader than that of Caribbean forms. Inasmuch as *G. oerstedii* has not been represented in any collection from Bermuda. I have assumed that records in the literature to Bermudan *G. oerstedii* actually should be referred to *G. bredini*. Dingle (1964) has commented on a marked color polymorphism in Bermudan specimens of *G. oerstedii;* none of the specimens from Bermuda available for study have exhibited any particular color pattern.

Specimens from the Carolinas, northeastern Florida, and the Gulf of Mexico appear to be much shorter and broader than typical Caribbean specimens; as in the case of the few specimens from Bermuda, the AWCLI for Gulf forms completely overlaps that of the Caribbean ones. The telson of the Gulf form is broader, with inflated carinae (fig. 88) which are often pitted, and the intermediate denticle is usually posterior to the apex of the intermediate tooth. The knob is usually completely obscured by the inflated median carina. The submedian teeth never have movable apices. Most specimens of this form can be distinguished at a glance from Caribbean specimens, but there appear to be intergrades between these two forms present from localities on the southwest coast of Florida. The color of these forms may be light green, with darker pigment along the posterior margin of each somite, or mottled pink as shown in fig. 88. The merus may be barred in some specimens; the dactylus is always pink.

Some small specimens from south Florida differ from all others on having an almost triangular telson, with sharp dorsal carinae, and with the intermediate denticles armed and set at or posterior to the apex of the intermediate tooth. In these forms the movable apices of the submedian teeth are present. As these forms are usually less than 5 mm in TL they may well be juveniles of the more typical Gulf form.

Finally, a series of small specimens taken by the FISH HAWK in

323

moderate depths off Puerto Rico have the carinae of the telson very inflated; this inflation is never seen in small specimens of the littoral form of *G. bredini*. The median carina is usually provided with an apical tubercle. The intermediate denticles are usually posterior to the apex of the intermediate tooth. The submedian carinae may be pitted as in the Gulf form. Even small specimens (TL less than 20mm) apparently lack the movable apices of the submedian teeth.

All of these forms seem to intergrade; although many specimens from

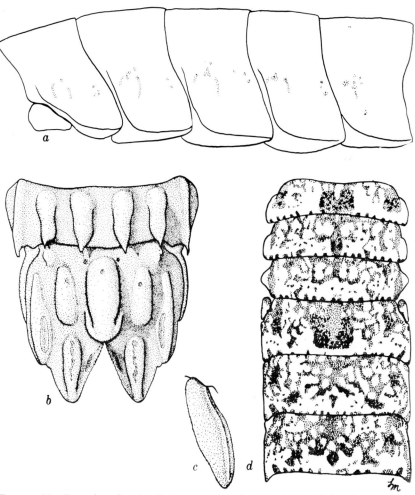

FIGURE 88. *Gonodactylus bredini*, n. sp. Female, TL 41.7, off South Carolina. *a*, abdomen, lateral view; *b*, sixth abdominal somite and telson; *c*, uropodal endopod (setae omitted). Female, TL 31.5, off Anna Maria Key, Florida; *d*, chromatophore pattern of body, dorsal view.

each of the areas listed above are immediately distinguishable there are also numerous specimens that fit none or several of those patterns.

*Etymology.*—This species is named in honor of Mr. J. Bruce Bredin, co-sponsor of the Smithsonian-Bredin expeditions to the Caribbean in 1956, 1958, 1959, and 1960.

*Ontogeny.*—The larval specimens reported from Bermuda by Gurney (1946) as *G. oerstedii* may be referable to *G. bredini.* Gurney's account deals mainly with composite early stages but some notes on the postlarva are included.

*Type.*—U. S. National Museum.

*Type-locality.*—Baradal, Tobago Cays, Grenadine Islands, Lesser Antilles.

*Distribution.*—Western Atlantic, from Bermuda, the Carolinas, the northern Gulf of Mexico through the Caribbean to Aruba, Bonaire, and Curaçao off the coast of South America; littoral to 55 m, in sponges, rocks, and on coral reefs.

*Gonodactylus oerstedii* Hansen, 1895

Figure 89

*Gonodactylus chiragra:* (?)Gibbes, 1845, p. 70 [listed].—(?)A. Milne-Edwards, 1868, p. 69.—(?)von Martens, 1872, p. 147.—(?)Neumann, 1878, p. 39.—(?)Miers, 1880, p. 118 [part].—(?)von Martens, 1881, p. 94 [discussion].—(?)Kingsley, 1884, p. 67 [discussion].—Brooks, 1892, pp. 337, 353, fig. on pl. 1, pls. 3, 14, 15 [biology and larvae; colored figure].—(?)Andrews, 1892, p. 72.—(?)Bigelow, 1893, p. 100; 1893a, p. 102; 1894, p. 495 [part].—(?) Nutting, 1895, p. 126 [discussion].—(?)Nobili, 1897, p. 6; (?)1898, p. 2.—(?)Rathbun, 1899, p. 628.—(?)Young, 1900, p. 500.—(?)Andrews, Bigelow, & Morgan, 1945, p. 340 [popular account].—(?) Holthuis, 1959, p. 173 [discussion] [not *G. chiragra* (Fabricius, 1781)].

*Gonodactylus chiragrus:* (?)Gibbes, 1850, p. 201 [p. 37 on separate].—(?)Guppy, 1894, p. 115 [listed] [not *G. chiragra* (Fabricius, 1781)].

*Gonodactylus oerstedti* (?)Thallwitz, 1893, p. 55 [*nomen nudum*].

*Gondactylus falcatus:* (?)Sharp, 1893, p. 105 [part] [not *G. falcatus* (Forskål, 1775)].

*Gonodactylus Oerstedii* Hansen, 1895, footnote, p. 65 [part]; (?)1921a, p. 79 [morphology].

*Gonodactylus oerstedi:* Borradaile, 1898, p. 35, pl. 5, fig. 3 [*G. chiragra* and *G. oerstedi* in text].

*Gonodactylus oerstedii:* (?)Rankin, 1898, p. 253.—(?)Thompson, 1901, p. 42.—Bigelow, 1901, p. 152, text-figs. 1-2 [part].—(?) McClendon, 1911, p. 57 [ecology].—(?)Kemp, 1913, p. 204 [listed].—(?)Nutting, 1919, p. 181 [discussion].—Rathbun, 1919, p. 348 [part].—(?)Kemp & Chopra, 1921, p. 309.—(?)Parisi, 1922, p. 111.—Schmitt, 1924, p. 80 [part]; 1924a, p. 96 [part].—Hansen, 1926, p. 27 [part].—(?)Pearse, 1929, p. 221 [ecology].—(?)Boone, 1930, p. 21, pl. 1.—Bigelow, 1931, p. 120 [part].—(?)Pearse, 1932, p. 121 [ecology].—(?)Rathbun, 1935, p. 113 [fossil].—(?)Pratt, 1935, p. 446.—(?)Lunz, 1937, p. 4.—(?)Berry, 1939, p. 466 [discussion, fossil].—(?)Schmitt, 1940, p. 211, text-figs. 26, 29

[part; text-figs. 27-28 and possibly 29 represent another species].—
Holthuis, 1941, p. 38.—(?)Steinbeck & Ricketts, 1941, p. 428.—(?)Pearse,
1950, p. 150 [ecology].—Chace, 1954, p. 449 [part].—Voss & Voss, 1955,
p. 227.—Chace, 1956, p. 162, upper fig. on second plate [part ?].—
Manning, 1959, p. 16 [part]; 1961, p. 43, pl. 11, figs. 5-6 [part]; 1963b,
p. 422 [embryology].—Manning & Provenzano, 1963, p. 467, text-
figs. 1-8 [larvae].—Bullis & Thompson, 1965, p. 13 [part].—(?)Elofsson,
1965, p. 8, text-fig. 5 [anatomy].—(?)Milne & Milne, 1965, p. 289
[listed; biology].—Manning, 1967a, p. 103.
    not *Gonodactylus chiragra* var. E *(G. oerstedi)* Borradaile, 1899, p. 402.
(?) not *Gonodactylus oerstedii:* Rankin, 1900, p. 545.—Verrill, 1923, p.
    189, pl. 50, figs. 3-4, pl. 51, figs. 2-2b, pl. 55, fig. 3 (larva), text-fig. 1.—
    Lunz, 1935, p. 152, text-fig. 1.—Gurney, 1946, p. 155, text-fig. 14.—
    Dennell, 1950, p. 63.—Pearse & Williams, 1951, p. 144.—Morrison &
    Morrison, 1952, p. 396.—Springer & Bullis, 1956, p. 22.—Menzel, 1956,
    p. 46 [listed].—Hazlett & Winn, 1962, p. 27 [biology].—Dingle, 1954,
    p. 236 [= *G. bredini,* n. sp.].
*Gonodactylus chiragra* var. O ?= *oerstedii:* Lanchester, 1903, p. 455 [listed].
*Gonodactylus OErstednii:* (?)Bouvier, 1918, p. 12 [listed; erroneous spelling
    of *G. oerstedii*].
*Gonodactylus OErstedti:* (?)de Boury, 1918, p. 17 [discussion; erroneous
    spelling].
not *Gonodactylus oerstedii:* (?) Andrade Ramos, 1951, p. 143 [listed].—
    (?)Lemos de Castro, 1955, p. 42, text-figs. 30-31, pl. 12, fig. 44, pl. 18,
    fig. 56 [?=*G. lacunatus* Manning; possibly more than one species].
(?) *Gonodactylus:* Milne & Milne, 1961, p. 422, figs. 7-9 [biology].
not *Gonodactylus oerstedii:* Manning, 1966, p. 371, text-fig. 6a [=*G.
    austrinus,* n. sp.].

*Previous records.*—BAHAMAS: Bimini (?Bigelow, 1893a, 1894; ?Boone,
1930; ?Pearse, 1950; ?Milne & Milne, 1961); Green Turtle Cay (?Bigelow,
1893a); Andros (?Parisi, 1922); Nassau (?Rankin, 1898); Cay Sal Bank
(?Boone, 1930); ?Andrews, Bigelow, & Morgan, 1945; Manning, 1967b;
no specific locality (Brooks, 1891; Borradaile, 1898; ?Lunz, 1937; ?Man-
ning, 1959).—FLORIDA: Soldier Key (Voss & Voss, 1955; Manning,
1959); Biscayne Bay, Plantation Key, Long Key, Bahia Honda, Sand Key,
Key West (all Manning, 1959); Florida Keys (?Bigelow, 1893, 1894;
?Boone, 1930; Manning 1967b); Key West (?Gibbes, 1850; ?Sharp, 1893;
?Chace, 1954); Dry Tortugas and environs (?Sharp, 1893; ?Nutting,
1895; McClendon, 1911; ?Pearse, 1929, 1932; Schmitt, 1940; ?Chace,
1954).—PANAMA: Caledonia Bay (Manning, 1961).—WEST INDIES:
?Neumann, 1878; ?Sharp, 1893; ?Hansen, 1926.—CUBA: no speci-
fic locality (?von Martens, 1872, 1881; ?de Boury, 1918; ?Bouvier, 1918);
Havana (?Parisi, 1922).—JAMAICA: Port Henderson (Andrews, 1892;
Bigelow, 1894); Port Royal Cays, Kingston Harbor (?Rathbun, 1899);
Port Antonio (?Rathbun, 1899; ?Boone, 1930; Manning, 1967b).—
PUERTO RICO: Bigelow, 1901, 1931.—VIRGIN ISLANDS: St. Thomas
(?Thallwitz, 1893; Brooks, 1886; ?Nobili, 1898; ?Thompson, 1901;
?Parisi, 1922).—LESSER ANTILLES: St. Eustatius (Rathbun, 1919;
Holthuis, 1941); Antigua (Schmitt, 1924a); Barbados (?Young, 1900;

Schmitt, 1924a); Testigos Ids. (?Holthuis, 1941).—COLOMBIA: Bahia Honda (?Manning, 1961); Gairaca, Santa Marta (?Holthuis, 1941).—VENEZUELA: Gulf of Paria (Guppy, 1894); Archipelago de los Roques (Chace, 1956); Cubagua Id. (Manning, 1959).—TOBAGO: Bucco Reef (?Manning, 1961).—ARUBA: ?Rathbun, 1919; Holthuis, 1941.—CURAÇAO: Rathbun, 1919; Schmitt, 1924; Holthuis, 1941; ?Manning, 1961. —BONAIRE: Rathbun, 1919; Holthuis, 1941.

Bullis & Thompson (1965) record this species from OREGON Sta. 3638; the remainder of their records are referred to G. austrinus, n. sp., G. bredini, n. sp., G. lacunatus, and G. torus, n. sp.

Material.—NO DATA: 1 ♀; paralectotype; UZM.—BAHAMAS: 1 ♂; Bimini; USNM 88689.—2 ♂, 1 ♀; same; USNM 88688.—1 ♂, 1 ♀; same; USNM 88690.—1 ♂; same; UMML.—3 ♀; Andros Bank; USNM 43043. —3 ♂, 3 ♀; Abaco; ALBATROSS; USNM.—1 juv.; New Providence; ALBATROSS; USNM 12328.—1 ♀; between Hog Id. and Athol Id.; RMNH.— 2 ♂, 1 ♀; Nassau; USNM 11545.—21 ♂, 25 ♀, 1 juv. (12 lots); Lyford Cay; USNM.—2 ♂; Powell's Point, Eleuthera Id.; USNM.—7 ♂, 1 ♀; Oyster Cay, Exuma Chain; USNM.—2 ♂, 1 ♀; NW of Little Major's Spot Cay, Exuma Chain; USNM.—2 ♂; Spanish Wells; USNM.—FLORIDA: 8 ♂, 16 ♀, 2 juv. (3 lots); Bear Cut, Key Biscayne, Miami; RMNH.—6 ♂, 18 ♀, 1 juv. (4 lots); same; USNM.—1 ♀; Cape Florida, Biscayne Bay; USNM 13571.—1 ♀; same; USNM 14005.—3 ♀; Biscayne Bay, near Cape Florida; USNM 21500.—1 ♂; same; USNM.—1 ♀; Cape Florida flats; UMML 32.173.—10 ♂, 11 ♀; Biscayne Bay off Key Biscayne; USNM.—1 ♂; Biscayne Bay; USNM 25649.—1 ♂, 3 ♀; same; UMML 32.185.—4 ♂, 8 ♀; same; UMML 32.186.—2 ♂; Soldier Key, Biscayne Bay; RMNH 391, 392.—4 ♂, 6 ♀; Sands Key Light; USNM. —5 ♂, 7 ♀ (2 lots); Harry Harris Park, Key Largo; USNM.—1 ♀; Key Largo; USNM 18005.—2 ♂, 1 ♀ (3 lots); off Alligator Reef Light; USNM.—3 ♂, 1 ♀; same; UMML.—11 ♂, 10 ♀ (3 lots); Lower Matecumbe Key; USNM.—1 ♂, 5 ♀ (2 lots); Long Key; UMML.—1 ♂; same; Tulane Univ.—1 ♂; Key Vacas; USNM.—3 ♂; Florida Bay off Marathon; UMML.—2 ♀; Hawk Channel; FISH HAWK Sta. 7427; USNM.—1 ♀; same; FISH HAWK Sta. 7428; USNM.—1 ♀; same; FISH HAWK Sta. 7463; USNM.—1 ♀; Channel Key Lake; FISH HAWK Sta. 7442; USNM.—1 ♀; Upper Jewfish Bush Lake; FISH HAWK Sta. 7445; USNM.—3 ♂, 2 ♀; Indian Key; USNM 14102.—1 ♂; same; USNM 13562.—1 ♂; Knight's Key; USNM 13563.—4 ♂, 1 ♀; same; USNM.—4 ♂, 3 ♀ (2 lots); same; FISH HAWK; USNM.—1 ♀; Little Torch Key; USNM.—1 ♀; Salt Pond Key; USNM 13577.—1 ♂, 2 ♀; West Summerland Key; UMML.—1 ♀; between Mangrove Key and Key West; USNM.—19 ♂, 18 ♀ (7 lots); Key West; USNM.—2 ♂, 1 ♀; Sand Key, Key West; UMML.—2 ♂; Tortugas; RMNH 159.—2 ♂, 6 ♀; same; RMNH 254.—159 ♂, 181 ♀, 35 juv. (100 lots); localities around Tortugas; W. L. Schmitt and C. R. Shoemaker; USNM.—MEXICO: 2 ♀; OREGON Sta. 3638; USNM.—1 ♂; Mujeres

Harbor, Quintana Roo; Smithsonian-Bredin Exped. Sta. 10-60; USNM.—
1 ♀; Mujeres Id., Quintana Roo; Bredin Sta. 29-60; USNM.—2 ♂, 1 ♀,
2 juv.; Nicchehabin Reef, Ascension Bay, Quintana Roo; Bredin Sta.
52-60; USNM.—1 ♀; same; Bredin Sta. 67-60; USNM.—3 ♂, 6 ♀; same;
Bredin Sta. 72-60; USNM.—1 ♂; same; Bredin Sta. 82-60; USNM.—1 ♂,
1 ♂, 1 juv.; same; Bredin Sta. 91-60; USNM.—2 ♂, 9 ♀, 1 juv.; Suliman
Point, Ascension Bay, Quintana Roo; Bredin Sta. 85-60; USNM.—5 ♂,
4 ♀, 3 juv.; same; Bredin Sta. 95-60; USNM.—1 juv.; N of Ascension
Bay, Quintana Roo; Bredin Sta. 87-60; USNM.—1 ♂; Cozumel Id.; Bredin
Sta. 47-60; USNM.—1 ♂, 1 ♀; 10 mi. SW of Miguel, Cozumel Id.; UMML.
—1 ♀; Cayo Norte, Banco Chinchorro, Yucatan; USNM.—BRITISH
HONDURAS: 1 ♀; S of Cay Bokel, Turneffe Id.; UMML.—1 ♂, 1 ♀;
Half Moon Cay, Lighthouse Reef; USNM.—1 ♀; W of Half Moon Cay;
USNM.—2 ♀; Spanish Cay; USNM 22598.—NICARAGUA: 1 ♂, 1 ♀;
Oregon Sta. 4836; USNM.—OLD PROVIDENCE ISLAND: 3 ♂,
4 ♀; USNM 78321.—PANAMA: 1 ♂, 3 ♀; Colon Reef, Colon; USNM
106048.—1 ♂; same; USNM 106051.—1 ♂; Colon; USNM 106049.
—2 ♂; same; USNM.—1 ♂; Margarita Id.; USNM 76429.—1 ♀; Coco
Solo; USNM.—3 ♂; Caledonia Bay; AHF Sta. A1-39; USNM—1 ♂;
same; AHF Sta. A2-39; USNM.—8 ♂, 5 ♀, 1 juv.; same; AHF
Sta. A50-39; USNM 106348.—WEST INDIES: 6 ♂, 8 ♀; Ørsted,
col.; paralectotypes; UZM.—2 ♀ (1 brk.); C. L. Thomas, col.;
paralectotypes; UZM.—CUBA: 1 ♂, 1 ♀; Tomas Barrera Exped.;
USNM 48513.—1 ♀; Cardenas Bay, Varadero, Matanzas Prov.; USNM
99975.—1 ♂, 1 ♀; Esperanza; USNM 48549.—1 ♂; Cabanas; USNM
48523.—JAMAICA: 1 ♀; Coral Key, E of Montego Bay; USNM.—1 ♀;
Montego Bay; USNM.—1 ♀; Port Henderson; USNM 64780.—2 ♂;
Northeast Cay, Pedro Cays; USNM.—1 ♀; Rocky Cay, Morant Cays;
UMML.—1 ♂; Northeast Cay, Morant Cays; USNM.—HAITI: 3 ♂, 1 ♀;
Tortuga Id.; Smithsonian-Hartford Exped. Sta. 12; USNM.—PUERTO
RICO: 1 ♂; Fish Hawk; USNM 64795.—1 juv.; W end of San Juan Id.;
Smithsonian-Hartford Exped. Sta. 17; USNM.—1 ♀; Caballo Blanco Reef;
USNM 64797.—2 ♂, 2 ♀; Rat Id., Joyuda; USNM.—3 ♂; Puerto Real;
Fish Hawk; USNM 64761.—1 ♂, 1 ♀; Mayaguez; Fish Hawk; USNM
64766.—1 ♂; Magimo, Parguera; RMNH.—1 ♂; Parguera; USNM.—
1 ♀; Guanica; Fish Hawk; USNM 64798.—2 ♂, 3 ♀; Hucares; Fish Hawk;
USNM 64763.—1 ♂, 5 ♀; Fayardo; Fish Hawk; USNM 64767.—13 ♂.
11 ♀ (11 lots); Culebra; Fish Hawk; USNM 64759, 64760, 64762,
64764, 64765, 64768, 64769, 64782, 64787, 64791.—VIRGIN IS-
LANDS: 1 ♂, 5 ♀; Danish West Indies; UZM.—3 ♂, 3 ♀; St. Thomas;
paralectotypes; UZM.—7 ♂, 15 ♀; St. Thomas; UZM.—9 ♂, 11 ♀; same;
Albatross; USNM 7658.—1 ♂; St. Thomas harbor; USNM.—1 ♂; same;
USNM.—4 ♂, 4 ♀, 5 juv.; same; Smithsonian-Hartford Exped. Sta. 23;
USNM.—1 ♂, 2 ♀; Banana Bay, Water Id., St. Thomas; Smithsonian-
Hartford Exped. Sta. 66; USNM.—3 ♂, 3 ♀; Smith Bay, St. Thomas;

Smithsonian-Hartford Exped. Sta. 68; USNM.—2 ♀; W side Haulover Cut, St. Thomas; Smithsonian-Hartford Exped. Sta. 70; USNM.—1 ♂, 1 ♀; French Bay, St. Thomas; USNM.—1 ♂; St. John; UZM.—1 ♂; Reef Bay, St. John; UMML.—3 ♂, 1 ♀; same; USNM.—2 ♂, 1 ♀; E shore Europa Bay, St. John; UMML.—1 ♂; Lesser Lameshur Bay, St. John; USNM.— 1 juv.; same; USNM.—3 ♂, 1 ♀, 3 postlarvae; Greater Lameshur Bay, St. John; UMML.—1 ♂, Fish Bay, St. John; USNM.—4 ♂, 4 ♀; off Lagoon Point, Coral Bay, St. John; USNM.—2 ♂, 2 ♀; Tortola; UMML.—2 ♂, 2 ♀; Burt Point, off Roadtown, Tortola Harbor; Bredin Sta. 5-58; USNM. —1 ♂, 3 ♀; Burt Point, Road Harbor, Tortola; Bredin Sta. 117-56; USNM. —2 ♂, 2 ♀; Soper's Hole, Tortola; Bredin Sta. 23-58; USNM.—2 ♀; E of Vixen Point, Gorda Sound, Virgin Gorda; Bredin Sta. 10-58; USNM.— 12 ♂, 19 ♀; Prickly Pear Id., Vixen Point, Gorda Sound; Bredin Sta. 111-56; USNM.—7 ♂, 10 ♀, 2 juv.; Colquhoun Reef, Gorda Sound; Bredin Sta. 112-56; USNM.—1 ♂, 2 ♀; Colquhoun Reef and Mosquito Id., Gorda Sound; Bredin Stats. 37, 38, 39-59; USNM.—2 ♂; Pomato Point, Anegada; Bredin Sta. 42-58; USNM.—1 ♂; White Bay, Guano Id.; Bredin Sta. 9-58; USNM.—1 ♂, 70.6; St. Croix; A. S. Ørsted, col.; lecto-type; UZM.—1 ♂; same; paralectotype; UZM.—1 ♂, 1 ♀; St. Croix; USNM 104853.—1 ♀; off Buck Id., N of St. Croix; UZM.—1 ♀; N side Buck Id.; Smithsonian-Hartford Exped. Sta. 38; USNM.—1 ♀; Judith Fancy Bay, St. Croix; USNM 105397.—2 ♂, 1 ♀; Salt River Reefs, Christiansted, St. Croix; USNM 71847.—1 ♂, Little Bay, Peter Id.; Bredin Sta. 22-58; USNM.—LESSER ANTILLES: 2 ♂, 2 ♀; NE corner Sandy Id., Anguila; Bredin Sta. 55-58; USNM.—1 ♂, 5 ♀, 2 juv. (4 lots); Great Bay, St. Martin; RMNH.—1 ♂, 1 ♀; lagoon near Oyster Pond Landing, Barbuda; Bredin Sta. 85-56; USNM.—2 ♀; Oyster Pond Landing, Barbuda; Bredin Sta. 92-56; USNM.—3 ♂, 7 ♀; Cocoa Pt., Barbuda; Bredin Sta. 98-59; USNM.—2 ♀; same; Bredin Sta. 102-59; USNM.—1 ♀; same; Bredin Sta. 102a-59; USNM.—5 ♂, 7 ♀; Gravenor Landing, Barbuda; Bredin Sta. 113a-58; USNM.—1 ♂; Spanish Point, Barbuda; Bredin Sta. 112c-58; USNM.—1 ♂; Saba Bank; Bredin Sta. 106-56; USNM.—1 ♂, 1 ♀; Dick Bay, St. Eustatius; RMNH 63.—3 ♂, 4 ♀, 1 juv.; Frigate Bay, St. Christopher; Bredin Sta. 103-56; USNM.—1 postlarva; Charlestown, Nevis; Bredin Sta. 69-58; USNM.—1 ♂; Nonsuch Bay, off Green Id., Antigua; Bredin Sta. 94-58; USNM.—3 ♀; N side Bird Id., Nonsuch Bay, Antigua; Bredin Sta. 96-58; USNM.—2 ♂, 2 ♀; Charlotte Harbor Point, English Harbor, Antigua; Bredin Sta. 73-56; USNM.—1 ♂; English Harbor, Antigua; USNM 68940.—1 ♀; Pillars of Hercules, English Harbor, Antigua; USNM 57980.—1 ♂; same; USNM 57981.—6 ♂, 9 ♀; N of Black's Point, Falmouth Bay, Antigua; Bredin Sta. 110-59; USNM.—3 ♀; same; Bredin Sta. 112-59; USNM.—6 ♂, 5 ♀, 2 juv.; Rat Ids., Pointe-à-Pitre, Guadeloupe; Bredin Sta. 69-56; USNM.—5 ♂, 4 ♀, 1 juv.; E of Cochons Id., Pointe-à-Pitre, Guadeloupe; Bredin Sta. 70-56; USNM.—1 ♀; Woodbridge Bay, Dominica; Bredin Sta. 58-56; USNM.—1 ♂; Aves Id., W of Dominica; RMNH.—1 ♂; Pigeon Id., St. Lucia; Bredin Sta. 47-56; USNM 107740.—1 ♀; same; Bredin Sta. 60-59; USNM 106062.—1 ♀; same; Bredin Sta. 68-59a; USNM 106063.—1 ♀; Barbados; USNM 57979.—1 ♂; same; USNM 68948.—1 ♂; same; USNM 68953.—1 ♀; off the Crane, Barbados; USNM 68947.—1 ♂; Worthing, Barbados; USNM 26407.— 1 ♀; same; USNM 26408.—1 ♀; Pelican Id., Barbados; USNM 68949.— 5 ♂, 6 ♀, 2 juv.; Carlide Bay, Pelican Id., Barbados; Smithsonian-Hartford Exped. Sta. 56; USNM.—2 ♂; Pelican Id., Barbados; USNM 68954.— 1 ♀; off Needham's Point, Barbados; USNM 68944.—6 ♂, 3 ♀; Tyrrell Bay, Carriacou Id., Grenadines; Bredin Sta. 16-56; USNM.—3 ♂, 3 ♀; same; Bredin Sta. 17-56; USNM.—3 ♂, 2 ♀; W side of Baradal, Tobago Cays, Grenadines; Bredin Sta. 22-56; USNM.—1 ♂, 1 ♀; same; Bredin Sta. 23-56; USNM.—1 ♂; Mustique, Grenadines; Bredin Sta. 34-56; USNM.—1 ♀; Grenada; UZM.—1 ♀; Point Salines, White Bay, Grenada; RMNH.—VENEZUELA: 2 ♂, 1 ♀; Cubagua Island; AHF Sta. A25-39; USNM 106345.—1 ♂, 2 ♀; 4 mi. N of Tortuga Id.; AHF Sta. A44-39; USNM.—1 ♂; Gran Roque, Los Roques Ids.; USNM 99098.—TOBAGO: 9 ♂, 11 ♀, 1 juv. (3 lots); no specific locality; UZM.—3 ♂, 5 ♀; W of Pigeon Point; Bredin Sta. 4-59; USNM 106057.—2 ♂, 3 ♀; same; Bredin Sta. 6-59; USNM 106058.—3 ♀; same; Bredin Sta. 31-59; USNM 106061. —2 juv.; same; Bredin Sta. 31a-59; USNM.—1 ♂, 1 ♀ (2 lots); Bucco Reef; RMNH.—6 ♂, 7 ♀; same; Bredin Sta. 8-59; USNM 106059.—1 ♂, 2 ♀; same; Bredin Sta. 26-59; USNM 106060.—ARUBA: 4 ♂, 8 ♀; no specific locality; RMNH 53.—2 ♂, 3 ♀; lagoon, SW coast; RMNH.—1 ♀; Rincon; RMNH.—4 ♂, 3 ♀, 2 juv.; Paarden Bay; RMNH.—1 ♂; Pta. Basara; AHF Sta. A16-39; USNM 106344.—3 ♂, 2 ♀, 3 juv.; Malmok, Aresji; RMNH.—5 ♂, 7 ♀ (6 lots); Boekoeti; RMNH.—1 ♀; Playa Master; RMNH.—9 ♂, 9 ♀ (2 lots); Poos Chikitoe, N of Oranjestad; RMNH. —1 ♂; lagoon near Poos Chikitoe; ZMA.—1 ♂; Punta Braboe; RMNH. —CURAÇAO: 1 ♂, 1 ♀; no specific locality; RMNH 65.—2 ♂, 1 ♀; same; ALBATROSS; USNM 7594.—4 ♂, 3 ♀, 5 juv. (2 lots); Plaja Hoeloe; RMNH.—1 ♀; Joris Bay; RMNH.—1 ♂; Caracas Bay; USNM 57529.— 1 ♀; Brakke Point, Spanish Water; RMNH.—2 ♂, 3 ♀; Plaja Frankie, Spanish Point; RMNH.—1 ♂, 1 ♀; Santa Martha Bay, nr. St. Nicolaas; RMNH.—2 ♂; rr. Willemstad; RMNH.—10 ♂, 6 ♀ (4 lots); Fuik Bay; RMNH.—8 ♂, 10 ♀ (8 lots); Piscadera Bay; RMNH.—BONAIRE: 1 ♂, flat at NE shore; RMNH.—1 ♂; off Bonaire; RMNH.—1 ♀; lake near Bonaire; ZMA.—1 ♂, 1 ♀ (2 lots); Lac Bonaire; RMNH.—1 ♂, 2 ♀ (2 lots); Paloe Lechi, S of Salinja; RMNH.—1 ♀; N of Point Vierkant; RMNH.—1 ♀, 2 juv.; Boca Washikemba; RMNH.—1 ♂, 4 ♀ (3 lots); Soerebon; RMNH.

*Description.*—Lower portion of posterior margin of pleura of first to fourth abdominal somites straight or slightly convex; pleura, in lateral view, particularly of first and second somites, with posterior margin concave above. convex or straight below; fifth abdominal somite rounded or angled pos-

terolaterally, unarmed; sixth abdominal somite with 6 carinae, variously
swollen (swelling more pronounced in adult males than females), apical
spines present in small specimens (TL 40 mm or less), absent or reduced
to tubercles in adults. A summary of abdominal width-carapace length
indices for specimens of various sizes follows:

| CL<br>in mm | n | AWCLI<br>mean | AWCLI<br>range |
|-------------|----|---------------|----------------|
| 3.0 | 1 | 880 | . . . . . . |
| 4.0 | 8 | 868 | 811-919 |
| 5.0 | 6 | 827 | 809-849 |
| 6.0 | 3 | 834 | 813-862 |
| 7.0 | 6 | 823 | 801-864 |
| 8.0 | 8 | 805 | 750-840 |
| 9.0 | 13 | 798 | 759-860 |
| 10.0 | 16 | 797 | 745-837 |
| 11.0 | 9 | 791 | 764-815 |
| 12.0 | 8 | 802 | 783-836 |
| 13.0 | 10 | 775 | 729-806 |
| 14.0 | 2 | 828 | 800-855 |
| 15.0 | 2 | 785 | 762-807 |
| 16.0 | 2 | 784 | 783-785 |
| 17.0 | 1 | 692 | . . . . . . |

Telson broader than long, of Oerstedii-type; median carina variously
swollen (usually more swollen in large males), if armed posteriorly, with
1 inconspicuous tubercle only; accessory median carinae short, unarmed,
usually fusing posteriorly with median carina to form "anchor;" knob
present, well-defined, unarmed; anterior submedian carinae short, swollen,
usually unarmed; carinae of submedian teeth well-defined, not armed, pitted or sulcate dorsally; submedian teeth with shelf on inner margin, shelf
with anterior angled prominence on large specimens; oblique swollen ridge
extending from base of each carina of submedian tooth towards knob low,
often absent; movable apices of submedian teeth always present; intermediate teeth with sharp apices; intermediate teeth separate from submedians, longitudinal axes subparallel; intermediate carina with accessory
carina on inner face; lateral carina distinct, apex blunt; submedian denticles
small, numerous; 2 intermediate denticles present, outer small, sharp, inner much larger, rounded, often with apical tubercle; intermediate denticles situated well anterior to apex of intermediate tooth.

Endopod of uropod with inner margin sinuous, convex proximally, concave distally.

Color.—Males and females dimorphic in color; males dark, body

331

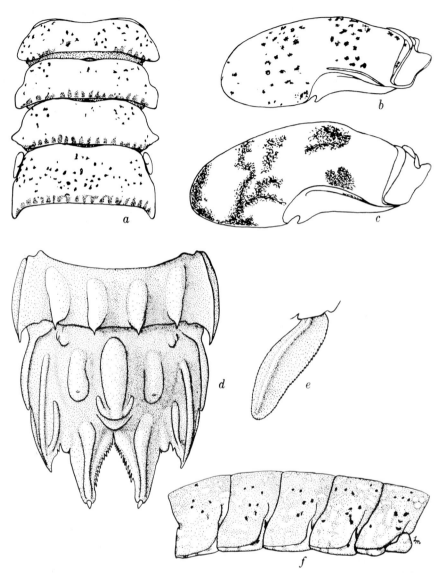

FIGURE 89. *Gonodactylus oerstedii* Hansen. Female, TL 62.8, Key Biscayne, Florida. *a,* chromatophore pattern of body; *b,* chromatophore pattern of merus of claw. Male, TL 35.5, Key Biscayne, Florida: *c,* chromatophore pattern of merus of claw. Female, TL 48.7, Tobago: *d,* sixth abdominal somite and telson; *e,* uropodal endopod (setae omitted); *f,* abdomen, lateral view.

bluish with scattered yellow pigment in no particular pattern; merus of chela usually with 2 broad dark bands, dactylus blue or light purple (fig. 89c); ventral portion of exposed thoracic somites and copulatory tubes dark; females light, body cream colored with many scattered black chromatophores, overall appearance speckled (fig. 89a); sixth thoracic and first abdominal somites with rectangular patches of black chromatophores; merus of chela (fig. 89b) with 2 broad bands of dark spots, dactylus as in male; carinae of sixth abdominal somite and telson often banded yellow and green; juveniles usually with light background and few scattered black chromatophores on body.

Color in life very variable in both males and females; either sex may be marbled, mottled, or with a uniform color. Brooks (1892) illustrated one of the more common color patterns.

*Size.*—Males, TL 10.3-76.0 mm; females, TL 8.0-68.3 mm; postlarvae, TL 7.5-10.0 mm. Other measurements of female, TL 67.3: carapace length, 15.1; fifth abdominal somite width, 11.5; telson length, 9.2, width, 10.0.

*Discussion.*—*G. oerstedii* is the most common littoral stomatopod in the western Atlantic. It can be distinguished from most other species by the broad telson lacking dorsal spinules, with the marginal teeth widely spaced, the submedian teeth always with movable apices, and the anteriorly-recessed intermediate denticles which are sharp and well-formed at all sizes. In addition to these features, the shape of the abdominal pleura is characteristic in *G. oerstedii;* none of the other species have as well-marked a double-inflection to the posterior margin of the pleura. Finally, the endopod of the uropod tapers more and is more sinuous in *G. oerstedii* than in any of the other western Atlantic species.

Three other species in the western Atlantic might be confused with *G. oerstedii, G. curacaoensis* Schmitt, *G. torus,* n. sp., and *G. austrinus,* n. sp. *G. curacaoensis* is a littoral species and may occur together with *G. oerstedii;* it differs in (a) having a small posterolateral spine on the fifth abdominal somite, (b) having a slenderer telson, with median and anterior submedian carinae sharp, spined posteriorly and the lateral tooth of the telson spined, and (c) in having a broad uropodal endopod, with the inner margin convex, not sinuous. *G. torus* lives in deeper water than either of those two species and probably would not be taken together with them. The telson of *G. torus* superficially resembles that of *G. oerstedii,* but *G. torus* matures at a much smaller size and this is reflected in the swollen carinae of the telson at a TL of 25 mm; specimens of *G. oerstedii* at this size never show signs of sexual maturity. Finally, *G. torus* is a slenderer species than is *G. oerstedii;* this is readily seen by a comparision of the AWCLI for each species. *G. austrinus* differs chiefly in lacking the double-inflection on the posterior margin of the abdominal pleura; other differences are described below.

*Remarks.*—*G. oerstedii* is extremely abundant in the Caribbean where it can be found in a wide variety of habitats, including coral reefs, *Porites* flats, *Phragmatopoma* clumps, sponges, and rocks. It may occur together with *G. bredini, G. curacaoensis,* and *G. spinulosus,* and it can be distinguished from all but *G. spinulosus* in the field wtihout difficulty. That species is similar in basic color pattern and its characteristic patches of black pigment on the sixth thoracic and first abdominal somites may be masked by the overall color; both *G. spinulosus* and *G. oerstedii* have blue dactyli on the raptorial claws.

The characters afforded by the marginal teeth of the telson, the position of the denticles, and the shape of the uropods can be used to distinguish *G. oerstedii* from other West Indian *Gonodactylus* even at a TL of around 10 mm.

*G. oerstedii* has been reported from several localities in the eastern Pacific by Schmitt (1940); it does occur there but its distribution there cannot be determined until the eastern Pacific species of the genus are studied in detail.

The record of *G. oerstedii* from the Miocene of North Carolina (Rathbun, 1935) is based on a portion of the basal prolongation of the uropod of a stomatopod, probably a *Gonodactylus* (USNM 166062). The fragment cannot be identified with certainty, for the basal prolongation has no diagnostic value in the American species.

*Ontogeny.*—The early larval stages of the species have been described by Brooks (1892) and Manning & Provenzano (1963). Manning (1963b) has also presented some superficial observations on embryological development. The species has been reared in the laboratory through metamorphosis to the juvenile and a report on the later stages is in preparation in collaboration with A. J. Provenzano, Jr.

*Type.*—Hansen did not specifically list any material in his original description of *G. oerstedii* which first appeared as a footnote in a report on the stomatopods of the Plankton Expedition in 1895. The West Indian *Gonodactylus* collection at the Copenhagen Museum was examined and was found to include numerous lots identified by Hansen as *Gonodactylus oerstedii* n. sp. All of these lots are considered to be syntypes; no specimens with dates after 1895 were labelled "n. sp.". A male, TL 70.6, from St. Croix, Virgin Islands, collected by A. S. Ørsted, is here selected as the lectotype. Included among the paralectotypes are specimens of *G. bredini, G. curacaoensis,* and *G. spinulosus* as well as a series of *G. oerstedii.*

*Type-locality.*—St. Croix, Virgin Islands.

*Distribution.*—Western Atlantic, from South Florida through the Caribbean to Curaçao, Bonaire and Aruba. Eastern Pacific (Schmitt, 1940). Littoral to 29 m, usually in 5 m or less.

## Gonodactylus torus, new species
### Figure 90

*Gonodactylus chiragra:* Bigelow, 1894, p. 495 [part; not *G. chiragra* (Fabricius, 1787)].
*Gonodactylus oerstedii:* Chace, 1954, p. 449 [part].—Bullis & Thompson, 1965, p. 13 [part] [not *G. oerstedii* Hansen, 1895].

*Previous records.*—CUBA: off Havana (Bigelow, 1894; Chace, 1954). Bullis & Thompson (1965) record this species from SILVER BAY Stats. 2361 and 2366 and OREGON Stats. 4227 and 4231.

*Holotype.*—1 ♂, 33.5; off Palm Beach, Florida; sand and rocky reef; 73-91 m; TRITON, Thompson-McGinty; 2 August 1950; USNM 119289.

*Paratypes.*—2 ♂, 32.5-32.6; off Palm Beach, Florida; rocky reef; 55-73 m; TRITON, Thompson-McGinty; August 1950; USNM 119291.—3 ♂, 26.0-31.0; 2 ♀, 26.0-30.8; off Palm Beach Florida; rocky reef; 55 m; TRITON; May 1951; USNM 119290.

*Other material.*—NORTH CAROLINA: 1 ♂, 13.7; PELICAN Sta. 192-7; USNM 119300.—2 ♂, 14.5-26.5; 33°55.5′N, 76°28.4′W; 63 m; Gray Sta. 1414; 21 May 1965; DUML.—FLORIDA: 2 ♂, 20.6-26.0; off Palm Beach; TRITON, McGinty; 1950; USNM 119294.—1 ♀, 26.5; same; rocky reef bottom; 64 m; 7 January 1950; USNM 119292.—1 ♂, 12.5; same; 73 m; February 1950; USNM 119293.—1 ♂, 30.6; 4 ♀, 22.5-33.0; same; 55-73 m; 20 April 1950; USNM 119296.—3 ♂, 24.7-33.7; 3 ♀, 20.3-27.4, same; 55-73 m; January 1951; USNM 119295.—2 ♂, 20.9-22.2; 1 ♀, 19.0; off Hollywood; 26°07-08′N, 80°04-05′W; 55 m; GERDA Sta. 423; 27 September 1964; UMML.—1 ♂, 12.9; off Miami; 25°47.6′N, 80°05′W; 55 m; GERDA Sta. 5; 4 May 1962; UMML.—1 ♂, 12.7; 3 mi. ENE Crocker Reef; 50 m; W. A. Starck, II, *et al.*; 27 August 1961; USNM 119299.—1 ♀, 17.1; 2 mi. NE of Alligator Reef Light; 40 m; R. A. Wade; 23 August 1961; USNM 119298.—1 ♀, 28.7; Straits of Florida; SILVER BAY Sta. 2361; USNM 119305.—2 ♀, 24.2-24.6; same; SILVER BAY Sta. 2366; USNM 119304.—1 ♀, 23.5; off Sand Key (Key West?); 10 m; J. B. Hendersen; May 1913; USNM 46062.—1 ♂, 28.0; western Dry Rocks, near Key West; 46 m; J. B. Henderson; USNM 110875.—1 ♀, 18.0; 11 mi. S of no. 2 buoy, off Dry Tortugas; 67 m; ANTON DOHRN; W. L. Schmitt; 10 June 1925; USNM 119308.—1 ♀, 25.1; 1 postlarva, 8.5; off Tortugas; 24°32′N, 83°18-20′W; 71 m; GERDA Sta. 562; 12 April 1965; UMML.—1 ♀, 19.6; same; 24°32-33′N, 83°09-15′W; 68 m; GERDA Sta. 564; 12 April 1965; USNM 119309.—1 ♀, 24.4; 7 mi. S of no. 2 red buoy, off Dry Tortugas; 36 m; ANTON DOHRN; W. L. Schmitt; 11 June 1925; USNM 119297.—1 ♂, 17.1; off Cape St. George; ALBATROSS Sta. 2406; USNM 110876.—MEXICO: 1 ♂, 25.8; 1 ♀, 23.1; off Yucatan; ALBATROSS Sta. 2363; USNM 119306.—1 ♀, 26.4; same; ALBATROSS Sta. 2365; USNM 21496.—PANAMA: 1 ♂, 22.5; ALBATROSS Sta. 2146; USNM 119307.—

CUBA: 1 ♀, 25.3; ALBATROSS Sta. 2323; USNM 9493.—BARBADOS: 1 ♀, 18.7; off Barbados; 91-364 m; found with eggs inside sponge cavity; J. B. Lewis; USNM 110257.—1 ♂, 18.3; off Needham's Point; 109-127 m; University of Iowa Barbados-Antigua Expedition, Sta. 11; 17 May 1918; USNM 57982.—BRAZIL: 2 ♂, 24.6-25.0; 1 ♀, 17.1; off Maranhao State; OREGON Sta. 4227; USNM 119301.—1 ♀, 20.1; same; OREGON Sta. 4231; USNM 119302.—3 ♂, 17.8-23.4; off Piaui State; OREGON Sta. 4243; USNM 119303.

*Diagnosis.*—Lower portion of posterior margin of pleura of first to fourth abdominal somites slightly convex; pleura, in lateral view, with dorsal portion concave, ventral portion convex; fifth abdominal somite angled posterolaterally, occasionally with posterolateral spine; sixth abdominal somite with 6 carinae, variously swollen, very inflated in males with TL greater than 25 mm, apical spines present. A summary of abdominal width-carapace length indices for specimens of various sizes follows:

| CL in mm | n | AWCLI mean | AWCLI range |
|---|---|---|---|
| 2.0 | 1 | 769 | . . . . . . |
| 3.0 | 3 | 787 | 727-820 |
| 4.0 | 4 | 767 | 721-789 |
| 5.0 | 16 | 767 | 733-796 |
| 6.0 | 8 | 738 | 716-781 |
| 7.0 | 2 | 765 | 757-772 |
| 8.0 | 6 | 714 | 680-733 |

Telson broader than long, of Oerstedii-type; median carina flask-shaped in juveniles and females, broadly inflated in males larger than TL 25 mm; median carina usually with apical spinule; accessory median carinae about one-third length of median, unarmed, fusing posteriorly with median to form anchor; anchor completely obscured by inflation of median carina in males; anterior submedian carinae variable in shape, slender, excavate posteriorly in juveniles and females, usually with apical tubercle, noticeably inflated in males, with apical tubercle absent; carinae of submedian teeth not inflated, sinuous, not pitted or spinulose dorsally, occasionally with accessory lateral carinae in females; submedian teeth with inconspicuous shelf on inner face, not angled anteriorly; swollen oblique ridge extending from anterior end of carinae of submedian teeth toward knob present; submedian teeth with subparallel apices, movable apices always present; intermediate marginal teeth widely separate from submedians, apices sharp, longitudinal axes of intermediates and submedians subparallel; accessory carina present on inner face of each intermediate carina; lateral tooth distinct from margin of intermediate tooth, not sharply spined;

submedian denticles small, numerous, distinct along entire face of tooth; intermediate denticles present, occasionally with short dorsal tubercle; intermediate denticles recessed as in *G. oerstedii,* larger always with apical spinule.

Endopod of uropod with inner margin straight or evenly convex.

*Color.*—Completely faded in most specimens. A few specimens are marked with black spots on the body, concentrated on the sixth thoracic and first abdominal somites, as well as a pair of large, black submedian spots on the telson.

*Size.*—Males, TL 12.5-33.7 mm; females, TL 17.1-33.0 mm; postlarva, TL 8.5 mm. Other measurements of female, TL 26.5: carapace length, 5.5; fifth abdominal somite width, 4.1; telson length, 3.5, width, 3.7.

*Discussion.*—G. torus closely resembles *G. oerstedii* in many features but is a much smaller species than is the latter. Males of *G. torus* show secondary sexual modifications at TL 25 mm; at this size males have the median and anterior submedian carinae of the telson very swollen, with the median almost completely obliterating the anchor. In *G. torus* the copulatory tubes of males are as well-developed at TL 20 mm as they are

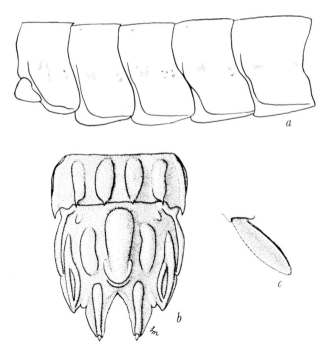

FIGURE 90. *Gonodactylus torus,* n. sp., male paratype, TL 32.5, off Palm Beach, Florida. *a,* abdomen, lateral view; *b,* sixth abdominal somite and telson; *c,* uropodal endopod (setae omitted).

in *G. oerstedii* at TL 40 mm. Finally, *G. torus* is a much narrower species than is *G. oerstedii*. At its maximum size, TL 33.7 mm, the AWCLI of *G. torus* (mean 714) shows no overlap with that found in *G. oerstedii* at the same size (mean 805). The carinae of the telson are never as well-developed in specimens of *G oerstedii* under TL 20 mm as they are in *G. torus.*

*G. torus* also resembles *G. curacaoensis* to some extent, particularly in having the fifth abdominal somite occasionally armed posteriorly, but in *G. torus* the lateral tooth is never as developed as in *G. curacaoensis* and the anterior submedian carinae never terminate in spines in the new species.

*G. torus* and *G. minutus* are the only two dwarf species of *Gonodactylus* to be recognized in the Americas. Both can occur off Brazil. *G. minutus* can immediately be distinguished from *G. torus* by the presence of dorsal spinules on some of the carinae of the telson.

*Remarks.*—The presence of eggs in a sponge cavity off Barbados along with a female of *G. torus* substantiates the view that this is a dwarf species. The female with eggs had a TL of 18.7 mm.

*Etymology.*—The name is from the Latin, "torus", meaning round elevation or protruberance, and alludes to the swollen median carina of the telson in the adult male.

*Ontogeny.*—Unknown.

*Type.*—U. S. National Museum.

*Type-locality.*—Off Palm Beach, Florida, in 73-91 m.

*Distribution.*—Western Atlantic, from North Carolina, southeast and northwest Florida, Yucatan, Panama, Cuba, Barbados, and northern Brazil, in depths between 10 and 364 m, usually in excess of 50 m.

**Gonodactylus austrinus,** new species
Figure 91

Gonodactylus chiragra: Pocock, 1890, p. 526 [not *G. chiragra* (Fabricius, 1781)].
Gonodactylus oerstedii: Bullis & Thompson, 1965, p. 13 [part].—Manning, 1966, p. 371, text-fig. 6a [not *G. oerstedii* Hansen, 1895].

*Previous records.*—BRAZIL: Fernando de Noronha (Pocock, 1890; Manning, 1966); Rocas Id., Abrolhos Ids. (Manning, 1966). Bullis & Thompson (1965) record this species (as *G. oerstedii*) from OREGON Stats. 1938, 3555, 3603, and 3605.

*Holotype.*—1 ♂, 30.0; Fernando de Noronha; 03°49.4'S, 32°24.4'W; 24 m; CALYPSO Sta. 16; 18 November 1961; MNHNP.

*Paratypes.*—4 ♂, 20.6-37.3; 8 ♀, 19.4-53.2; Fernando de Noronha; H. H.

338

Ridley; BMNH.—9 ♂, 17.8-48.0; 12 ♀ (6 brk.), 28.4-42.9; Fernando de Noronha; 10564-96; April 1926; USNM 119264.

*Other material.*—HONDURAS: 1 ♀, 35.8; OREGON Sta. 1938; USNM 119158.—NICARAGUA: 4 ♂, 19.6-38.5; 6 ♀, 21.7-29.2; OREGON Sta. 3603; USNM 119139.—1 ♀, 22.7; OREGON Sta. 2605; USNM.—1 ♀, 22.7; OREGON Sta. 3605; USNM 119263.—PANAMA: 5 ♂, 25.4-36.0; 8 ♀, 21.2-36.5; OREGON Sta. 3555; USNM 119262.—BRAZIL: 1 ♂, 47.8; Fernando de Noronha; 10215; April 1926; USNM 119267.—1 ♂, 21.7; same; 10244; April 1926; USNM.—1 ♀, 18.7; same; 10838; April 1926; G. Finlay Simmons; USNM 119265.—2 ♂, 11.5-27.0; 2 ♀, 11.6-13.5; Lagon Id., Rocas Id.; 2 m; CALYPSO Sta. 5; 17 November 1961; USNM 113246.—2 ♂, 9.6-25.1; W of Rocas Id.; 7-8 m; CALYPSO Sta. 6; 17 November 1961; MNHNP.—1 ♂, 12.8; 1 ♀, 27.5; Rocas Id.; 03°50.5'S, 33°50.7'W; 28 m; CALYPSO Sta. 9; 17 November 1961; MNHNP.—1 ♀, 15.9; Abrolhos Ids.; 2-5 m; CALYPSO Sta. 85; 28 November 1961; MNHNP.

*Diagnosis.*—Lower portion of posterior margin of pleura of first to fourth abdominal somites straight or slightly convex; pleura, in lateral view, at most faintly sinuous, usually evenly concave; fifth abdominal somite rounded posterolaterally; sixth abdominal somite with 6 swollen carinae, spined except in large males (TL in excess of 25 mm). A summary of abdominal width-carapace length indices for specimens of various sizes follow:

| CL in mm | n | AWCLI mean | AWCLI range |
|---|---|---|---|
| 3.0 | 1 | 853 | . . . . . . |
| 4.0 | 3 | 836 | 825-846 |
| 5.0 | 9 | 804 | 761-827 |
| 6.0 | 11 | 806 | 778-828 |
| 7.0 | 5 | 773 | 764-783 |
| 8.0 | 9 | 769 | 702-803 |
| 9.0 | 1 | 788 | . . . . . . |
| 10.0 | 2 | 778 | 745-810 |
| 11.0 | 1 | 772 | . . . . . . |
| 12.0 | 1 | 714 | . . . . . . |
| 13.0 | 1 | 754 | . . . . . . |

Telson length and width subequal, of Oerstedii-type; median carina swollen in adults, usually with apical tubercle; accessory median carinae of moderate length, with apical protruberance in juveniles, fused with median carina to form anchor in adults; knob present, well-formed, unarmed; anterior submedian carinae short, swollen, unarmed, excavate posteriorly; carinae of submedian teeth neither armed nor pitted or sulcate dorsally; oblique swollen ridge present between knob and carinae of submedian teeth; inner margin of submedian teeth swollen; movable apices of submedian teeth always present, apices of teeth posteriorly convergent; intermediate teeth separate from submedians, longitudinal axes subparallel; intermediate carinae each with accessory carina on inner face; apex of lateral carina blunt; submedian denticles small, numerous; 2 intermediate denticles present (usually), outer small, sharp, inner rounded, with apical spinule; all specimens with intermediate denticles situated anterior to apex of intermediate tooth.

Endopod of uropod with inner margin convex or straight.

*Color.*—Largely faded in all specimens, although some have greenish cast with a few scattered dark chromatophores on the body; dactylus pink.

*Size.*—Males, TL 9.6-48.0 mm; females, TL 11.6-53.2 mm; other measurements of female, TL 53.2: carapace length, 11.4; fifth abdominal somite width, 8.8; telson length, 6.6, width, 7.3.

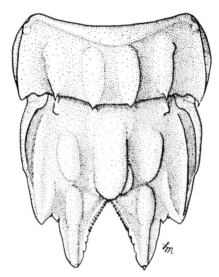

FIGURE 91. *Gonodactylus austrinus,* n. sp., male, TL 27.0, CALYPSO Sta. 5, sixth abdominal somite and telson (from Manning, 1966).

*Discussion.*—*G. austrinus* may be considered the southern counterpart of *G. oerstedii.* It is a smaller species, for males at TL 25 mm exhibit secondary sexual characters, including the swollen dorsal carinae on the telson. As in *G. oerstedii,* in *G. austrinus* the apices of the submedian teeth of the telson are provided with movable apices, the intermediate marginal tooth is separate from the submedian, and the intermediate

340

denticles are recessed anteriorly. *G. austrinus* differs from *G. oerstedii* in lacking the strong double inflection of the posterolateral margin of each abdominal somite, in having the apices of the submedian teeth of the telson convergent rather than subparallel, and in lacking the obtusely-angled anterior projection on the inner shelf of each submedian tooth. The marginal teeth of the telson are blunter in *G. austrinus* than in *G. oerstedii*.

In having the apices of the submedian teeth of the telson posteriorly convergent, the telson of *G. austrinus* resembles that of *G. bredini* and *G. lacunatus*. *G. austrinus* differs from both of these species in having spined, anteriorly-recessed intermediate denticles and in having minute movable apices on the submedian marginal teeth.

*Etymology.*—The name is derived from the Latin, "austrinus", southern, and alludes to the distribution of this species.

*Ontogeny.*—Unknown.

*Type.*—Muséum National d'Histoire Naturelle, Paris. A series of paratypes is in the U.S. National Museum and the British Museum (Natural History).

*Type-locality.*—Fernando de Noronha, Brazil.

*Distribution.*—Southern Caribbean Sea, from off Honduras, Nicaragua, and Panama, and off Brazil, from Fernando de Noronha, Rocas Id., and the Abrolhos Ids., sublittoral to 73 m.

APPENDIX

STATION DATA FOR U. S. F. C. STEAMER *ALBATROSS*
(Complete data for all stations are on file, U. S. National Museum)

| Sta. | Latitude | Longitude | Depth in m | Date |
|---|---|---|---|---|
| 2120 | 11°07″N | 62°14′30″W | 134 | 30 January 1884 |
| 2121- | 10°37′00-40″N | 61°42′40″- | | |
| 2122 | | 61°44′22″W | 62 | 3 February 1884 |
| 2143 | 09°30′45″N | 76°25′30″W | 282 | 23 March 1884 |
| 2146 | 09°32″N | 79°54′30″W | 62 | 2 April 1884 |
| 2248 | 40°07″N | 69°57′W | 122 | 27 September 1884 |
| 2282 | 35°21′10″N | 75°22′40″W | 25 | 19 October 1884 |
| 2283 | 35°21′15″N | 75°23′15″W | 25 | 19 October 1884 |
| 2285 | 35°21′25″N | 75°24′25″W | 24 | 19 October 1884 |
| 2296 | 35°35′20″N | 74°58′45″W | 49 | 20 October 1884 |
| 2323 | 23°10′51″N | 82°19′03″W | 297 | 17 January 1885 |
| 2363 | 22°07′30″N | 87°06′W | 38 | 30 January 1885 |
| 2365 | 22°18′N | 87°04′W | 44 | 30 January 1885 |
| 2373 | 29°14′N | 85°29′15″W | 47 | 7 February 1885 |
| 2378 | 29°14′30″N | 88°09′30″N | 124 | 11 February 1885 |
| 2402 | 28°36′N | 85°33′30″W | 202 | 14 March 1885 |
| 2405 | 28°45′N | 85°02′W | 55 | 15 March 1885 |
| 2406 | 28°46′N | 84°49′W | 47 | 15 March 1885 |
| 2407 | 28°47′30″N | 84°37′W | 44 | 15 March 1885 |
| 2409 | 27°04′N | 83°21′15″W | 47 | 18 March 1885 |

| Sta. | Latitude | Longitude | Depth in m | Date |
|------|----------|-----------|------------|------|
| 2411 | 26°33′30″N | 83°15′30″W | 49 | 18 March 1885 |
| 2412 | 26°18′30″N | 83°08′45″W | 49 | 19 March 1885 |
| 2413 | 26°00′N | 82°57′30″W | 44 | 19 March 1885 |
| 2596 | 35°08′30″N | 75°10′W | 89 | 17 October 1885 |
| 2623 | 33°38′N | 77°36′W | 27 | 20 October 1885 |
| 2640 | 25°05′N | 80°15′W | 102 | 9 April 1886 |
| 2655 | 27°22′N | 78°07′30″W | 615 | 2 May 1886 |
| 2758 | 06°59′30″S | 34°47′W | 36 | 16 December 1887 |
| 2762 | 23°08′S | 41°34′W | 107 | 30 December 1887 |
| 2769 | 45°22′S | 64°20′W | 94 | 15 January 1888 |

STATION DATA FOR M/V *COMBAT*
(See also Bullis & Thompson, 1965)

| Sta. | Latitude | Longitude | Depth in m | Date |
|------|----------|-----------|------------|------|
| 203 | 30°01′N | 80°32′W | 40 | 14 January 1957 |
| 235 | 27°27′N | 78°58′W | 328 | 2 February 1957 |
| 236 | 27°29′N | 78°58′W | 364 | 2 February 1957 |
| 237 | 27°28′N | 78°44′W | 393 | 3 February 1957 |
| 238 | 27°30′N | 78°52′W | 419 | 3 February 1957 |
| 334 | 29°15′N | 80°13′W | 55 | 1 June 1957 |
| 375 | 35°06′N | 75°18′W | 55 | 16 June 1957 |
| 378 | 34°58′N | 75°32′W | 55 | 17 June 1957 |
| 445 | 25°15′N | 79°13′W | 364 | 23 July 1957 |

STATION DATA FOR TRAWLER *COQUETTE*
(Complete data on file in U. S. National Museum)

| Sta. | Latitude | Longitude | Depth in m | Date |
|------|----------|-----------|------------|------|
| 1 | 06°22′N | 55°06′W | 26 | 11 May 1957 |
| 2 | 06°23′N | 55°05.5′W | 27 | 11 May 1957 |
| 4 | 06°25′N | 55°05′W | 29 | 11 May 1957 |
| 6 | 06°24.5′N | 55°03′W | 27 | 11 May 1957 |
| 15 | 06°24.5′N | 54°59.5′W | 29 | 11 May 1957 |
| 20 | 06°28′N | 54°57.5′W | 31 | 11 May 1957 |
| 21 | 06°21′N | 55°00′W | 26 | 12 May 1957 |
| 33 | 06°52′N | 54°53′W | 51 | 12 May 1957 |
| 36 | 06°55′N | 54°54′W | 55 | 12 May 1957 |
| 140 | 06°22-24′N | 54°55-59′W | 25 | 30 May 1957 |
| 144 | 06°22.5′N | 54°58′W | 26 | 30 May 1957 |
| 159 | 06°22′N | 55°02.5′W | 26 | 4 June 1957 |
| 209 | 06°41′N | 54°33′W | 40 | 14 June 1957 |
| 218 | 06°42′N | 54°13.5′W | 44 | 14 June 1957 |
| 280 | 06°45′N | 55°35′W | 47 | 26 June 1957 |
| 281-282 | 06°46-46.5′N | 55°36.5-38.0′W | 36 | 26 June 1957 |
| 283 | 06°47′N | 55°40′W | 46 | 26 June 1957 |
| 284 | 06°49′N | 55°42′W | 46 | 26 June 1957 |
| 297 | 06°45-50 5′N | 55°17-27′W | 44 | 28 June 1957 |

(Complete data on file in U. S. National Museum. Original data for many
stations consisted of bearings. In such cases only general locality is given
here. Puerto Rico stations were assigned two numbers, as noted below.)

| Sta. | Latitude | Longitude | Depth in m | Date |
|---|---|---|---|---|
| 820 | Narragansett Bay | | 10 | 23 August 1880 |
| 865- | 40°05′00″N- | 70°22′06″W- | | |
| 867 | 40°05′42″N | 70°23′W | 116-118 | 4 September 1880 |
| 876 | 39°57′N | 70°56′W | 218 | 13 September 1880 |
| 882 | Narragansett Bay | | 23 | 17 September 1880 |
| 883 | Narragansett Bay | | 24 | 17 September 1880 |
| 949 | 40°03′N | 70°31′W | 182 | 23 August 1881 |
| 955 | Buzzard's Bay | | 13 | 26 August 1881 |
| 961 | Buzzard's Bay | | 15 | 26 August 1881 |
| 1058 | Chesapeake Bay | | 5-46 | 28 February 1882 |
| 1075 | Chesapeake Bay | | 20-25 | 13 March 1882 |
| 1076 | Chesapeake Bay | | 27 | 13 March 1882 |
| 1077 | Chesapeake Bay | | 20-22 | 13 March 1882 |
| 1110 | 40°02′N | 70°35′W | 182 | 22 August 1882 |
| 1247- | 41°00′45″- | 70°59′15″- | | |
| 1251 | 41°10′N | 71°00′05″W | 31-49 | 6 September 1887 |
| 1555 | Long Island Sound | | 15 | 25 September 1890 |
| 1560 | Long Island Sound | | 22 | 25 September 1890 |
| 1727 | Long Island Sound | | 27 | 21 September 1892 |
| 1755 | Long Island Sound | | 15 | 23 September 1892 |
| 1768 | Long Island Sound | | 30 | 29 September 1892 |
| 6053 | San Juan Harbor, Puerto Rico | | 7-14 | 16 January 1899 |
| [P. R. 125] | | | | |
| 6058 | Harbor of Mayaguez, P. R. | | 13 | 18 January 1899 |
| [P. R. 130] | | | | |
| 6059 | Harbor of Mayaguez, P. R. | | 22 | 19 January 1899 |
| [P. R. 131] | | | | |
| 6061 | Harbor of Mayaguez, P. R. | | 33 | 20 January 1899 |
| [P. R. 133] | | | | |
| 6062 | Harbor of Mayaguez, P. R. | | 46-55 | 20 January 1899 |
| [P. R. 134] | | | | |
| 6066 | Harbor of Mayaguez, P. R. | | 313 | 20 January 1899 |
| [P. R. 138] | | | | |
| 6072 | off Point Molines, P. R. | | 13 | 25 January 1899 |
| [P. R. 144] | | | | |
| 6075 | off Boca Prieta, P. R. | | 15 | 25 January 1899 |
| [P. R. 147] | | | | |
| 6079 | off St. Thomas, Virgin Islands | | 36-42 | 6 February 1899 |
| [P. R. 156] | | | | |
| 6084 | off Vieques Id., P. R. | | 20 | 8 February 1899 |
| [P. R. 151] | | | | |
| 6085 | off Vieques Id., P. R. | | 25 | 8 February 1899 |
| [P. R. 157] | | | | |
| 6086 | off Culebra Id., P. R. | | 27 | 8 February 1899 |
| [P. R. 158] | | | | |
| 6089 | off Vieques Id., P. R. | | 38 | 10 February 1899 |
| [P. R. 161] | | | | |
| 6091 | off Vieques Id., P. R. | | 27 | 10 February 1899 |
| [P. R. 163] | | | | |

| Sta. | Latitude | Longitude | Depth in m | Date |
|------|----------|-----------|------------|------|
| 6092 | off Vieques Id., P. R. | | 29 | 10 February 1899 |
| [P. R. 164] | | | | |
| 6093 | off Culebra Id., P. R. | | 27 | 10 February 1899 |
| [P. R. 165] | | | | |
| 6096 | off Vieques Id., P. R. | | 11 | 14 February 1899 |
| [P. R. 168] | | | | |
| 6098 | off Humacao, Hucares, P. R. | | 23 | 14 February 1899 |
| [P. R. 170] | | | | |
| 6099 | off Humacao, Hucares, P. R. | | 17 | 14 February 1899 |
| [P. R. 171] | | | | |
| 7109 | Tampa Bay | | 12 | 29 March 1901 |
| 7151 | 29°43′40″N | 83°49′45″W | 10 | 7 November 1901 |
| 7152 | 29°39′30″N | 83°53′10″W | 14 | 7 November 1901 |
| 7153 | 29°53′20″N | 83°56′00″W | 17 | 7 November 1901 |
| 7158 | 29°19′30″N | 83°46′00″W | 18 | 20 November 1901 |
| 7160 | 20°21′N | 83°32′W | 12 | 21 November 1901 |
| 7161 | 29°18′N | 83°37′W | 15 | 21 November 1901 |
| 7171 | 29°06′15″N | 83°33′W | 15 | 27 November 1901 |
| 7181 | 29°02′30″N | 83°14′00″W | 8 | 28 November 1901 |
| 7184 | 28°55′00″N | 83°28′10″W | 19 | 28 November 1901 |
| 7190 | Apalachee Bay, Florida | | 6 | 5 December 1901 |
| 7191 | 29°54′N | 84°06′W | 8 | 5 December 1901 |
| 7192 | 29°49′N | 84°06′15″W | 11 | 5 December 1901 |
| 7195 | 29°34′N | 84°07′20″W | 19 | 5 December 1901 |
| 7201 | 29°32′30″N | 83°50′00″W | 16 | 6 December 1901 |
| 7207 | 28°57′30″N | 82°58′00″W | 5 | 9 December 1901 |
| 7208 | 28°55′30″N | 83°02′00″W | 7 | 9 December 1901 |
| 7209 | 28°52′45″N | 83°07′00″W | 10 | 9 December 1901 |
| 7217 | 28°27′00″N | 83°13′00″W | 20 | 15 January 1902 |
| 7231 | 28°08′30″N | 83°10′00″W | 18 | 23 January 1902 |
| 7234 | 28°01′30″N | 83°08′00″W | 20 | 23 January 1902 |
| 7272 | off Key West, Florida | | 13 | 13 February 1902 |
| 7273 | off Key West, Florida | | 14 | 13 February 1902 |
| 7277 | off Key West, Florida | | 10 | 13 February 1902 |
| 7278 | off Key West, Florida | | 10 | 13 February 1902 |
| 7293 | 24°42′30″N | 81°55′52″W | 13 | 24 February 1902 |
| 7361 | 25°07′10″N | 81°29′00″W | 9 | 18 December 1902 |
| 7364 | 25°03′50″N | 81°20′30″W | 6 | 18 December 1902 |
| 7367 | 25°03′15″N | 81°08′38″W | 4 | 18 December 1902 |
| 7370 | 25°00′40″N | 81°15′37″W | 5 | 19 December 1902 |
| 7372 | 25°00′55″N | 81°22′15″W | 7 | 19 December 1902 |
| 7375 | 24°58′05″N | 81°28′30″W | 9 | 19 December 1902 |
| 7379 | 24°57′35″N | 81°13′25″W | 5 | 19 December 1902 |
| 7387 | 24°54′55″N | 81°23′45″W | 5 | 20 December 1902 |
| 7390 | off Cape Sable, Florida | | 9 | 22 December 1902 |
| 7391 | off Cape Sable, Florida | | 9 | 22 December 1902 |
| 7392 | 24°52′10″N | 81°30′20″W | 9 | 22 December 1902 |
| 7410 | N. of Knight's Key Channel, Florida | | 16 | 8 January 1903 |
| 7427 | Hawk Channel, Florida | | 5 | 27 January 1903 |
| 7428 | Hawk Channel, Florida | | 5 | 27 January 1903 |
| 7431 | Grassey Key Lake, Florida | | 15 | 28 January 1903 |
| 7442 | Channel Key Lake, Florida | | 3 | 29 January 1903 |

| Sta. | Latitude | Longitude | Depth in m | Date |
|------|----------|-----------|------------|------|
| 7445 | Upper Jewfish Bush Lake, Florida | | 3 | 30 January 1903 |
| 7463 | Hawk Channel, Florida | | 5 | 18 February 1903 |
| 7464 | Hawk Channel, Florida | | 5 | 18 February 1903 |
| 7469 | Hawk Channel, Florida | | 4 | 19 February 1903 |
| 7482 | Biscayne Bay, Florida | | 3 | 7 March 1903 |
| 7493 | N. of Pumpkin Key, Card Sound, | | | |
| | Biscayne Bay, Florida | | 18 | 10 March 1903 |
| 7795 | Sanibel Id., Florida | | 4 | 31 December 1912 |
| 8278- | | | | |
| 8280 | S. Carolina | | 11-13 | 13 July 1915 |
| 8341 | 37°22'12"N | 76°10'25"W | 17 | 22 October 1915 |
| 8342 | 37°35'25"N | 76°11'20"W | 14 | 23 October 1915 |
| 8345 | 37°48'50"N | 75°58'05"W | 20 | 23 October 1915 |
| 8353 | 38°20'25"N | 76°18'15"W | 46 | 25 October 1915 |
| 8359 | 38°50'24"N | 76°24'00"W | 49 | 26 October 1915 |
| 8361 | 38°55'09"N | 76°23'12"W | 42 | 27 October 1915 |
| 8366 | 37°00'40"N | 76°14'55"W | 29 | 2 December 1915 |
| 8372 | 37°13'12"N | 76°04'30"W | 33 | 3 December 1915 |
| 8388 | 38°07'12"N | 76°13'30"W | 38 | 6 December 1915 |
| 8394 | 38°20'12"N | 76°17'54"W | 49 | 7 December 1915 |
| 8395 | 38°31'18"N | 76°24'54"W | 20 | 8 December 1915 |
| 8396 | 38°40'18"N | 76°25'06"W | 38 | 9 December 1915 |
| 8511 | 37°35'48"N | 76°10'24"W | 13 | 23 April 1916 |
| 8523 | 38°20'24"N | 76°18'14"W | 47 | 25 April 1916 |
| 8599 | 37°15'07"N | 76°04'44"W | 46 | 26 July 1916 |
| 8603 | 37°23'30"N | 76°04'50"W | 24 | 26 July 1916 |

## Station Data for M/V *OREGON*
(See also Springer & Bullis, 1956; Bullis & Thompson, 1965)

| Sta. | Latitude | Longitude | Depth in m | Date |
|------|----------|-----------|------------|------|
| 27 | 29°47'N | 86°58'W | 191 | 14 June 1950 |
| 36 | 28°30'N | 85°36'W | 218 | 27 June 1950 |
| 60 | 29°09'N | 88°33'W | 200 | 2 August 1950 |
| 83 | 29°20.5'N | 88°08'W | 82 | 9 August 1950 |
| 88 | 29°14'N | 88°35'W | 67 | 23 August 1950 |
| 90 | 29°14'N | 88°35'W | 73 | 24 August 1950 |
| 101 | 29°07.5'N | 88°50.5'W | 84 | 12 September 1950 |
| 103 | 29°12'N | 88°50'W | 55 | 12 September 1950 |
| 107 | 28°53.3'N | 89°36.5'W | 67 | 13 September 1950 |
| 110 | 28°57'N | 89°36'W | 60 | 14 September 1950 |
| 123 | 28°46.5'N | 89°45'W | 73 | 18 September 1950 |
| 156 | 27°22'N | 96°08'W | 187 | 27 November 1950 |
| 158 | 27°27'N | 96°17'W | 118 | 27 November 1950 |
| 159 | 27°29'N | 96°16'W | 106 | 27 November 1950 |
| 162 | 27°18'N | 96°02'W | 364 | 28 November 1950 |
| 185 | 30°13'N | 88°08'W | 12 | 14 December 1950 |
| 188 | 29°52.5'N | 88°04'W | 35 | 14 December 1950 |
| 273 | 29°09'N | 85°59'W | 200 | 18 February 1951 |
| 274 | 29°35'N | 85°16.5'W | 11 | 19 February 1951 |
| 275 | 29°29'N | 85°10'W | 22 | 19 February 1951 |
| 276 | 29°03'N | 84°24'W | 27 | 23 February 1951 |

| Sta. | Latitude | Longitude | Depth in m | Date |
|------|----------|-----------|------------|------|
| 326 | 29°57′N | 86°57.5′W | 149 | 30 April 1951 |
| 332 | 30°02.5′N | 86°53′W | 109 | 5 May 1951 |
| 340 | 28°50′N | 89°33′W | 78 | 7 May 1951 |
| 342 | 28°50′N | 89°33′W | 78 | 7 May 1951 |
| 382 | 29°11.5′N | 88°07.5′W | 346-382 | 21 June 1951 |
| 411 | 21°15′N | 92°16′W | 62-66 | 17 August 1951 |
| 449 | 20°52.5′N | 90°51′W | 18 | 27 August 1951 |
| 562 | 24°45.5′N | 82°30′W | 27 | 28 May 1952 |
| 662 | 24°12′N | 97°17′W | 73 | 13 October 1952 |
| 713 | 20°05.6′N | 91°13′W | 22 | 8 December 1952 |
| 720 | 20°12′N | 91°40′W | 36 | 11 December 1952 |
| 843 | 29°22′N | 88°40′W | 51 | 22 October 1953 |
| 845 | 28°55′N | 89°15′W | 60 | 23 October 1953 |
| 896 | 28°50′N | 85°06′W | 64 | 7 March 1954 |
| 907 | 29°04′N | 83°51′W | 22 | 10 March 1954 |
| 920 | 28°09′N | 84°54′W | 146 | 11 March 1954 |
| 944 | 29°50′N | 86°30′W | 91 | 21 March 1954 |
| 945 | 29°48′N | 86°37′W | 122 | 21 March 1954 |
| 982 | 26°16′N | 82°21′W | 15 | 6 April 1954 |
| 993 | 25°50′N | 81°52′W | 11 | 7 April 1954 |
| 1082 | 26°11′N | 96°52′W | 38 | 2 June 1954 |
| 1084 | 26°15′N | 96°38′W | 47 | 2 June 1954 |
| 1215 | 25°45′N | 82°03′W | 18 | 19 November 1954 |
| 1340 | 22°55′N | 79°27′W | 437 | 15 July 1955 |
| 1341 | 22°55′N | 79°16′W | 437 | 16 July 1955 |
| 1343 | 22°59′N | 79°17′W | 455 | 16 July 1955 |
| 1344 | 22°50′N | 79°08′W | 364-410 | 16 July 1955 |
| 1382 | 29°20′N | 86°10′W | 186 | 7 September 1955 |
| 1719 | 29°00′N | 83°32′W | 15 | 16 February 1957 |
| 1864 | 16°23′N | 83°20′W | 100 | 21 August 1957 |
| 1870 | 16°39′N | 82°29′W | 410 | 21 August 1957 |
| 1879 | 16°38′N | 81°39′W | 273 | 22 August 1957 |
| 1883 | 16°52′N | 81°30′W | 364 | 23 August 1957 |
| 1902 | 11°27′N | 83°11′W | 246 | 9 September 1957 |
| 1938 | 16°05′N | 82°05′W | 44 | 15 September 1957 |
| 1983 | 09°53′N | 59°53′W | 228 | 3 November 1957 |
| 1985 | 09°41′N | 59°47′W | 273 | 3 November 1957 |
| 2021 | 07°18′N | 53°32′W | 182 | 8 November 1957 |
| 2052 | 04°03′N | 50°25′W | 91 | 13 November 1957 |
| 2203 | 29°13.5′N | 88°12′W | 228 | 26 June 1958 |
| 2207 | 09°58′N | 61°11′W | 36-40 | 26 August 1958 |
| 2208 | 09°55′N | 60°53′W | 62-64 | 26 August 1958 |
| 2209 | 09°45′N | 60°47′W | 36-40 | 26 August 1958 |
| 2221 | 09°22′N | 59°43′W | 91 | 28 August 1958 |
| 2226 | 08°32′N | 59°05′W | 51-60 | 28 August 1958 |
| 2228 | 08°30′N | 58°56′W | 67 | 28 August 1958 |
| 2231 | 08°32′N | 58°42′W | 82-87 | 29 August 1958 |
| 2236 | 08°09′N | 58°23′W | 42 | 29 August 1958 |
| 2244 | 08°12′N | 58°21′W | 56-71 | 31 August 1958 |
| 2249 | 07°40′N | 57°34′W | 49-55 | 31 August 1958 |
| 2250 | 07°38′N | 57°34′W | 47-49 | 31 August 1958 |
| 2261 | 07°20′N | 56°49′W | 60 | 1 September 1958 |

| Sta. | Latitude | Longitude | Depth in m | Date |
|------|----------|-----------|------------|------|
| 2262 | 07°18'N | 56°49'W | 55-60 | 1 September 1958 |
| 2267 | 06°58'N | 56°02'W | 46 | 2 September 1958 |
| 2272 | 06°30'N | 55°52'W | 31 | 3 September 1958 |
| 2276 | 06°42'N | 55°37'W | 40-42 | 3 September 1958 |
| 2279 | 06°20'N | 55°34'W | 13-27 | 3 September 1958 |
| 2285 | 07°27'N | 54°54'W | 246-273 | 8 September 1958 |
| 2286 | 07°26'N | 54°49'W | 191-218 | 8 September 1958 |
| 2288 | 07°26'N | 54°40'W | 173 | 8 September 1958 |
| 2295 | 07°27'N | 53°47'W | 218-228 | 9 September 1958 |
| 2306 | 05°58'N | 52°24'W | 55-57 | 11 September 1958 |
| 2307A-B | 05°56-57'N | 52°20'W | 51-56 | 11 September 1958 |
| 2327 | 06°26'N | 54°20'W | 31 | 15 September 1958 |
| 2351 | 11°31'N | 62°24'W | 337-364 | 23 September 1958 |
| 2355 | 17°23'N | 63°22'W | 20-22 | 25 September 1958 |
| 2606 | 18°37.5'N | 64°57'W | 400 | 26 September 1959 |
| 2639 | 17°47'N | 66°04'W | 410-437 | 4 October 1959 |
| 2648 | 18°12'N | 64°18'W | 273 | 6 October 1959 |
| 2649 | 18°13'N | 64°14.5'W | 228 | 6 October 1959 |
| 2652 | 18°18'N | 67°18.5'W | 546 | 6 October 1959 |
| 2658 | 18°26'N | 67°11.5'W | 319 | 7 October 1959 |
| 2664 | 18°31.5'N | 66°50'W | 328 | 7 October 1959 |
| 2782 | 11°33'N | 62°30'W | 364 | 20 April 1960 |
| 2799 | 29°12'N | 88°26'W | 142 | 29 May 1960 |
| 2827 | 29°16.5'N | 88°04.5'W | 146-164 | 17 July 1960 |
| 2862 | 28°37'N | 91°51.5'W | 33-40 | 4 August 1960 |
| 3201 | 29°19'N | 88°04'W | 118 | 2 February 1961 |
| 3537 | 17°37'N | 77°03'W | 27 | 14 May 1962 |
| 3555 | 17°07'N | 78°23'W | 27 | 17 May 1962 |
| 3574 | 12°31'N | 82°21'W | 364 | 23 May 1962 |
| 3577 | 12°32'N | 82°25'W | 155 | 23 May 1962 |
| 3585 | 09°12'N | 81°30'W | 246-255 | 25 May 1962 |
| 3587 | 09°18'N | 80°25'W | 137 | 29 May 1962 |
| 3588 | 09°18'N | 80°27'W | 182 | 29 May 1962 |
| 3590 | 09°18'N | 80°22'W | 228 | 29 May 1962 |
| 3595 | 09°02'N | 81°26'W | 182 | 30 May 1962 |
| 3597 | 09°04'N | 81°25'W | 273-291 | 31 May 1962 |
| 3603 | 12°16'N | 82°54'W | 27-36 | 2 June 1962 |
| 3604 | 12°15'N | 82°57'W | 31-40 | 2 June 1962 |
| 3605 | 12°16'N | 82°53'W | 40-73 | 2 June 1962 |
| 3620 | 15°58'N | 81°12'W | 27 | 6 June 1962 |
| 3626 | 16°45'N | 81°27'W | 273 | 7 June 1962 |
| 3627 | 16°50'N | 81°21'W | 364 | 7 June 1962 |
| 3638 | 21°10'N | 86°26'W | 29 | 11 June 1962 |
| 3649 | 29°13'N | 88°07.5'W | 273 | 24 July 1962 |
| 3713 | 29°38.5'N | 88°30'W | 40 | 21 August 1962 |
| 3768 | 28°26.5'N | 91°00.5'W | 36-40 | 12 September 1962 |
| 3769 | 28°22.5'N | 91°04'W | 53-58 | 12 September 1962 |
| 3776 | 28°25.5'N | 91°45.5'W | 51-55 | 13 September 1962 |
| 3968 | 27°32'N | 96°40'W | 51-53 | 28 September 1962 |
| 4002 | 29°11.5'N | 88°11.5'W | 273 | 23 October 1962 |
| 4089 | 27°43'N | 83°34'W | 40 | 4 December 1962 |
| 4090 | 27°41'N | 83°24'W | 29 | 5 December 1962 |

| Sta. | Latitude | Longitude | Depth in m | Date |
|------|----------|-----------|------------|------|
| 4170 | 06°23'N | 56°05'W | 31 | 19 February 1963 |
| 4171 | 06°16'N | 55°56'W | 27 | 19 February 1963 |
| 4179 | 06°46'N | 54°44'W | 46 | 20 February 1963 |
| 4209 | 04°30'N | 50°52'W | 66 | 27 February 1963 |
| 4211 | 02°50'N | 49°15'W | 58 | 27 February 1963 |
| 4215 | 00°27'S | 47°09'W | 22 | 8 March 1963 |
| 4227 | 01°24'S | 43°11'W | 73 | 10 March 1963 |
| 4229 | 01°50'S | 43°06'W | 66 | 10 March 1963 |
| 4231 | 01°50'S | 42°43'W | 66 | 10 March 1963 |
| 4243 | 02°10'S | 41°33'W | 66 | 11 March 1963 |
| 4302 | 07°35'N | 54°25'W | 273 | 24 March 1963 |
| 4306 | 06°54'N | 57°47'W | 18 | 25 March 1963 |
| 4392 | 12°32'N | 71°05'W | 73 | 25 September 1963 |
| 4393 | 12°32'N | 71°04'W | 84 | 25 September 1963 |
| 4394 | 12°37'N | 71°10'W | 118 | 25 September 1963 |
| 4405 | 11°53'N | 69°28'W | 391 | 27 September 1963 |
| 4434 | 11°11'N | 68°11'W | 364 | 8 October 1963 |
| 4446 | 10°54'N | 68°01'W | 182 | 10 October 1963 |
| 4467 | 10°25'N | 65°42'W | 91 | 17 October 1963 |
| 4492 | 10°29'N | 62°00'W | 33 | 24 October 1963 |
| 4493 | 10°29'N | 62°14'W | 18 | 24 October 1963 |
| 4583 | 29°13'N | 88°11'W | 182 | 12 December 1963 |
| 4836 | 14°21'N | 80°15.5'W | 0-2 | 12 May 1964 |
| 4838 | 11°09.5'N | 74°24.5'W | 309-328 | 16 May 1964 |
| 4844 | 11°06.7'N | 74°30'W | 182 | 17 May 1964 |
| 4846 | 11°06.8'N | 74°23.6'W | 91 | 17 May 1964 |
| 4847 | 11°06'N | 74°29'W | 64-73 | 17 May 1964 |
| 4850 | 11°04.1'N | 74°23'W | 18-19 | 18 May 1964 |
| 4851 | 11°06.7'N | 74°25.6'W | 73-82 | 18 May 1964 |
| 4852 | 11°07'N | 74°26'W | 109-118 | 18 May 1964 |
| 4856 | 11°08'N | 74°23.8'W | 182 | 19 May 1964 |
| 4857 | 11°07'N | 74°21.5'W | 73-91 | 19 May 1964 |
| 4858 | 11°09.5'N | 74°25'W | 291 | 19 May 1964 |
| 4860 | 11°09'N | 74°26'W | 282-291 | 19 May 1964 |
| 4864 | 10°59.3'N | 75°05.5'W | 55 | 22 May 1964 |
| 4866 | 10°57.6'N | 75°10'W | 27 | 23 May 1964 |
| 4880 | 10°24'N | 75°50'W | 346-355 | 24 May 1964 |
| 4886 | 09°38'N | 75°54.5'W | 42 | 25 May 1964 |
| 4898 | 08°53.8'N | 76°51.6'W | 182 | 27 May 1964 |
| 4906 | 10°05'N | 75°53'W | 73-109 | 28 May 1964 |
| 4911 | 11°50'N | 73°05'W | 319-346 | 31 May 1964 |
| 4930 | 15°46'N | 81°37'W | 55 | 8 June 1964 |
| 4943 | 29°42'N | 86°45'W | 182 | 14 June 1964 |
| 5018 | 13°00'N | 59°33'W | 319 | 20 September 1964 |
| 5025 | 11°08'N | 60°58'W | 73-87 | 21 September 1964 |
| 5028 | 11°30'N | 60°46'W | 364-437 | 22 September 1964 |
| 5050 | 12°38'N | 61°24.5'W | 18 | 27 September 1964 |
| 5058 | 14°02'N | 61°00'W | 16-24 | 29 September 1964 |
| 5059 | 14°02.5'N | 61°01'W | 24-29 | 29 September 1964 |

## STATION DATA FOR M/V *PELICAN*
(Data on file in U. S. National Museum)

| Sta. | Latitude | Longitude | Depth in m | Date |
|------|----------|-----------|------------|------|
| 1 | 29°51.5'N | 88°42.5'W | 16 | 22 January 1938 |
| 4 | SE Pass, Louisiana | | | 4 February 1938 |
| 9 | 29°02'N | 88°41.5'W | 218-308 | 4 February 1938 |
| 11 | 29°11'N | 88°30'W | 161 | 5 February 1938 |
| 20 | Grand Isle, Louisiana | | 7 | 16 February 1938 |
| 39 | 27°46'N | 96°51.5'W | 24 | 20 April 1938 |
| 42 | ?27°41'N | 96°36.5'W | 44 | 22 April 1938 |
| 54-1 | 28°18'N | 96°17'W | 19 | 2 May 1938 |
| 69-6 | 28°48'N | 89°51'W | 55 | 13 May 1938 |
| 81-3 | 28°41.5'N | 91°08.5'W | 15 | 11 July 1938 |
| 84-3 | 28°14'N | 91°41.0'W | 71 | 12 July 1938 |
| 85-6 | 28°19.5'N | 91°24'W | 64 | 12 July 1938 |
| 96-1 | 28°39.5'N | 89°58.5'W | 146 | 14 November 1938 |
| 156-2 | 29°08'N | 85°00'W | 31 | 11 March 1939 |
| 169-1 | 28°22.5'N | 80°32'W | 13 | 18 January 1940 |
| 181-13 | 32°03'N | 79°49.5'W | 25 | 3 February 1940 |
| 182-8 | 32°09'N | 79°02'W | 182-319 | 4 February 1940 |
| 182-11 | 32°30'N | 79°16'W | 33 | 4 February 1940 |
| 182-16 | 32°53'N | 79°30'W | 9 | 12 February 1940 |
| 188-8 | 35°02.5'N | 76°00'W | 13 | 1 March 1940 |
| 192-1 | 34°36'N | 76°35'W | 15 | 6 March 1940 |
| 192-7 | 33°59.5'N | 76°29.5'W | 46 | 6 March 1940 |
| 194-10 | 32°34'N | 79°05'W | 35 | 9 March 1940 |
| 202-6 | 29°28.5'N | 81°04.5'W | 15 | 28 March 1940 |
| 203-1 | 29°25.5'N | 81°00.5'W | 16 | 28 March 1940 |
| 204-4 | 28°56'N | 80°01.5'W | 182 | 29 March 1940 |
| 205-5 | 28°08'N | 79°54'W | 182 | 30 March 1940 |
| 206A-3 | 28°12'N | 80°31.5'W | 18 | 3 April 1940 |
| 206A-5 | 28°20'N | 80°34'W | 13 | 3 April 1940 |

## STATION DATA FOR M/V *SILVER BAY*
(See also Bullis & Thompson, 1965)

| Sta. | Latitude | Longitude | Depth in m | Date |
|------|----------|-----------|------------|------|
| 53 | 28°33'N | 84°24'W | 36-40 | 16 July 1957 |
| 54 | 28°09'N | 83°50'W | 36 | 16 July 1957 |
| 67 | 26°20'N | 82°03'W | . . . . . . | 18 July 1957 |
| 69 | 26°13'N | 82°05'W | 13 | 18 July 1957 |
| 71 | 25°26'N | 81°40'W | 8 | 18 July 1957 |
| 100 | 29°10'N | 85°48'W | 100-129 | 26 July 1957 |
| 181 | 28°04'N | 92°05'W | 78 | 22 September 1957 |
| 182 | 28°07'N | 92°37'W | 89 | 22 September 1957 |
| 445 | 28°03'N | 78°44'W | 910-946 | 9 June 1958 |
| 581 | 29°31'N | 84°44'W | 18 | 25 July 1958 |
| 1196 | 24°11'N | 83°21.5'W | 728 | 8 June 1959 |
| 1695 | 33°57'N | 77°01'W | 35 | 29 February 1960 |
| 1704 | 34°13'N | 76°48'W | 29-31 | 29 February 1960 |
| 1967 | 27°35'N | 79°55'W | 135-137 | 22 April 1960 |
| 1968 | 27°39'N | 79°54'W | 182 | 22 April 1960 |
| 1992 | 28°37'N | 80°13'W | 33-36 | 24 April 1960 |
| 2039 | 28°23'N | 79°59'W | 116-137 | 27 April 1960 |

| Sta. | Latitude | Longitude | Depth in m | Date |
|------|----------|-----------|------------|------|
| 2142 | 29°59'N | 80°34'W | 35 | 13 June 1960 |
| 2165 | 34°39'N | 76°56'W | 11-13 | 17 July 1960 |
| 2361 | 24°58'N | 80°22'W | 47-55 | 25 October 1960 |
| 2366 | 24°59'N | 80°20'W | 56-64 | 25 October 1960 |
| 2390 | 24°42'N | 80°44'W | 91 | 26 October 1960 |
| 2402 | 24°35'N | 81°09'W | 46-55 | 27 October 1960 |
| 2445 | 24°08'N | 80°08'W | 251 | 3 November 1960 |
| 2458 | 23°40'N | 79°18'W | 528 | 5 November 1960 |
| 2460 | 23°55'N | 79°34'W | 182-237 | 6 November 1960 |
| 2468 | 23°52'N | 79°11'W | 364-382 | 6 November 1960 |
| 2470 | 24°25'N | 79°13'W | 228 | 7 November 1960 |
| 2480 | 26°06'N | 79°10'W | 222-228 | 9 November 1960 |
| 2725 | 28°23'N | 79°56'W | 171 | 1 February 1961 |
| 2731 | 28°46'N | 80°01'W | 146 | 2 February 1961 |
| 2732 | 28°41'N | 80°01'W | 146 | 2 February 1961 |
| 3279 | 28°03'N | 80°00.5'W | 73 | 14 July 1961 |
| 3466 | 27°55'N | 79°05'W | 182 | 25 October 1961 |
| 3468 | 27°26'N | 78°57'W | 137 | 25 October 1961 |
| 3516 | 24°24'N | 80°00'W | 728-855 | 9 November 1961 |
| 3648 | 33°20'N | 77°41'W | 29 | 11 December 1961 |
| 3655 | 32°43'N | 78°34'W | 40 | 14 December 1961 |
| 5096 | 27°55'N | 80°07'W | 35-38 | 26 September 1963 |
| 5130 | 25°48'N | 77°06'W | 20 | 1 October 1963 |
| 5149 | 20°01'N | 71°39'W | 44-53 | 13 October 1963 |
| 5161 | 19°57.5'N | 71°05'W | 273-346 | 14 October 1963 |
| 5166 | 19°48.5'N | 70°30.5'W | 400-546 | 15 October 1963 |

# BIBLIOGRAPHY

ALIKUNHI, K. H.
  1944. Final pelagic larva of *Squilla hieroglyphica* Kemp. Current Sci., Bangalore, *13* (9): 237-238, fig. 1.
  1947. *Squilla hieroglyphica* Kemp. Current Sci., Bangalore, *16* (9): 289.
  1952. An account of the stomatopod larvae of the Madras plankton. Rec. Indian Mus., *49* (3-4): 239-319, fig. 1-25.
  1958. Notes on a collection of stomatopod larvae from the Bay of Bengal, off the Mahanadi estuary. Journ. Zool. Soc. India, *10* (2): 120-147, figs. 1-23.

ALLEE, W. C.
  1923. The distribution of common littoral invertebrates of the Woods Hole region. Studies in marine ecology, I. Biol. Bull., Woods Hole, *44*: 167-191, pl. 1.

ALLEN, D. M., AND A. INGLIS
  1958. A pushnet for quantitative sampling of shrimp in shallow estuaries. Limn. and Oceanogr., *3* (2): 239-241, figs. 1-2.

AMOS, WILLIAM H.
  1959. The living wreck. Estuarine Bull., *4* (1): 8-10, figs.

ANDRADE RAMOS, F. DE P.
  1951. Nota preliminar sôbre alguns Stomatopoda de costa brasileira. Bol. Inst. oceanogr., São Paulo, *2* (1): 139-150, pl. 1.

350

ANDREWS, E. A.
1892. Notes on the fauna of Jamaica. John Hopkins Univ. Circ., *11*: 72-77, 2 figs.

ANDREWS, E. A., R. P. BIGELOW, AND T. H. MORGAN
1945. Three at Bimini. Sci. Monthly, *61*: 333-344, figs. 1-9.

ANONYMOUS
1942. Annotated list of the fauna of the Grand Isle region, 1928-1941. Mar. Lab. Louisiana State Univ.: 1-20 Baton Rouge, Louisiana.

ANONYMOUS
1952. Mantis shrimp. Maryland Tidewater News, *9* (7): 2.

ARCHER, A. A.
1948. The shrimp surveys. Alabama Conservation., *20* (2): 10, fig.

ARNOLD, AUGUSTA FOOTE
1901. The sea-beach at ebb-tide: xii + 490, unnumbered figs. The Century Co., N. Y.

BAHAMONDE N., NIBALDO
1957. Sobre la distribución geográfica de *Lysiosquilla polydactyla* von Martens, 1881 (Crustacea, Stomatopoda). Invest. Zool. Chil., *3* (5-7): 119-121, figs. 1-3.

BALSS, H.
1916. Stomatopoda. Crustacea III. Beiträge zur Kenntnis des Meeresfauna Westafrikas, herausgegeben von W. Michaelsen (Hamburg), *2:* 47-52.
1938. Stomatopoda. *In*: Bronn, H. G., Klassen u. Ordnungen des Tierreichs, Bd. 5, Abt. 1, Buch 6, Teil 2: 1-173, figs. 1-114. Akademische Verlagsgesellschaft, Leipzig.

BARNARD, K. H.
1950. Descriptive list of South African stomatopod Crustacea (mantis shrimps). Ann. South African Mus., *38*: 838-864, figs. 1-4.

BATE, C. S.
1868. Carcinological gleanings. No. III. Ann. Mag. nat. Hist., ser. 4, *1*: 442-448, pl. 21.

BEHRE, E. H.
1950. Annotated list of the fauna of the Grand Isle region, 1928-1946. Occ. Pap. Mar. Lab. Louisiana State Univ., no. 6: 1-66.

BERG, C.
1900. Datos sobre algunos crustáceos nuevos para la fauna Argentina. Comm. Mus. Nac. Buenos Aires, *1* (7): 223-235, 1 fig.

BERRY, CHARLES T.
1939. A summary of the fossil Crustacea of the Order Stomatopoda and a description of a new species from Angola. Amer. midl. Nat., *21* (2): 461-471, 2 figs.

BERTHOLD, A. A.
1827. Natürliche Familien des Thierreichs, aus dem Französischen mit Anmerkungen und Zusätzen: ix + 606. Weimar.

BIGELOW, R. P.
1891. Preliminary notes on some new species of *Squilla*. Johns Hopkins Univ. Circ., *10*: 93-94.

1893. Preliminary notes on the Stomatopoda of the Albatross collections and on other specimens in the National Museum. Johns Hopkins Univ. Circ., *12* (106): 100-102.

1893a. The Stomatopoda of Bimini. Johns Hopkins Univ. Circ., *12* (106): 102-103.

1894. Report on the Crustacea of the Order Stomatopoda collected by the steamer Albatross between 1885 and 1891 and on other specimens in the U. S. National Museum. Proc. U. S. Nat. Mus., *17:* 489-550, pls. 20-22, text-figs. 1-28.

1901. The Stomatopoda of Porto Rico. Bull. U. S. Fish. Comm., *20* (2): 149-160, figs. 1-13 (bound volume issued in 1902).

1931. Stomatopoda of the southern and eastern Pacific Ocean and the Hawaiian Islands. Bull. Mus. comp. Zool. Harvard, *72* (4): 105-191, text-figs. 1-10, pls. 1-2.

1941. Notes on *Squilla empusa* Say. J. Wash. Acad. Sci., *31* (9): 399-403, fig. 1.

BLAKE, S. F.
1953. The Pleistocene fauna of Wailes Bluff and Langleys Bluff, Maryland. Smithsonian Misc. Coll., *121* (12): 1-32, fig. 1, pl. 1.

BOONE, L.
1927. Crustacea from tropical east American seas. Bull. Bingham oceanogr. Coll., *1* (2): 1-147, figs. 1-33.

1930. Crustacea: Stomatopoda and Brachyura. Scientific results of the cruises of the yachts "Eagle" and "Ara," 1921-1928, Wm. K. Vanderbilt, commanding. Bull. Vanderbilt Mar. Mus., *2:* 228 p., 74 pls.

1934. Crustacea: Stomatopoda and Brachyura. Scientific results of the world cruise of the yacht "Alva," 1931, William K. Vanderbilt, commanding. Bull. Vanderbilt Mar. Mus., *5:* 210 p., 109 pls.

BORRADAILE, L. A.
1898. On some crustaceans from the South Pacific. Part I. Stomatopoda. Proc. Zool. Soc. London, 1898: 32-38, pls. 5-6.

1899. On the Stomatopoda and Macrura brought by Dr. Willey from the South Seas. *In*: Willey, A., Zoological Results based on material from New Britain, New Guinea, Loyalty Islands and elsewhere, collected during the years 1895, 1896, and 1897, *4:* 395-428, pls. 36-39.

1907. Stomatopoda from the western Indian Ocean. The Percy Sladen Trust Expedition to the Indian Ocean in 1905, under the leadership of Mr. J. Stanley Gardiner. Trans. Linn. Soc. London, Zool., ser. 2, *12:* 209-216, pl. 22.

BOSS, KENNETH J.
1965. A new mollusk (Bivalvia, Erycinidae) commensal on the stomatopod crustacean *Lysiosquilla*. Amer. Mus. Novitates, no. 2215: 1-11, figs. 1-3.

DE BOURY, M. E.
1918. Quelques observations sur la moeurs et sur l'habitat des crustacés à l'île de Cuba. Bull. Mus. Hist. Nat. Paris, *24*: 16-18.

BOUVIER, E. L.
1918. Sur une petit collection de crustacés de Cuba offerte au Museum par M. de Boury. Bull. Mus. Hist. Nat. Paris, *24:* 6-15, figs. 1-7.

BOUVIER, E. L., AND P. LESNE
1901. Sur les arthropodes du Mozambique et de San Thome offerts au Muséum par M. Almada Negreiros. Bull. Mus. Hist. Nat. Paris, *7:* 12-15.

Brooks, W. K.
1878. The larval stages of *Squilla empusa* Say. Johns Hopkins Univ., Chesapeake Zool. Lab. Sci. Res., 1878: 143-170, pls. 9-12.
1885. Notes on the Stomatopoda. Johns Hopkins Univ. Circ., 5 (43): 10-11.
1886. Report on the Stomatopoda collected by H.M.S. Challenger during the years 1873-76. Rep. Sci. Res. Challenger, Zool., *16:* 1-116, pls. 1-16.
1886a. Notes on Stomatopoda. Ann. Mag. nat. Hist., ser. 5, *17:* 166-168.
1886b. The Stomatopoda of the "Challenger" collection. Johns Hopk. Univ. Circ., 5 (49): 83-85.
1892. The habits and metamorphosis of *Gonodactylus chiragra.* Chap. III. *In:* Brooks, W. K., and H. F. Herrick, The embryology and metamorphosis of the Macrura. Mem. Nat. Acad. Sci., 5 (4): 353-360, pls. 1, 3, 14, 15.

Brown Frank A., Jr.
1948. Color changes in the stomatopod crustacean, *Chloridella empusa.* Anat. Record, *101:* 732-733 (abstract).

Brullé, M.
1836-1844. Crustacés. *In:* Barker-Webb, P., and S. Berthelot, Histoire Naturelle des Iles Canaries. Zool., 2 (2) Entomologie: 13-18, atlas, pl. unique.

Bullis, Harvey R., Jr., and John R. Thompson
1965. Collections by the exploratory fishing vessels *Oregon, Silver Bay, Combat,* and *Pelican* made during 1956-1960 in the southwestern North Atlantic. Spec. Sci. Rep., U. S. Fish Wildl. Serv., no. 510: 1-130.

Burdon-Jones, C.
1962. The feeding mechanism of *Balanoglossus gigas.* Fac. Filos. Cienc. Letr. Univ. São Paulo, Bol. no. 261, Zoologia no. 24: 255-280, figs. 1-9.

Büttikofer, J.
1890. Reisbilder aus Liberia. Resultate geographischer, natuurwissenschaftlicher und ethnographischer Unterschungen wahrend der Jahre 1879-1882 und 1886-1887, 2: viii + 510. E. J. Brill, Leiden.

Cadenat, J.
1950. Sur quelques espèces des squilles des côtes du Sénégal. C. R. Prem. Conf. Internat. Afr. Ouest, Dakar, *1:* 192-193.
1957. *Lysiosquilla aulacorhynchus,* espèce nouvelle de stomatopode de la cote occidentale d'Afrique. Diagnose preliminaire. Bull. Inst. Français Afrique Noire, ser. A, *19* (1): 126-133, figs. 1-8.

Calman, W. T.
1917. Crustacea. Part IV. Stomatopoda, Cumacea, Phyllocarida, and Cladocera. Brit. Antarct. Terra Nova Exped., 1910. Nat. Hist. Rep., Zool., *3* (5): 137-162, figs. 1-9.

Cary, L. R. and H. M. Spaulding
1909. Further contributions to the marine fauna of the Louisiana coast. Publ. Gulf Biol. Sta., Cameron: 1-21.

Chace, Fenner A., Jr.
1939. Preliminary descriptions of one new genus and seventeen new species of decapod and stomatopod Crustacea. Reports on the scientific results of the first Atlantis expedition to the West Indies, under the joint auspices of the University of Havana and Harvard University. Mem. Soc. Cubana Hist. nat., *13* (1): 31-54.

1942. Six new species of decapod and stomatopod Crustacea from the Gulf of Mexico. Proc. New England Zool. Club, *19*: 79-92, pls. 23-28.

1954. Stomatopoda. *In:* Galtsoff, P. S., ed., Gulf of Mexico, its origin, waters, and marine life. Fish. Bull., U. S. Fish Wildl. Serv., no. 89: 449-450.

1956. Crustaceos decapodos y stomatopodos del Archipielago de los Roques e Isla de la Orchila, pp. 145-168, 4 pls. *In:* Mendez, A., *et al.,* El Archipielago de los Roques y la Orchila, 257 pp., figs. Soc. Cienc. Nat. LaSalle, Caracas.

1958. A new stomatopod crustacean of the genus *Lysiosquilla* from Cape Cod, Massachusetts. Biol. Bull., Woods Hole, *114* (2): 141-145, pl. 1.

CHOPRA, B.
1939. Stomatopoda. The JOHN MURRAY Expedition, Sci. Reps., *6* (3): 137-181, figs. 1-13.

CLAUS, C.
1871. Die Metamorphose der Squilliden. Abh. Ges. Wiss. Göttingen, *16* (1): 111-163, pls. 1-8 (pp. 1-55 on separate).

COUES, ELLIOTT AND H. C. YARROW
1878. Notes on the natural history of Fort Macon, N. C., and vicinity. (No. 5). Proc. Acad. nat. Sci. Philadelphia, 1878: 297-315.

COVENTRY, G. AYRES
1944. The Crustacea. Results of the Fifth George Vanderbilt Expedition, 1941 (Bahamas, Caribbean Sea, Panama, Galapagos Archipelago and Mexican Pacific Islands). Monogr. Philadelphia Acad. Sci., *6*: 531-544.

COWLES, R. P.
1930. A biological study of the offshore waters of Chesapeake Bay. Bull. U. S. Bur. Fish., *48:* 277-381, 16 figs.

CUNNINGHAM, ROBERT O.
1870. Notes on the reptiles, Amphibia, fishes, Mollusca, and Crustacea obtained during the voyage of H. M. S. 'NASSAU' in the years 1866-69. Trans. Linn. Soc. London, *27:* 465-502, pls. 58-59.

DAHL, ERIC
1954. Stomatopoda. Rep. Lund Univ. Chile Exp. 1948-49, No. 15. Acta Univ. Lund, N. F., Avd 2, Bd. 49, Nr. 17: 1-12, 1 text-fig.

DANA, J. D.
1852. Crustacea. Part I. United States Exploring Expedition during the years 1838, 1839, 1840, 1841, 1842 under the command of Charles Wilkes, U.S.N., *13:* 1-685. Atlas, 1855, pp. 1-27, pls. 1-96. C. Sherman, Philadelphia.

DARTEVELLE, EDM.
1951. Stomatopodes de la côte du Congo. Bull. Inst. Roy. col. Belge, *22* (4): 1020-1038, text-figs. 1-7, 1 pl.

DAWSON, C. E.
1963. Progress report. Gulf Coast Research Laboratory Museum, 1960-1962: iii + 34, mimeo. Ocean Springs, Mississippi.

1965. Progress report, Gulf Coast Research Laboratory Museum, 1963-1964: 57 pp., mimeo. Ocean Springs, Mississippi.

DEEVEY, GEORGIANA B.
1952. A survey of the zooplankton of Block Island Sound, 1943-1946. Bull. Bing. oceanogr. Coll., *13* (3): 65-119, figs. 1-18.

DeKay, J. E.
1844. Zoology of New York . . . Part 6. Crustacea: 1-70, pls. 1-13. Carroll and Cook, Albany.

Dennell, R.
1950. The occurrence of *Lysiosquilla scabricauda* (Lamarck) in Bermuda. Proc. Linn. Soc. London, *162* (1): 63-64.

Desmarest, A. G.
1823. Malacostracés. Dictionnaire des Sciences Naturelles . . . , *28:* 138-425. F. G. Levrault, Paris, Strasbourg.
1825. Considérations générales sur la classe des Crustacés, et descriptions des espèces de ces animaux, qui vivent dans la mer, sur les côtes ou dans les eaux douces de la France: xix + 446, pls. 1-56. F. G. Levrault, Paris.

Desmarest, E.
1858. Crustacés — Mollusques — Zoophytes. *In:* Chenu, J., Encyclopédie d'Histoire Naturelle . . . : 1-312, pls. 1-40, text-figs. 1-320. Marescq et Companie, Paris.

Dingle, Hugh
1964. A colour polymorphism in *Gonodactylus oerstedi* Hansen, 1895 (Stomatopoda). Crustaceana, *7* (3): 236-240.

Doflein, F., and H. Balss
1912. Die Dekapoden und Stomatopoden der Hamburger Magalhaenischen Sammelreise, 1892-1893. Mitteil. a.d. Naturhist. Mus., *29:* 25-44, figs. 1-4.

Edmondson, C. H.
1921. Stomatopoda in the Bernice P. Bishop Museum. Occ. Pap. Bishop Mus., *7* (13): 281-302, figs. 1-2.

Elofsson, Rolf
1965. The nauplius eye and frontal organs in Malacostraca (Crustacea). Sarsia, no. 19: 1-54, figs. 1-31.

Ernst, A.
1870. A species of *Squilla*. Proc. Zool. Soc. London, 1870: 3.
1880. On Dana's *Lysiosquilla inornata*. Ann. Mag. nat. Hist., ser. 5, *5:* 436.

Eydoux, A. M. and L. Souleyet
1842. Crustacés. Voyage autour du monde exécuté pendant les années 1836 et 1837 sur la corvette, La Bonite commandée par M. Vaillant, Capitaine de Vaisseau. Zoologie, *1:* 219-272, pl. 5. Arthus Bertrand, Paris.

Fabricius, J. C.
1775. Systema entomologiae, sistens insectorum classes, ordines, genera, species, adiectis synonymis, locis, descriptionibus, observationibus: 832 pp. Libraria Kortii (Flensburgi et Lipsiae).
1787. Mantissa insectorum sistens eorum species nuper detectas adjectis characteribus genericis, differentiis specificis, emendationibus, observationibus, *1:* xx + 348 pp. Christ. Gottl. Proft, Hafniae.

Faxon, Walter
1882. Crustacea. *In:* Agassiz, A., W. Faxon, and E. L. Mark, Selections from embryological monographs, I. Mem. Mus. comp. Zool. Harvard, *9* (1): pls. 1-14.

355

1895. The stalk-eyed Crustacea. Reports on an exploration off the west coasts of Mexico, Central and South America, and off the Galapagos Islands . . . IV. Mem. Mus. comp. Zool. Harvard, *18:* 1-292, pls. A-K, 1-56.

1896. Supplementary notes on the Crustacea. Reports on the results of dredging, under the supervision of Alexander Agassiz . . . XXXVII. Bull. Mus. comp. Zool. Harvard, *30* (3): 153-166, pls. 1-2.

FIGUEIREDO, M. J.
1962. Un stomatopode nouveau pour la faune Portugaise et pour l'ocean Atlantique, *Pseudosquilla ferussaci* (Roux). Not. Estud. Inst. Biol. Mar., no. 25: 5-9, pls. 1-5.

FISH, CHARLES J.
1925. Seasonal distribution of the plankton of the Woods Hole region. Bull. U. S. Bur. Fish., *41:* 91-179, figs. 1-81.

FOWLER, H. W.
1912. The Crustacea of New Jersey. Ann. Rep. New Jersey State Museum, Part II, 1911: 29-650, pls. 1-150.

1913. Notes on the fishes of the Chincoteague region of Virginia. Proc. Acad. nat. Sci., Philadelphia, *LXV:* 61-65.

FOXON, G. E. H.
1932. Report on the stomatopod larvae, Cumacea and Cladocera. Sci. Rep. Great Barrier Reef Exped., *4* (11): 375-398, figs. 1-10.

FUKUDA, T.
1910. Report on Japanese Stomatopoda, with descriptions of two new species. Annot. Zool. Japon., *7:* 139-152, pl. 4.

GIBBES, L. R.
1845. Catalogue of the crustaceans in the cabinet of the Boston Society of Natural History, September 1, 1845. Proc. Boston Soc., *2:* 69-70.

1848. Catalogue of the fauna of South Carolina, pp. i-xxiv, appendix. *In:* Tuomey, M., Report on the geology of South Carolina: vi + 293, text-figs.

1850. On the carcinological collections of the cabinets of natural history in the United States, with an enumeration of the species contained therein, and descriptions of new species. Proc. Amer. Assoc. Adv. Sci., 3rd meeting: 167-201 (pp. 1-37 on separate).

1850a. Catalogue of the Crustacea in the Cabinet of the Academy of Natural Sciences of Philadelphia, August 20th, 1847, with notes on the most remarkable. (With additions and observations by the committee). Proc. Acad. nat. Sci. Philadelphia, 1850: 22-30.

GIESBRECHT, W.
1910. Stomatopoden. Erster Theil. Fauna u. Flora Neapel, Monogr. 33: vii + 239, pls. 1-11.

GLASSELL, STEVE A.
1934. Some corrections needed in recent carcinological literature. Trans. San Diego Soc. nat. Hist., 7 (38): 453-454.

GOODE, G. BROWN
1879. The voices of crustaceans. Proc. U. S. nat. Mus., *1:* 7-8.

GORDON, ISABELLA
1929. Two rare stomatopods of the genus *Lysiosquilla*, Dana. Ann. Mag. nat. Hist., ser. 10, *4:* 460-462, figs. 1-2.

GOULD, A. A.
1833. Crustacea, pp. 563-564. *In:* Hitchcock, E., Report on the geology, mineralogy, botany, and zoology of Massachusetts: xii + 692. J. S. and C. Adams, Amherst.

GUÉRIN-MÉNEVILLE, F. E.
1829-1844. Iconographie de Règne Animal de G. Cuvier, ou représentation d'après nature de l'une des espèces les plus remarquables et souvent non encore figurées, de chaque genre d'animaux. Avec une texte descriptif mis au courant de la science. Ouvrage pour servir d'atlas à tous les traités de Zoologie, *2:* pls. 1-104. J. B. Baillière, Paris.
1855. Crustaceos. *In:* de la Sagra, R., Historia fisica politica y natural de la isla de Cuba, Historia Natural, *7:* v-xxxii; atlas, *8:* pls. 1-3. Arthus Bertrand, Paris.
1857. Crustacés. *In:* de la Sagra, R., Histoire physique, politique et naturelle de l'ile de Cuba: xiii-lxxxvii, pls. 1-3. Arthus Bertrand, Paris.

GUNTER, G.
1950. Seasonal population changes and distributions as related to salinity, of certain invertebrates of the Texas coast, including the commercial shrimp. Publ. Inst. Mar. Sci., Univ. Texas, *1* (2): 7-51, figs. 1-8.

GUPPY, R. J. L.
1894. Observations upon the physical conditions and fauna of the Gulf of Paria. Proc. Victoria Inst. Trinidad, pt. 2, 1894: 105-115.

GURNEY, R.
1946. Notes on stomatopod larvae. Proc. zool. Soc. London, *116* (1): 133-175, figs. 1-14.
1946a. The larval stomatopod *Alima lebouri* Gurney, identified with *Squilla tricarinata* Holthuis. Proc. zool. Soc. London, *116* (3-4): 734.

DE HAAN, W.
1833-1850. Crustacea. *In:* de Siebold, P. F., Fauna Japonica, sive descriptio animalium, quae in itinere per Japoniam, jussu et auspiciis superiorum, qui summum in India Batavia Imperium tenent, suscepto, annis 1823-1830 collegit, notis, observationibus et adumbrationibus illustravit: ix-xvi, i-xxxi, vii-xvii, 1-243, pl. A-Q, 1-55, circ. 2. A. Arnz (Lugdunum Batavorum).

HANSEN, H. J.
1895. Isopoden, Cumaceen und Stomatopoden der Planktonexpedition. Ergbn. Planktonexped. Humboldt-Stiftung, *2* (Gc): 1-105, pls. 1-8.
1921. On some malacostracous Crustacea (Mysidacea, Euphausiacea, and Stomatopoda) collected by Swedish Antarctic expeditions. Arch. Zool., *13* (20): 1-7.
1921a. Studies on Arthropoda, I, pp. 1-80, pls. 1-4. Gyldendalske Boghandel, Copenhagen.
1926. The Stomatopoda of the SIBOGA Expedition. SIBOGA Exped., monogr. 35: 1-48, pls. 1-2.

HAZLETT, BRIAN A., AND H. E. WINN
1962. Sound production and associated behavior of Bermuda crustaceans (*Panulirus, Gonodactylus, Alpheus,* and *Synalpheus*). Crustaceana, *4* (1): 25-38, text-figs. 1-2, pl. 1.

HEDGPETH, JOEL W.
1950. Annotated list of certain marine invertebrates found on Texas jetties, pp. 72-85. *In:* Whitten, H. L., H. F. Rosene, and Joel W. Hedgpeth, The invertebrate fauna of Texas coast jetties; a preliminary survey. Publ. Inst. Mar. Sci., Univ. Texas, *1* (2): 53-87, figs. 1-4, pl. 1.

357

1953. An introduction to the zoogeography of the northwestern Gulf of Mexico with reference to the invertebrate fauna. Publ. Inst. Mar. Sci., Univ. Texas, *3* (1): 107-224, figs. 1-46.

1954. Bottom communities of the Gulf of Mexico. *In:* Galtsoff, P. S., ed., Gulf of Mexico, its origin, waters ,and marine life. Fish. Bull., U. S. Fish Wildl. Serv., no. 89: 203-214, figs. 51-54.

HEILPRIN, ANGELO

1888. Contributions to the natural history of the Bermuda Islands. Proc. Acad. nat. Sci. Philadelphia, *40:* 302-328, pls. 14-16.

1889. The Bermuda Islands. A contribution to the physical history and geology of the Somers Archipelago. With an examination of the structure of coral reefs, pp. 1-231, pls. 1-17, 11 pls. in text. Acad. nat Sci., Philadelphia.

HEMMING, FRANCIS

1945. Suspension of the rules for *Squilla* Fabricius (J. C.), 1787 (Class Crustacea, Order Stomatopoda). Opinion 186. Opin. Decl. Int. Comm. Zool. Nomencl., *3* (5): 53-64.

1954. Validation, under the Plenary Powers, of the generic name *Lysiosquilla* Dana, 1852 (Class Crustacea, Order Stomatopoda). Opinion 294. Opin. Decl. Int. Comm. Zool. Nomencl., *8* (11): 143-154.

HERKLOTS, J. A.

1851. Addimenta ad Faunam Carcinologicam Africae Occidentalis . . . : 1-31, pls. 1-2. Lugduni-Batavorum.

1861. Symbolae Carcinologicae. I. Catalogue de crustacés qui ont servi de base au système carcinologique de M. W. de Haan, rédigé d'après la collection du Musée des Pays-Bas et les crustacés de la Faune du Japon. Tijdschr. Entomol., *4:* 116-156 (pp. 1-43 on separate).

HESSLER, ROBERT R.

1964. The Cephalocarida, comparative skeletomusculature. Mem. Connecticut Acad. Sci., *16:* 1-97, figs. 1-47.

HILDEBRAND, H. H.

1954. A study of the fauna of the brown shrimp *(Penaeus aztecus* Ives) grounds in the western Gulf of Mexico. Publ. Inst. Mar. Sci., Univ. Texas, *3* (2): 233-366, 7 figs.

1955. A study of the fauna of the pink shrimp (*Penaeus duorarum* Burkenroad) grounds in the Gulf of Campeche. Publ. Inst. Mar. Sci., Univ. Texas, *4* (1): 169-232, figs. 1-2.

HILGENDORF, F.

1890. Eine neue Stomatopoden-Gattung Pterygosquilla. Sber. Ges. naturforsch. Freunde: 172-177, 1 fig.

HOLTHUIS, L. B.

1941. Note on some Stomatopoda from the Atlantic coasts of Africa and America, with the description of a new species. Zool. Meded., Leiden, no. 23: 31-43, 1 text-fig.

1941a. The Stomatopoda of the Snellius Expedition. Biological results of the Snellius Expedition, XII. Temminckia, *6:* 241-294, figs. 1-9.

1951. Proposed use of the Plenary Powers to validate the generic name *Odontodactylus* Bigelow, 1893 (Class Crustacea, Order Stomatopoda). Bull. Zool. Nomencl., *2:* 86-87.

1951a. Proposed use of the Plenary Powers to render the generic name *Lysiosquilla* Dana, 1852 (Class Crustacea, Order Stomatopoda) the oldest available name for the species currently referred thereto. Bull. Zool. Nomencl., *2:* 83-84.

1959. Stomatopod Crustacea of Suriname. Stud. Fauna Suriname, *3* (10): 173-191, pls. VIII-IX, text-fig. 76.
1964. Preliminary note on two new genera of Stomatopoda. Crustaceana, *7* (2): 140-141.

HOLTHUIS, L. B., AND RAYMOND B. MANNING
1964. Proposed use of the Plenary Powers (A) to designate a type-species for the genera *Pseudosquilla* Dana, 1852, and *Gonodactylus* Berthold, 1827, and (B) for the suppression of the generic name *Smerdis* Leach, 1817 (Crustacea, Stomatopoda). Z. N. (s) 1609. Bull. Zool. Nomencl., *21* (2): 137-143.
In press. Stomatopoda. *In:* Moore, Raymond C., ed., Treatise on Invertebrate Paleontology,

HOWARD, L. O.
1883. A list of the invertebrate fauna of South Carolina. Chap. 11, pp. 265-310. *In:* South Carolina. Resources and population, institutions and industries, viii + 726 pp. S. Carolina State Board of Agriculture.

HUEBNER, J.
1816-1826. Verzeichniss bekannter Schmetterlinge, pp. 1-432. Augsburg (not seen).

VON IHERING, H.
1897. A ilha de São Sebastião. Rev. Mus. Paulista, *2:* 129-171, 5 figs.

INGLE, R. W.
1958. On *Squilla alba* Bigelow, a rare stomatopod crustacean new to St. Helena and South Africa. Ann. Mag. nat. Hist., ser. 13, *1* (1): 49-56, figs. 1-6.
1960. *Squilla labadiensis* n. sp. and *Squilla intermedia* Bigelow, two stomatopod crustaceans new to the west African coast. Ann. Mag. nat. Hist., ser. 13, *2* (2): 565-576, figs. 1-14.
1963. Crustacea Stomatopoda from the Red Sea and Gulf of Aden. Contributions to the knowledge of the Red Sea, no. 26. Bull. Sea Fish. Res. Sta., Haifa, no. 33: 1-69, figs. 1-73.

IVES, J. E.
1891. Crustacea from the northern coast of Yucatan, the harbor at Vera Cruz, the west coast of Florida and the Bermuda Islands. Proc. Acad. nat. Sci., Philadelphia, *43* (1): 176-207, pls. 5-6.

JOHNSTON, H.
1906. List of invertebrate animals of Liberia founded on the collections of Büttikofer, Reynolds, Whicker, H. H. Johnston, A. White, etc., pp. 860-883. *In:* Johnston, H., Liberia, *2:* xvi + 1183, text-figs. Hutchinson and Co., London.

JURICH, BRUNO
1904. Die Stomatopoden der deutschen Tiefsee-Expedition. Wiss. Ergbn. Deutschen Tiefsee-Exped. VALDIVIA, *7:* 361-408, pls. 25-30 (separate pp. 1-51, pls. 1-6).

KEMP, S.
1911. Preliminary descriptions of new species and varieties of Crustacea Stomatopoda in the Indian Museum. Rec. Indian Mus., *6:* 93-100.
1913. An account of the Crustacea Stomatopoda of the Indo-Pacific region based on the collection in the Indian Museum. Mem. Indian Mus., *4:* 1-217, pls. 1-10, 10 text-figs.
1915. On a collection of stomatopod Crustacea from the Philippine Islands. Philippine Journ. Sci., sec. D, *10* (3): 169-187, pl. I.

KEMP, S. AND B. CHOPRA
1921. Notes on Stomatopoda. Rec. Indian Mus., *22* (4): 297-311, figs. 1-4.

KINGSLEY, J. S., ED.
1884. Crustacea and Insects. The Standard Natural History, *2:* vi + 555, 666 figs., 20 pls. S. E. Cassino and Co., Boston.

KOELBEL, C.
1892. Beiträge zur Kenntnis der Crustaceen der Canarischen Inseln. Ann. naturh. Hotmus. Wien, *7:* 105-116, pl. 10.

KOMAI, T.
1914. On some species of Japanese Stomatopoda (in Japanese). Dobuts. Z. Tokyo, *26:* 459-468, 1 pl., 1 text-fig.
1927. Stomatopoda of Japan and adjacent localities. Mem. Coll. Sci. Kyoto Imp. Univ., ser. B, *3* (3): 307-354, pls. xiii-xiv, text-figs. 1-2.
1938. Stomatopoda occurring in the vicinity of Kii Peninsula. Annot. Zool. Japon., *17* (3-4): 264-275, figs. 1-3.

KURIAN, C. V.
1947. On the occurrence of *Squilla hieroglyphica* Kemp (Crustacea Stomatopoda) in the coastal waters of Travancore. Current Sci. Bangalore, *16* (4): 124.
1954. Contributions to the study of the crustacean fauna of Travancore. Bull. Centr. Res. Inst., Univ. Travancore, Trivandrum, ser. C, *3* (1): 69-91, 3 figs.

DE LAMARCK, J. B. P. A.
1818. Histoire naturelle des animaux sans vertèbres, présentant les caractéres généraux et particuliers de ces animaux, leur distribution, leur classes, leurs familles, leurs genres, et la citation des principales espèces qui s'y rapportent; précédée d'une introduction offrant la détermination des caractères essentiels, de l'animal, sa distinction du végétal et des autres corps naturelles, enfin, l'exposition des principes fondamentaux de la zoologie, *5:* 612 pp. Deterville, Paris.

LANCHESTER, W. F.
1903. Marine crustaceans. VIII. Stomatopoda, with an account of the varieties of *Gonodactylus chiragra. In:* Gardiner, J. S., The fauna and geography of the Maldive and Laccadive Archipelagoes. Being an account of the work carried on and of the collections made by an expedition during the years 1899 and 1900, *1:* 444-459, pl. 23.

LATREILLE, P. A.
1802-1803. Histoire naturelle, générale et particulière, des crustacés et des insectes, *3:* 1-468. F. Dufart, Paris.
1817. Les crustacés, les arachnides et les insectes. *In:* Cuvier, G., La règne animal distribué d'après son organisation, pour servir de base á l'histoire naturelle des animaux et d' introduction á l'anatomie comparée, ed. 1, *3:* xxix + 653. Deterville, Paris.
1818. Crustacés, arachnides et insectes. Tableau Encyclopédique et méthodique de trois règnes de la nature, *24:* 1-38, pls. 269-397. Mme. Veuve Agasse, Paris.
1825. Familles naturelles du règne animal, exposées succinctement et dans un ordre analytique, avec l'indication de leurs genres: 570 pp. J. B. Baillière, Paris.
1828. Squille, *Squilla.* Encyclopédie Methodique Histoire Naturelle . . ., *10:* 467-475. Agasse, Paris.

LEACH, W. E.
1817-1818. A general notice of the animals taken by Mr. John Cranch, during the expedition to explore the source of the River Zaire. Appendix no. IV. *In:* Tuckey, J. K., Narrative of an expedition to explore the River Zaire, usually called the Congo, in South Africa, in 1816, under the direction of Captain J. K. Tuckey, R. N. To which is added the journal of Professor Smith; Some general observations on the country and its inhabitants; and an appendix: Containing the natural history of that part of the Kingdom of Congo through which the Zaire flows: 407-419 (1818), 1 pl. (1817). John Murray, London.
1818. Sur quelques genres nouveaux de crustacés. Journ. phys. Chim. Hist. nat., *86:* 304-307, figs. 4-11.

LEBOUR, M. V.
1934. Stomatopod larvae. Résultats scientifiques du voyage aux Indes Orientales Néerlandaises de LL. AA. RR. le Prince et la Princesse Léopold de Belgique. Mem. Mus. Roy. Hist. Nat. Belgique, *3* (16): 9-17, figs. 1-5.
1954. The planktonic decapod Crustacea and Stomatopoda of the Benguela Current. Part I. First survey, R.R.S. 'WILLIAM SCORESBY,' March, 1950. Discovery Rep., 27: 219-233, figs. 1-6.

LEE, B. D., AND W. N. MCFARLAND
1963. Osmotic and ionic concentrations in the mantis shrimp *Squilla empusa* Say. Publ. Inst. Mar. Sci., Univ. Texas, *8:* 126-142, figs. 1-10.

LEIDY, J.
1855. Contributions towards a knowledge of the marine invertebrate fauna of the coasts of Rhode Island and New Jersey. Journ. Acad. nat. Sci. Philadelphia, ser. 2, *3:* 1-20, pls. 10-11.

LEMOS DE CASTRO, ALCEU
1945. A tamburutaca. Rev. Mus. Nac., Rio de Janeiro, *2* (5): 29-30, figs.
1955. Contribuição ao conhecimento dos crustáceos do ordem Stomatopoda do litoral brasileiro: (Crustacea, Hoplocarida). Boll. Mus. Nac., Rio de Janeiro, n. s., Zool., no. 128: 1-68, pls. I-XVIII, figs. 1-31.
1962. Sôbre os crustáceos referidos por Marcgrave em sua "História Naturalis Brasiliae" (1648). Ann. Mus. Nac., Rio de Janeiro, *52:* 37-51, pls. 1-4.

LINNAEUS, CAROLUS
1758. Systema naturae per regna tria naturae, secundum classes, ordines, genera, species, cum characteribus, differentiis, synonymis, locis, ed. 10, *1:* iii + 824. Holmiae.

LONGHURST, A. R.
1958. An ecological survey of the west African marine benthos. Brit. Col. Off., Fish. Pub., no. 11: 1-102, figs. 1-11.

LUCAS, H.
1840. Histoire naturelle des crustacés, des arachnides et des myriapodes: 1-601, pls. 1, 1-7, 1-20, 1-13, 1-3, 1, 1. P. Duménil, Paris.

LUEDERWALDT, HERMANN
1919. Lista dos crustaceos superiores (Thoracostraca) do Museu Paulista, que foram encontrados no Estado de S. Paulo. Rev. Mus. Paulista, *11:* 427-435 (pp. 1-9 on separate).
1929. Resultados de uma excursão scientifica á Ilha de São Sebastião no litoral de Estado de São Paulo e em 1925. Rev. Mus. Paulista, *16:* 1-79, 3 pls.

361

Lunz, G. Robert, Jr.
  1933. The rediscovery of *Squilla neglecta* Gibbes. Charleston (S.C.) Mus. Leaf., no. 5: 1-8, pl. 1.
  1935. The stomatopods (mantis shrimps) of the Carolinas. J. Elisha Mitchell Sci. Soc., *51* (1): 151-159, figs. 1-6.
  1937. Stomatopoda of the Bingham Oceanographic Collection. Bull. Bingham oceanogr. Coll., *5* (5): 1-19, figs. 1-10.

McLendon, J. F.
  1911. On adaptations in structure and habits of some marine animals of Tortugas, Florida. Pap. Tortugas Lab., Carnegie Inst. Washington, 3: 55-62, text-figs. 1, pls. 1-2.

Manning, Raymond B.
  1959. A checklist of the stomatopod crustaceans of the Florida-Gulf of Mexico area. Quart. J. Fla. Acad. Sci., *22* (1): 14-24.
  1960. A useful method for collecting Crustacea. Crustaceana, 1 (4): 372-373.
  1961. Stomatopod Crustacea from the Atlantic coast of northern South America. Allan Hancock Atlantic Exped., Rep. no. 9: 1-46, pls. 1-11.
  1961a. A new deep-water species of *Lysiosquilla* (Crustacea, Stomatopoda) from the Gulf of Mexico. Ann. Mag. nat. Hist., ser. 13, 3: 693-697, pl. 10, figs. 1-2, pl. 11, figs. 3-4, text-fig. 5.
  1961b. Sexual dimorphism in *Lysiosquilla scabricauda* (Lamarck) a stomatopod crustacean. Quart. J. Fla. Acad. Sci., *24* (2): 101-107, figs. 1-2.
  1962. A new species of *Parasquilla* (Stomatopoda) from the Gulf of Mexico, with a redescription of *Squilla ferussaci* Roux. Crustaceana, *4* (3): 180-190, figs. 1-2.
  1962a. A striking abnormality in *Squilla bigelowi* Schmitt (Stomatopoda). Crustaceana, *4* (3): 243-244, fig. 1.
  1962b. *Alima hyalina* Leach, the pelagic larva of the stomatopod crustacean *Squilla alba* Bigelow. Bull. Mar. Sci. Gulf & Carib., *12* (3): 496-507, figs. 1-4.
  1962c. Seven new species of stomatopod crustaceans from the northwestern Atlantic. Proc. Biol. Soc. Washington, 7ɛ: 215-222.
  1962d. A redescription of *Lysiosquilla biminiensis pacificus* Borradaile (Stomatopoda). Crustaceana, *4* (4): 301-306, fig. 1.
  1963. A new species of *Lysiosquilla* (Crustacea, Stomatopoda) from the northern Straits of Florida. Bull. Mar. Sci. Gulf & Carib., *13* (1): 54-57, fig. 1.
  1963a. Preliminary revision of the genera *Pseudosquilla* and *Lysiosquilla* with descriptions of six new genera (Crustacea: Stomatopoda). Bull. Mar. Sci. Gulf & Carib., *13* (2): 308-328.
  1963b. Notes on the embryology of the stomatopod crustacean *Gonodactylus oerstedii* Hansen. Bull. Mar. Sci. Gulf & Carib., *13* (3): 422-432.
  1963c. *Hemisquilla ensigera* (Owen, 1832) an earlier name for *H. bigelowi* (Rathbun, 1910) (Stomatopoda). Crustaceana, 5 (4): 315-317.
  1964. A new west American species of *Pseudosquilla* (Stomatopoda). Crustaceana, *6* (4): 303-308, fig. 1.
  1966. Stomatopod Crustacea. 3. Campagne de la "Calypso" au large des cotes Atlantiques de l'Amerique du Sud (1961-1962). I. Ann. Inst. océanogr. Monaco, *44:* 359-384, figs. 1-8.
  1966a. Notes on some Australian and New Zealand stomatopod Crustacea, with an account of the species collected by the Fisheries Investigation Ship Endeavour. Rec. Australian Mus., *27* (4): 79-137, figs. 1-10.

1967.  Preliminary account of a new genus and a new family of Stomatopoda. Crustaceana, *13* (2): 238-239.
1967a. Review of the genus *Odontodactylus* (Crustacea: Stomatopoda). Proc. U. S. Nat. Mus., *123* (3606): 1-35, figs. 1-8.
1967b. Stomatopoda in the Vanderbilt Marine Museum. Crustaceana, *12* (1): 102-106.
1968.  A revision of the family Squillidae (Crustacea, Stomatopoda), with the description of eight new genera. Bull. Mar. Sci., *18* (1): 105-142, figs. 1-10.
1968a. Correction of the type locality of *Squilla mantoidea* Bigelow (Stomatopoda). Crustaceana, *14* (1): 107.

MANNING, RAYMOND B., AND A. J. PROVENZANO, JR.
1963.  Studies on development of stomatopod Crustacea. I. Early larval stages of *Gonodactylus oerstedii* Hansen. Bull. Mar. Sci. Gulf & Carib., *13* (3): 467-487, figs. 1-8.

MARCGRAVE, G.
1648.  Historia Naturalis Brasiliae [História Natural do Brasil]. Translation by Dr. José Procopio de Magalhães. Museum Paulista, São Paulo, 1942: 1-293; Comentarios, p. I-CIV.

VON MARTENS, E.
1872.  Ueber Cubanische Crustaceen nach den Sammlungen Dr. J. Gundlach's. Arch. Naturgesch., *38* (1): 77-147, pls. 4-5.
1881.  Squilliden aus dem Zoologischen Museum in Berlin. S. B. Ges. naturf. Fr. Berlin, 1881: 91-94.

MAYER, A. G.
1905.  Sea shore life. The invertebrates of the New York coast. N. Y. Aquarium Series, no. 1: 1-181, 120 figs., N. Y. Zool. Soc.

MELLIS, JOHN C.
1875.  St. Helena. A physical, historical, and topographical description of the island, including its geology, fauna, flora, and meteorology, xiv + 426, pls. 1-56, 1 fig. L. Reeve & Co., London.

MENZEL, R. WINSTON, ED.
1956.  Annotated check-list of the marine fauna and flora of the St. George's Sound-Apalachee Bay region, Florida Gulf coast: iv + 78, 1 fig. Mimeo, Oceanographic Inst., Florida St. Univ., Tallahassee.

MIERS, E. J.
1880.  On the Squillidae. Ann. Mag. nat. Hist., ser. 5, *5:* 1-30, 108-127, pls. 1-3.
1884.  Crustacea. Report on the zoological collections made in the Indo-Pacific Ocean during the voyage of H.M.S. 'ALERT,' 1881-2: 178-322, 513-575, pls. 18-35, 46-52. British Museum, London.
1891.  Crustacea. Podophthalmia. *In:* Whymper, E., Supplementary appendix to Travels Amongst the Great Andes of the Equator: 121-124, figs. A-B, 1 pl. John Murray, London.

MILNE, LORUS J., AND M. MILNE
1961.  Scanning movements of the stalked compound eyes in crustaceans of the order Stomatopoda, pp. 422-426, figs. 1-11. *In:* Christensen, B. Chr., and B. Buchmann, eds., Progress in photobiology. Proc. 3rd Internat. Congr. Photobiology. Elsevier Publ. Co., Amsterdam.
1965.  Stabilization of the visual field. Biol. Bull., Woods Hole, *128* (2): 285-296, figs. 1-4.

MILNE-EDWARDS, A.
  1868. Observations sur la faune carcinologique de îles du Cap Vert. Nouv. Arch. Mus. Hist. nat. Paris, *4:* 49-68, pls. 16-18.
  1878. Description de quelques espèces nouvelles de crustacés provenant du voyage aux iles du Cap Vert de MM Bouvier et de Cessac. Bull. Soc. Phil. Paris, *2:* 225-232 (pp. 6-13 on separate).
  1891. Crustacés, Mission Scientifique du Cap Horn, 1882-83, Zool., *6* (2F): 1-76, pls. 1-7.

MILNE-EDWARDS, H.
  1837. Histoire naturelle des crustacés, comprenant l'anatomie, la physiologie et la classification de ces animaux, *2:* 1-532, Atlas, pp. 1-32, pls. 1-42. Roret, Paris.
  1837. Les Crustacés. *In:* Cuvier, G., La règne animal . . . , ed. 4, *17:* 1-278; *18:* pls. 1-80. Fortin, Masson et Cie, Paris.
  1838. Arachnides, crustacés, annélides, cirrhipèdes. *In:* Lamarck, J. B. P. A. de, Histoire naturelle des animaux sans vertèbres . . . , ed. 2, *5:* 1-699. J. B. Baillière, Paris.
  1839. Les crustacés, pp. 316-436. *In:* Lamarck, J. B. P. A. de, Histoire naturelle des animaux sans vertèbres . . . , ed. 3, *2:* 1-436. Meline, Cans et Compagnie, Bruxelles.

MINER, ROY WALDO
  1950. Field book of seashore life: xv + 888, 251 pls, 34 colored pls. G. P. Putnam's Sons, N. Y.

MONOD, TH.
  1925. Sur les stomatopodes de la cote occidentale d'Afrique. Bull. Soc. Sci. Nat. Maroc, *5* (3): 86-93, pls. 20-21.
  1932. Sur quelques crustacés de l'Afrique Occidentale (Liste des décapodes Mauritaniens et des xanthides Ouest-Africans). Bull. Comm. Etud. sci. Afr. occ. Franç., *15:* 456-548, figs. 1-26 (pp. 1-93 on separate).
  1939. Sur quelques crustacés de la Guadeloupe (Mission P. Allorge, 1936). Bull. Mus. Hist. nat. Paris, ser. 2, *11* (6): 557-568, figs. 1-11.
  1951. Sur quelques stomatopodes ouest-africains. Bull. Inst. Franc. Afr. Noire, *13* (1): 139-144, figs. 1-5.

MOREIRA, CARLOS
  1901. Crustaceos do Brazil. Arch. Mus. Nac., Rio de Janeiro, *11:* 1-151, pls. 1-5.
  1903. Campanhas de pesca de hiate "Annie," dos Srs. Bandeira e Bravo. Estudos preliminares. Crustaceos. Lavoura, *7* (1-3): 60-67 (Jan.-Mar. 1903).
  1903a. Campanhas de pesca do hiate "Annie," dos Srs. Bandeira e Bravo. Estudos preliminares. Crustaceos. Lavoura, *7* (1-3): 1-14, 5 figs. (15 May 1903).
  1903b. Crustaceos de Ponta do Pharol em S. Francisco do Sul, no Estado de Santa Catharina. Arch. Mus. Nac. Rio de Janeiro, *12:* 119-123.
  1905. Campanhas de pesca do "Annie." Crustaceos. Arch. Mus. Nac. Rio de Janeiro, *13:* 123-145, pls. I-IV (reprint with separate pagination dated 1906).

MORRISON, PETER R., AND K. C. MORRISON
  1952. Bleeding and coagulation in some Bermudan Crustacea. Biol. Bull., Woods Hole, *103* (3): 395-406.

NEUMANN, R.
  1878. Systematische Uebersicht der Gattungen der Oxyrhynchen, Catalog der Podophthalmen Crustaceen des Heidelberger Museums. Beschreibung einiger neuer Arten: 1-39.

NOBILI, G.
1897. Decapodi e stomatopodi raccolti dal Dr. Enrico Festa nel Darien, a Curaçâo, La Guayra, Porto Cabello, Colon, Panama, ecc. Boll. Mus. Zool. Anat. comp. Torino, *12* (280): 1-8.
1898. Crostacei decapodi e stomatopodi di St. Thomas (Antille). Boll. Mus. Zool. Anat. comp. Torino, *13* (314): 1-3.
1903. Crostacei di Singapore. Boll. Mus. Zool. Anat. comp. Torino, *18* (455): 1-39, 1 text-fig., 1 pl.

NORMAN, A. M.
1886. Crustacea. III. Museum Normanianum, or a catalogue of the Invertebrata of Europe, and the Arctic and the North Atlantic Oceans, which are contained in the collection of the Rev. Canon A. M. Norman, M. A., D.C.L., F.L.S., 1-26. Morton, Houghton-le-Spring.
1905. Crustacea. III. Museum Normanianum, or a catalogue of the Invertebrata of the Artic and North Atlantic Temperate Ocean and Palaearctic Region, which are contained in the collection of the Rev. Canon A. M. Norman, M.A., D.C.L., LL.D, F.R.S., F.L.S., etc., 2nd ed.: vi + 47. Thomas Caldcleugh, Durham.

NUTTING, C. C.
1895. Narrative and preliminary report of Bahama Expedition. Nat. Hist. Bull., St. Univ. Iowa, *3* (1-2): vi + 251, figs.
1919. Barbados-Antigua Expedition. Univ. Iowa Stud. Nat. Hist., *8* (3): 1-274, pls. 1-48.

DE OLIVEIRA, L. P. H.
1940. Contribuiçao ao conhecimento dos crustaceos do Rio de Janeiro. Catalogo dos crustaceos da Baia de Guanabara. Mem. Inst. Oswaldo Cruz, *35* (1): 137-151.
1944. Sobra uma nova especie de Crustacea Stomatopoda: *Squillerichthus Aragoi*. Mem. Inst. Oswaldo Cruz, *41:* 335-337, figs. 1-7.

OSORIO, B.
1887. Liste de crustacés des possessions Portugaises d'Afrique occidentale, dans les collections du Muséum d'Histoire Naturelle de Lisbonne. Jorn. Sci. Math. phys. nat. Lisboa, *11:* 220-231.
1888. Liste des crustacés des possessions Portugaises d'Afrique occidentale, dans les collections du Muséum d'Histoire Naturelle de Lisbonne. Jorn. Sci. Math. phys. nat. Lisboa, *12:* 186-191.
1889. Nouvelle contribution pour la connaissance de la faune carcinologique des iles Saint Thomé et du Prince. Jorn. Sci. Math. Phys. Nat. Lisboa, ser. 2, *1:* 129-139.
1898. Da distribuição geographica dos peixes e crustaceos colhidos nas possessões Portuguezas d'Afrique occidentale e existentes no Museu Nacional de Lisboa. Jorn. Sci. Math. phys. nat. Lisboa, ser. 2, *5:* 185-202.

PARISI, B.
1922. Elenco degli stomatopodi del Museo di Milano. Atti. Soc. Ital. sci. nat., *61:* 91-114, figs. 1-7.

PARKER, ROBERT H.
1956. Macroinvertebrate assemblages as indicators of sedimentary environments in East Mississippi Delta region. Bull. Amer. Assoc. Petrol. Geol., *40* (2): 295-376, figs. 1-32, pls. 1-8.
1960. Ecology and distributional patterns of marine macroinvertebrates, northern Gulf of Mexico, pp. 302-381, figs. 1-17, pls. 1-6. *In:* Recent sediments, northwest Gulf of Mexico, 1951-1958. Amer. Assoc. Petrol. Geol., Tulsa.

PAULMEIER, F. C.
 1905. Higher Crustacea of New York City. Bull. New York State Mus.,
    Zool., *12:* 117-189, figs. 1-59.

PEARSE, A. S.
 1929. Observations on certain littoral and terrestrial animals at Tortugas,
    Florida, with special reference to migrations from marine to terrestrial
    habitats. Carnegie Inst. Washington, Publ. no. 391: 205-223, figs. 1-3.
 1932. Inhabitants of certain sponges at Dry Tortugas. Pap. Tortugas Lab.,
    Carnegie Inst. Washington, *28:* 117-124, pls. 1-2, text-fig. 1.
 1950. Notes on the inhabitants of certain sponges at Bimini. Ecology, *31*
    (1): 149-151.

PEARSE, A. S., H. J. HUMM, AND G. W. WHARTON
 1942. Ecology of sand beaches at Beaufort, North Carolina. Ecol. Monogr.,
    *12* (2): 135-190, figs. 1-24.

PEARSE, A. S., AND LOUIS J. WILLIAMS
 1951. The biota of the reefs off the Carolinas. Journ. Elisha Mitchell Sci.
    Soc., *67* (1): 133-161, figs. 1-5.

PEREZ-FARFANTE, I.
 1954. The discovery of a new shrimp bank at Golfo de Batabano, Cuba.
    Proc. Gulf & Carib. Fish. Inst., 6th Ann. Sess., Miami Beach, 1953:
    97-98.

PETIVER, J.
 1712. Pterigraphia Americana. Icones continens plusquam 400 filicum
    variarum specierum . . . : 3 pp., pls. 1-20 (not seen).

POCOCK, R. I.
 1890. Crustacea. *In:* Ridley, H. N., Notes on the zoology of Fernando
    Noronha. Journ. Linn. Soc. London, Zool., *20:* 506-526.
 1893. Report upon the stomatopod crustaceans obtained by P. W. Basset-
    Smith, Esq., Surgeon R. N., during the cruise, in the Australian and
    China seas, of H.M.S. 'Penguin,' Commander W. U. Moore. Ann.
    Mag. nat. Hist., ser. 6, *11:* 473-479, pl. XXB.

PRATT, H. S.
 1916. A manual of the common invertebrate animals, exclusive of insects.
    737 pp., 1017 figs. A. C. McClurg and Co., Chicago.
 1935. A manual of the common invertebrate animals, exclusive of insects.
    xviii + 854, 974 figs. P. Blakiston's Son & Co., Philadelphia.

PREUDHOMME DE BORRE, A.
 1882. Liste des squillides du Musée Royal d'Histoire Naturelle de Belgique.
    C. r. Soc. entom. Belge, ser. 3, no. 20: cxi-cxv.

RANKIN, W. M.
 1898. The Northrop collection of Crustacea from the Bahamas. Ann. New
    York Acad. Sci., *11* (12): 225-258, pls. 29-30.
 1900. The Crustacea of the Bermuda Islands, with notes on the collections
    made by the New York University expeditions in 1897 and 1898.
    Ann. New York Acad. Sci., *12* (12): 521-548, pl. 17.

RATHBUN, MARY J.
 1892. List of Crustacea collected. Report of the Commissioner of Fish and
    Fisheries respecting the establishment of fish-cultural stations in the
    Rocky Mountain regions and Gulf states. Senate Misc. Doc. No. 65,
    52nd Congr., First Session, 1892: 87-88.
 1892. List of Crustacea collected, pp. 89-90. *In:* Evermann, B. W., A report
    upon investigations made in Texas in 1891. Bull. U. S. Fish. Comm.,
    *11:* 61-90, pls. 28-36.

1899. Jamaica Crustacea. Journ. Inst. Jamaica, 2 (6): 628-629.
1900. The decapod and stomatopod Crustacea. I. Results of the Branner-Agassiz Expedition to Brazil. Proc. Washington Acad. Sci., 2: 133-156, pl. 8.
1905. List of the Crustacea. Fauna of New England, 5. Occ. Pap. Boston Soc. nat. Hist., VII: 1-117.
1919. Stalk-eyed crustaceans of the Dutch West Indies. In: Boeke, J., Rapport betreffende een vorloopig onderzoek naar den toestand van der Visscherij en de Industrie van Zeeproducten in de Kolonie Curaçao, ingevolge het Ministerieel Besluit van 22 November 1904, 2: 317-349, figs. 1-5.
1935. Fossil Crustacea of the Atlantic and Gulf coastal plain. Geol. Soc. America, Spec. Pap. no. 2: vii, 1-160, pls. 1-26, text-figs. 1-2.

RATHBUN, RICHARD
1883. Collection of economic crustaceans, worms, echinoderms, and sponges. B. Great International Fisheries Exhibition, London, 1883. Bull. U. S. Nat. Mus., 27: 109-137.
1884. Crustaceans, worms, radiates, and sponges. Natural history of useful aquatic animals. Part V. In: Goode, G. Brown, The Fisheries and fishery industries of the United States, 1: 759-850, atlas, pls. 260-277 (also published in: 47th Congress, 1st Session, Senate Misc. Doc. No. 124, pts 1-2).
1893. Natural history of economic crustaceans of the United States. Reprinted from: Goode, G. Brown, The fisheries and fishery industries of the United States, 1: 763-830, pls. 260-275, pl. CXXI.

REED, CLYDE THEODORE
1941. Marine life in Texas waters. Texas Acad. Pub. Nat. Hist., non-tech ser.: xii + 88, figs. Anson Jones Press, Houston.

RICHARDS, HORACE G.
1938. Animals of the seashore: 273 pp., frontis. + 28 pls., 45 figs. Bruce Humphries Inc., Boston.

RICHARDSON, L. R.
1953. Variation in Squilla armata M. Edw. (Stomatopoda) suggesting a distinct form in New Zealand waters. Trans. Roy. Soc. New Zealand, 81 (2): 315-317, figs. 1-3.

RODRIGUEZ, GILBERTO
1959. The marine communities of Margarita Island, Venezuela. Bull. Mar. Sci. Gulf & Carib., 9 (3): 237-280, figs. 1-26.

ROXAS, H. A., AND E. ESTAMPADOR
1930. Stomatopoda of the Philippines. Nat. app. Sci. Bull., Manila, 1: 93-131, pls. 1-6.

DE SAUSSURE, H.
1853. Descriptions de quelques crustacés nouveaux de la côte occidentale du Mexique. Rev. Mag. Zool., no. 8: 354-368, pls. 12-13 (pp. 1-15 on separate).

SAWAYA, PAULO
1942. Comentários sobre crustáceos. Chap. XIX, p. LXI-LXV. In: Marcgrave, G., 1648 (1942 translation), História natural do Brasil, Comentários (appendix), 104 pp.

SAY, THOMAS
1818. An account of the Crustacea of the United States. Jour. Acad. nat. Sci., Philadelphia, 1: 235-253.

367

SCHINZ, H. K.
 1823. Krebse, Spinnen, Fusetten. Dritter Bd. *In:* von Cuvier, R., Das Thierreich eingeheilt nach dem Bau der Thiere als Grundlage ihrer Natur-
 geschichte un der vergleichenden Anatomie von dem Herrn Ritter
 von Cuvier: xviii + 932. Stuttgartt und Tubingen.

SCHMITT, WALDO L.
 1924. The macruran, anomuran, and stomatopod Crustacea. Bijd. Dierk.,
 *23:* 61-81, pl. 8, text-figs. 1-7.
 1924a. Report on the Macrura, Anomura, and Stomatopoda collected by
 the Barbados-Antigua Expedition from the University of Iowa in
 1918. Univ. Iowa Stud. Nat. Hist., *10* (4): 65-99, pls. 1-5.
 1940. The stomatopods of the west coast of America based on collections
 made by the Allan Hancock Expeditions, 1933-1938. Allan Hancock
 Found. Pacif. Exped., *5* (4): 129-225, figs. 1-33.

SERÈNE, R.
 1951. Observations sur deux *Pseudosquilla* d'Indochine. Treubia, *21* (1):
 11-25, figs. 1-8.
 1954. Observations biologiques sur les stomatopodes. Mém. Inst. océanogr.
 Nhatrang, Viet Nam, no. 8: 1-93, pls. 1-10, text-figs. 1-15 (also
 published with same pagination in Ann. Inst. océanogr. Monaco, *29).*
 1962. Révision du genre *Pseudosquilla* (Stomatopoda) et définition de
 genres nouveaux. Bull. Inst. océanogr. Monaco, no. 1241: 1-27,
 figs. 1-5.

SHARP, B.
 1893. Catalogue of the crustaceans in the museum of the Academy of Natural Sciences of Philadelphia. Proc. Acad. nat. Sci. Philadelphia,
 1893: 104-127.

SHELDON, J. M. ARMS
 1905. Guide to the invertebrates of the synoptic collection in the museum
 of the Boston Society of Natural History: v + 505. Boston.

SMITH, RALPH I., ED.
 1964. Keys to marine invretebrates of the Woods Hole region. Mar. Biol.
 Lab., Woods Hole: x + 208, pls. 1-28. Mimeo.

SMITH, SYDNEY I.
 1869. Notice of the Crustacea collected by Prof. C. F. Hartt on the coast
 of Brazil in 1867, together with a list of the described species of
 Brazilian Podophthalmia. Trans. Connecticut Acad. Arts Sci., *2:*
 1-41, pl. 1.
 1869a. Abstract of a notice of the Crustacea collected by Prof. C. F. Hartt,
 on the coast of Brazil in 1867. Contributions to Zoology from the
 Museum of Yale College. IV. Amer. Journ. Arts Sci., ser. 2, *48:*
 388-391.
 1873. The metamorphoses of the lobster, and other Crustacea, pp. 522-537,
 text-fig. 4 (pp. 228-243 in 1874 reprint). *In:* Verrill, A. E., Report
 upon the invertebrate animals of Vineyard Sound and the adjacent
 waters, with an account of the physical characters of the region. Rep.
 U. S. Fish Comm., *1:* 295-778, pls. 1-39 (reprint by Verrill and
 Smith, 1874: vi + 478, pls. 1-38).
 1881. Preliminary notice of the Crustacea dredged, in 64 to 325 fathoms,
 off the south coast of New England, by the U. S. Fish Commission
 in 1880. Proc. U. S. Nat. Mus., *3:* 413-452.

SPRINGER, STEWART, AND HARVEY R. BULLIS, JR.
1956. Collections by the *Oregon* in the Gulf of Mexico. Spec. Sci. Rep., U. S. Fish Wildl. Serv., no. 196: 1-134.

STEBBING, THOMAS R. R.
1893. A History of Crustacea. Recent Malacostraca. Internat. Sci. Ser. LXXI: xvii + 466, 32 figs., 19 pls. D. Appleton and Co., N. Y.
1904. South African Crustacea. Part II. *In:* Marine Investigations in South Africa, *2:* 1-92, pls. 5-16. The Cape Times Ltd., Cape Town.

STEINBECK, JOHN, AND E. F. RICKETTS
1941. Sea of Cortez: x + 598 pp., figs. The Viking Press, N. Y.

STEPHENSON, W.
1967. A comparison of Australasian and American specimens of *Hemisquilla ensigera* (Owen, 1832) (Crustacea: Stomatopoda). Proc. U. S. Nat. Mus., *120* (3564): 1-18, figs. 1-3.

STRAWN, KIRK
1954. The pushnet, a one-man net for collecting in attached vegetation. Copeia, 1954, no. 3: 195-197, pl. 1, text-figs. 1-2.

SUMNER, F. B., R. C. OSBURN, AND L. J. COLE
1913. A biological survey of the waters of Woods Hole and vicinity. Part I. Physical and zoological. Bull. U. S. Bur. Fish., *31:* 3-442, 274 charts.

TABB, DURBIN C., AND RAYMOND B. MANNING
1962. A checklist of the flora and fauna of northern Florida Bay and adjacent brackish waters of the Florida mainland collected during the period July, 1957 through September, 1960. Bull. Mar. Sci. Gulf & Carib., *11* (4): 552-649, figs. 1-8.

THALLWITZ, J.
1893. Dekapoden-studien, insbesondere basiert auf A. B. Meyer's Sammlungen im ostindischen Archipel, nebst einer Aufzählung der Dekapoden und Stomatopoden des Dresdener Museums. Abh. zool. -anthrop. Mus. Dresden, 1890/91, pt. 3: 1-55, pl. 1.

THOMPSON, D'ARCY W.
1901. A catalogue of Crustacea and of Pycnogonida contained in the Museum of University College, Dundee. Univ. of St. Andrews, Dundee, printed for the Museum: v + 56 pp.

TORRALBAS, F.
1917. Contribución al estudio de los crustáceos de Cuba. Notas del Dr. Juan Gundlach † 1886 compiladas y completadas por el Dr. José I. Torralbas † 1903. An. Acad. Ci. med. fis. nat. Habana, *53:* 543-624, figs. 1-73 (pp. 1-92 on separate with index to plates; portions originally published in An. Acad. Ci. Habana, *36* [1900], *37* [1900-1901]).

TOWNES, H. K.
1939. Ecological studies on the Long Island marine invertebrates of importance as fish food or as bait. Ann. Rep. New York Conserv. Dept., *28*, suppl.: 163-176, figs. 34-39.

TOWNSLEY, SIDNEY J.
1953. Adult and larval stomatopod crustaceans occurring in Hawaiian waters. Pacific Sci., *7* (4): 399-437, figs. 1-28.

VERRILL, A. E.
  1873. Report upon the invertebrate animals of Vineyard Sound and the adjacent waters, with an account of the physical characters of the region. Rep. U. S. Fish Comm., *1:* 295-778, pls. 1-39 (reprint by Verrill and Smith, 1874: vi + 478, pls. 1-38).
  1885. Results of the explorations made by the steamer "Albatross" off the northern coast of the United States, in 1883. Rep. U. S. Fish. Comm., *11:* 503-699, pls. 1-44.
  1901. Additions to the fauna of the Bermudas from the Yale Expedition of 1901, with notes on other species. Trans. Connecticut Acad. Arts Sci., *11:* 15-62, pls. 1-9, text-figs. 1-6.
  1923. Crustacea of Bermuda. Schizopoda, Cumacea, Stomatopoda and Phyllocarida. Trans. Connecticut Acad. Arts Sci., *26:* 181-211, pls. 49-56, text-figs. 1-3.

VERRILL, A. E., S. I. SMITH, AND O. HARGER
  1873. Catalogue of the marine invertebrate animals of the southern coast of New England, and adjacent waters, pp. 537-747 (pp. 243-453 in 1874 reprint). *In:* Verrill, A. E., Report upon the invertebrate animals of Vineyard Sound and the adjacent waters, with an account of the physical characters of the region. Rep. U. S. Fish Comm., *1:* 295-778, pls. 1-39 (reprint by Verrill and Smith, 1874: vi + 478, pls. 1-38).

VILELA, HERCULANO
  1949. Crustaceos decapodes e estomatópodes da Guiné Portugesa. Ann. Jta. Invest. colon. Lisboa, *4:* 47-70, figs. 1-17.

VOIGT, F. S.
  1836. Die Anneliden, Crustaceen, Arachniden und die ungeflügelten Insekten. *In:* Cuvier, G., Das Thierreich . . . , *4:* xiv + 516 pp. Brodhaus, Leipzig.

VOSS, GILBERT L., AND NANCY VOSS
  1955. An ecological survey of Soldier Key, Biscayne Bay, Florida. Bull. Mar. Sci. Gulf & Carib., *5* (3): 203-229, figs. 1-4.

WASS, MARVIN L.
  1961. A revised preliminary check list of the invertebrate fauna of marine and brackish waters of Virginia. Virginia Inst. Mar. Sci., Spec. Sci. Rep., no. 24: 67 pp. Mimeo.

WHITE, A.
  1847. List of the specimens of Crustacea in the collection of the British Museum: viii + 141. London.

WILSON, H. V.
  1900. Marine biology at Beaufort. Amer. Nat., *34:* 339-360, figs. 1-5.

YOUNG, G. C.
  1900. The stalk-eyed Crustacea of British Guiana, West Indies, and Bermuda: xix + 514, text-figs., pls. 1-7. John M. Watkins, London.

## ADDENDUM

    Since this manuscript was submitted for publication a number of references on stomatopods from the western Atlantic have come to my attention. These have been listed below with a few annotations. These references have not been cited in the synonymies in the present paper.

370

ALMEIDA RODRIGUES, SERGIO DE
1966. Ocorrência de *Acanthosquilla floridensis* Manning na Costa de São Paulo. Ciencia e Cultura, *18* (2): 142.

DINGLE, HUGH, AND ROY L. CALDWELL
1966. Aggressive, territorial, and reproductive behavior of the mantis shrimp *Gonodactylus oerstedi* (Stomatopoda). Abstract. American Zool., *6* (4): 503 [*G. bredini* Manning, Bermuda].

EVANS, G. OWEN, AND W. E. CHINA
1966. Opinion 785. *Pseudosquilla* Dana, 1852, and *Gonodactylus* Berthold, 1827 (Crustacea, Stomatopoda): Designation of type-species under the Plenary Powers. Bull. zool. nomencl., *23* (5): 204-206 [*Coronida* Brooks, 1886 (Name no. 1729), *Coronis* Desmarest, 1823 (Name no. 1730), *Gonodactylus* Berthold, 1827 (Name no. 1731), *Hemisquilla* Hansen, 1895 (Name no. 1732), and *Pseudosquilla* Dana, 1852 (Name no. 1733) placed on Official List; Squillidae Latreille, 1802-03 (Name no. 416), placed on Official List; *Coroniderichthus* Hansen, 1895 (Name no. 1816), *Coronis* Hübner, 1823 (Name no. 1817), *Gonerichthus* Brooks, 1886 (Name no. 1818), *Pseuderichthus* Brooks, 1886 (Name no. 1819), *Pseudosquille* Eydoux and Souleyet, 1842 (Name no. 1820), *Pseuderichthus* Dames, 1886 (Name no. 1821), *Alimerichthus* Claus, 1871 (Name no. 1822), *Smerdis* Leach, 1817 (Name no. 1823), and *Coronis* Gloger, 1827 (Name no. 1824) placed on Official Index].

FAUSTO FILHO, JOSÉ
1966. Sôbre a ocorrência de *Squilla lijdingi* Holthuis, 1959 no litoral Brasileiro (Crustacea Stomatopoda). Arq. Est. Biol. Mar. Univ. Fed. Ceará, *6* (2): 139-141, figs. 1-2.

HOLTHUIS, L. B.
1967. Fam. Lysiosquillidae et Bathysquillidae. Stomatopoda I. *In:* Gruner, H.-E., and L. B. Holthuis, eds., Crustaceorum Catalogus, ed. a, Pars 1: 1-28 [Complete synonymies for all species then known].

MOORE, D. R., AND K. J. BOSS
1966. Records for *Parabornia squillina*. Nautilus, *80* (1): 34-35 [Commensals found with *Lysiosquilla scabricauda* (Lamarck) from Mississippi].

RODRIGUES DA COSTA, H.
1964. Crustacea coletados numa excursão ao estado do Espirito Santo. I. Bol. Mus. Nac. Rio de Janeiro, n. s., zool., no. 250: 1-14 [*G. oerstedii* var. *spinulosus*, p. 10 ( = *G. minutus* Manning ?)].

# INDEX *

*Italics* indicate current names and page numbers of principal accounts. LARGE AND SMALL
CAPITALS indicate new taxa.

373

375

378

379